DRAINING NEW ORLEANS

DRAINING NEW ORLEANS

THE 300-YEAR QUEST TO DEWATER THE CRESCENT CITY

Richard Campanella

LOUISIANA STATE UNIVERSITY PRESS

BATON ROUGE

Published with the support of the Tulane University School of Architecture

Published by Louisiana State University Press
lsupress.org

Manufactured in the United States of America
First printing

DESIGNER: Michelle A. Neustrom
TYPEFACE: Calluna
PRINTER AND BINDER: Sheridan Books, Inc.

COVER IMAGES: *Top:* Sketch of wood screw pump in patent application filed in 1913
(courtesy of American Society of Mechanical Engineers and New Orleans Sewerage & Water
Board); *center:* George Hero at work with his engineers installing the West Bank drainage
system (courtesy of Hero family records); *bottom:* Linus W. Brown's 1895 contour map
of New Orleans (courtesy of New Orleans Sewerage & Water Board).

LIBRARY OF CONGRESS CATALOGING-IN-PUBLICATION DATA

Names: Campanella, Richard, author.
Title: Draining New Orleans : the 300-year quest to dewater the Crescent City /
  Richard Campanella.
Description: Baton Rouge : Louisiana State University Press, [2023] | Includes
  bibliographical references and index.
Identifiers: LCCN 2022031890 (print) | LCCN 2022031891 (ebook) | ISBN 978-0-8071-
  7854-6 (cloth) | ISBN 978-0-8071-7942-0 (pdf) | ISBN 978-0-8071-7941-3 (epub)
Subjects: LCSH: Drainage—Louisiana—New Orleans—History. | Reclamation of
  land—Louisiana—New Orleans—History. | New Orleans (La.)—History.
Classification: LCC TC977.L8 C36 2023 (print) | LCC TC977.L8 (ebook) |
  DDC 631.609763/35—dc23/eng/20221012
LC record available at https://lccn.loc.gov/2022031890
LC ebook record available at https://lccn.loc.gov/2022031891

To my wife Marina,
and to our young son Jason—
who can dewater (and rewater)
a batture lagoon in
minutes flat.

CONTENTS

Images follow page 118.

DRAINING NEW ORLEANS

Prologue

Dawn broke electric on Saturday, February 13, 1915—no ordinary time, in no ordinary place. It was the weekend before Mardi Gras, the crescendo of Carnival, in a rambunctious metropolis advocates cheerfully called the City That Care Forgot. More so, it was Hero Day, and as every Mardi Gras has its royalty, this one had its own special monarch: the "Drainage King," George Alfred Hero, a dapper sixty-one-year-old polymath who, word had it, was about to conquer the biggest, oldest problem of the Queen City of the South. And he was going to do it New Orleans–style, extravagantly, starting with a parade and ending with a yacht ride out to a backswamp fete—the very morass the Drainage King was about to dewater, eliminating what for centuries had been considered a nuisance and a nemesis.[1]

To be "king" resonates in Louisiana. Anachronistic as it may be, the honorific comes untinged with the sarcasm or irony it might bear when used elsewhere in this nation founded against monarchism. Perhaps this is because of the Bourbon origins of Louisiana, the only American state named for a French king. More likely it reflects a penchant for make-believe Carnival royalty, in which being crowned *rex* commands genuine deference across social strata. So when "George Hero [became] known in Louisiana as the 'drainage king,'" the sobriquet signaled an esteem not doled out liberally.[2] Here was a hero *and* a king—a King named Hero!—with his own parade on his own special day, during Carnival, in the City of Dreams.

Hero Day began with a sumptuous banquet at the Grunewald Hotel on Baronne Street, across from the French Quarter. Chauffeured automobiles then carried the entourage to the Garden District to join the parade assembling on St. Charles Avenue. First came two marching bands, followed by police on foot and horseback, then "mounted heralds wearing shields with letters spelling 'HERO.'"

Next were open-top luxury cars bearing Louisiana Governor Luther Hall, New Orleans Mayor Martin Behrman, City Chief Engineer John Coleman, and the Drainage King himself, all beaming and waving to appreciative crowds.

The cavalcade rolled under balmy midwinter skies down St. Charles Avenue and onto Canal Street, passing its famed venues and elegant emporia. It pulled up to the banks of the continent's largest river, upon the wharves of the South's busiest port, where three hundred additional guests awaited aboard the *Hanover.* They came from the apex of local society, and if any two professions prevailed, it was real estate agent and politician.[3]

The *Hanover* steamed across the Mississippi and into the Harvey Canal, where guests transferred to the yacht *Daisy* and proceeded southward toward Bayou Barataria. They glided past bucolic enclaves and across former plantations to the ragged backswamp, where thin soils gave way to stagnant water. Soon the *Daisy* arrived upon a clearing where twelve hundred members of *haute société* gathered around a gigantic ultra-modern machine, strikingly incongruous amid the dense thicket. Everyone eagerly anticipated the activation of that pumping station, and none other than President Woodrow Wilson would do the honors, via telephone from the White House.

The audience settled. Journalists and photographers staked out positions. Dignitaries gave soaring speeches. "King" George Hero got the loudest cheer, as he orated on draining a million more acres, making Louisiana a billion dollars richer.

Finally came the moment. Engineers spun the turbines. Electricity flowed. Aids telephoned the White House. The crowd held its breath.

"Then came the President's message," wrote a journalist, "handed up by a hastened messenger. And then the pumps began to whirr."[4]

INTRODUCTION

These Oozy and Muddy Lands

George Hero first became fascinated with draining swamps as a boy mucking around on a relative's Assumption Parish plantation. No wonder: children seem drawn to manipulating water. Put a child on a beach or by a stream, and little hands immediately get to work ditching, damming, and directing water against gravity and common sense. Perhaps this is because of drainage's instantaneous yoking of cause and effect, the pat clarity of the challenge, or the visibly satisfying transformation of the realm.

Drainage is hydraulics applied. When rain falls on land, some of the water gets absorbed into the soil body or retained on the surface, a portion of which evaporates. The rest, called runoff or stormwater, flows across the surface depending on its hydraulic head, namely the gradient from source to destination (called an outfall), which may be a stream, bayou, river, canal, lake, bay, or ocean. If the hydraulic head is strong, the runoff flows swiftly and the land surface drains effectively. If weak, runoff flows slowly and may back up on itself. If it's extremely weak, as in southern Louisiana, the runoff ponds to various depths, and may become permanently impounded. If that ponded area is forested, it's called a swamp; if it's grassy, it's a marsh; if it merges with tidal zones, it's a saline marsh; and if it's open water, it's a lake or bay.[1]

Drainage becomes complicated when humans enter the picture. Runoff that does not drain swiftly can become inconvenient, even intolerable—a "flood." Ponding that had naturally formed a swamp or marsh now becomes deemed a "disaster." The ensuing engineering interventions aimed to prevent a future flooding disaster drastically alter the value of land to humans, and once that new valuation is built upon, those engineered drainage systems become absolutely essential for any future value to be added. Lives and livelihoods come to depend utterly on that initial decision to muck around with muck, by inserting humans and machines in the role that gravity once played.

Anthropogenic drainage is simple in theory yet complex in execution. It has many motivations but one goal. The word itself defines succinctly, yet has many applications. And while its intention is always envisaged as beneficial, its consequences are often deleterious—in ways often incomprehensible to its advocates. Drainage is one of humanity's biggest, oldest acts of natural defiance, and our cities and farmlands would all be configured very differently, if at all, had we just let dirt and water alone. Drainage is a paradox, and it is key to understanding the paradoxical city of New Orleans.

At its most fundamental level, drainage means the removal of a liquid. To a physician, the liquid is a body fluid; to a plumber, it's wastewater; to a civil engineer, it's urban runoff; to a hydrological engineer, it's standing water; and to subsurface hydrologists, it's groundwater. For the urbanist, drainage renders habitability, and for the agriculturalist, it spells the difference between useless mud and rich, loamy soil.

Mud, on the contrary, is worse than useless; it is hazardous. It precludes humans and mires their motion; it could immobilize even the greatest army. Ergo the human predisposition to intervene, like that child at the stream, and fix the intolerable. In the heroic age of engineering, drainage was called "reclamation," a term plainly insinuating that hydric soils are a mistake waiting to be corrected, and that humans, through brain and brawn, can claim back that which had been rightfully intended for their use all along. Over the course of a century, waterfront cities like Boston, New York, Chicago, Miami, San Francisco, and Seattle joined New Orleans in exerting herculean engineering efforts to "reclaim" for themselves a new morphology, topography, and hydrology. But only New Orleans sat entirely upon an idiosyncratic land form known as a fluvial delta; only New Orleans would sink massively below sea level; and only New Orleans would end up ranking as "the world's toughest drainage problem."[2]

In this book, the term "reclamation" is used to mean the mechanized removal of standing water from swamps, marshes, or open-water bodies. Reclamation turns a waist-deep morass into mere mud, or a lake or bay into artificial land. "Subsurface drainage" implies the installation of subterranean pipes to draw down surface water and soil moisture. It turns muddy fields into dry ground, and with the help of some additional dirt ("artificial fill") brought in from external sources and graded into streets and blocks, the former morass becomes developable land.[3] Economic uses such as agriculture, forestry, and mineral extraction may proceed with reclamation alone. But for urbanization to occur, reclamation must be followed by subsurface drainage and artificial fill.

Understanding how this dewatering happened in New Orleans, who engineered it and why, and what consequences it had, is the goal of this book. The story is organized along the lines of historical geography—the transformation of place over time—and while drainage is a social project, its actualization in New Orleans is unusually biographical. This is a story of people and power, not just soil and water, and readers will meet a procession of drainage "kings"—lordly colonials, brilliant engineers, *noblesse oblige* elites, savvy entrepreneurs, hardworking contractors, and dedicated public servants, as well as the occasional rogue, huckster, and fraud. It is a story of determination and perseverance, but also of misguided policies, unintended consequences, and "path dependency"— that proclivity to make decisions based not on best practices or enlightened understanding, but on prior investments ("sunk costs"), however faulty they proved to be. The study area spans the greater New Orleans metropolitan area south of Lake Pontchartrain, and while the focus is on drainage, parallel storylines intertwine, such as levee systems, potable water, and sewerage, as well as transportation, urban expansion, and neighborhood formation. Together they limn the larger story of the historical geography of New Orleans; that's how intrinsic water is to this place. What the book does not attempt to be is a bureaucratic history of the New Orleans Sewerage & Water Board (S&WB), an inventory of its equipment, or an advocacy for a certain water-management approach, though all those angles are germane to our three-hundred-year history.

Drainage in New Orleans has become politically fraught in recent decades. Go to any public meeting on the topic, and the default tone is anger, as residents fume over potholed streets and persistent floods. My larger goal is to understand the backstory to these problems and the engineering decisions behind them, in the context of their era. My motivation is not anger but curiosity and concern; my objective is not to rebuke but to explain; and my hope is that an understanding of our drainage history can help calibrate our expectations and improve our future relationship with water—paradoxes and all.

THE DRAINAGE IMPERATIVE

Drainage is nearly as old as agriculture, its original motivation, and it laid the groundwork for the creation of cities. Farmers of the Tepe Pardis and Choga Mami regions excavated irrigation ditches as early as 5000 to 8000 years BCE, directing surplus water to arid areas in what are now Iran and Iraq. It worked: farms flourished, food supplies, and people had more options to live their lives. Irrigation-driven drainage spread into the Indus Valley region of today's Pakistan/India border around 4500 BCE, and elsewhere in Mesopotamia, Babylonia, and China to 1200 BCE, during which time the motivations for moving water expanded to hydropower, water storage, and the farming of drained wetlands.

By 400 BCE, Egyptians were digging networks of ditches to drain land for crops, aided by what would later be described as the Archimedean screw, the world's first positive displacement pump. This device had a screw-like thread spiraling within a pipe which "traps a fixed amount of water at the intake [and] transports it to a discharge elevation."[4] Here was an efficient tool to lift water into impoundments with either natural banks or artificial berms—like the levees of future Louisiana.

The Greeks, the Romans, and other ancient civilizations deployed comparable efforts during 200 BCE to 700 AD; collectively, they enhanced agricultural productivity toward the creation of food surpluses, which in turn abetted the emergence of towns and cities. People living in higher densities, assured of their sustenance thanks to the stockpiling of grains grown via drainage and irrigation, could now begin to build structures and infrastructure. These built environments would soon develop their own drainage motivations—for the provision of potable water, the removal of wastewater, the outflow of runoff, and in time, for the expansion of urban living space.[5]

Drainage as well as irrigation became facets of imperialism, serving to put conquered lands into productive use and, along with fortifications and urbanization, bringing "political control [through] spatial control."[6] The Romans were the premier drainage engineers of the ancient world, for their roads, aqueducts, and cities with latrines and public baths. "To meet the urban drainage needs," wrote two engineers in a review of drainage history, the Romans built "an intricate network of open channels and underground sewers, or *cloacae* . . . which drained the lowest parts of Roman . . . into the Tiber River." The constant flow allowed residents to use the channels to carry away domestic waste, making the *cloacae* an early example of a combined stormwater, wastewater, and marsh-drainage system.[7] This fused approach would remain the norm for centuries. Well into the nineteenth century, "drainage for the removal of rainwater and groundwater in urban settings was uniformly viewed as part of the more general issue of sewerage, [such that] urban drainage and sewerage were synonymous."[8]

Rome's most famous edifice, the Colosseum, was built on a drained lake, which itself was artificial.

Drainage was costly, and the Romans deployed it selectively. They did not introduce it to the Netherlands, for example, on account of the enormity of the challenge and its peripheral position. "The Romans had no interest whatsoever in . . . the Dutch lowlands[,] an uninhabitable marsh where water and wind had free reign," and instead of draining them, instead only erected "fortresses and defensive works . . . on the northern boundary of the empire."[9]

That task instead fell to the locals, whose limited living space and tillable land compelled them to reimagine their wetlands. Clustered in nascent cities upon slightly elevated sand ridges, the Dutch starting in the thirteenth century began building dykes around their communities, impounding water for wind-driven screw pumps to extract through parallel ditches for expulsion beyond the dykes. What resulted they called "polders," small topographical subbasins—mini-watersheds, or catchments—in which a natural hydrosphere became an anthropogenic lithosphere. While the now-dried soils in polders tended to consolidate and sink, the "bowls" also became highly fertile new farmland. Food production increased, and Dutch cities grew, thus necessitating further polderization and pumping, which called for greater commitments of resources and interregional policies. In time, nationhood developed around water management, and distinctive cultural landscapes as well as efficient administrative jurisdictions arose from hydrological engineering. Few other places had such an intertwined array of life-critical motivations for drainage, and such a sophisticated slate of tools and techniques for its execution. It would take many centuries and countless floods, from overtopping rivers and a stormy North Sea, but by the time Louisiana came upon the colonial docket in the New World, the Dutch had become the undisputed "drainage kings" of the Old World.[10] "God created the world," goes the saying, "but the Dutch created the Netherlands."[11]

THE FLUVIAL DELTA

Coastal Louisiana and the Netherlands are both deltas, and their peoples and environs are often viewed as colleagues and counterparts. Yet their distinctions are worth pointing out. Put plainly, the two deltas have different "genders," and the male is more difficult. Coastal Netherlands is a deltaic plain formed by the Alps-born Rhine River as it twice bifurcates into subchannels, each delivering sediment upon the shore of the North Sea. Unlike in the Gulf of Mexico, offshore currents flow strongly here, and daily tidal regimes span fully six feet, compared to only a foot or so off Louisiana. This constant wave and tidal action had the effect of sweeping away much of the deposited alluvium, leaving behind only a necklace of barrier islands and sandy dunes fronting sporadic alluvial marshes pocked by shallow bays, an area that collectively lay generally flush with Europe's North Sea bight. It is these lagoon-strewn interior wetlands of the Rhine-Meuse-Scheldt Delta that the Dutch, starting in the 1300s, had polderized, drained, and barricaded from the sea.

Whereas the Rhine-Meuse-Scheldt Delta is a sea-dominated delta system with an interior (female) structure, that of the Mississippi is a river-dominated (fluvial) delta with an exterior (male) morphology. Its formative sources of fresh water and sediment—the meandering, channel-shifting Mississippi River and its distributaries, bearing over seven times' the Rhine's water volume, and a commensurate amount of its sediment—overwhelmed the ability of the Gulf of Mexico's meager tides and weak longshore currents to sweep away its alluvial deposits.[12] The river thus effectively won the battle, depositing soil particles faster than Gulf waves could dislodge them. (How much? "The river-water is remarkably muddy," wrote an English captain of the lower Mississippi in 1770. "I have filled a half-pint tumbler with it, and have found a sediment of two inches of slime.")[13] What resulted, quite different from the inward incursions of the Rhine-Meuse-Scheldt Delta, was a prograded delta extending outwardly from the Gulf Coast, like "a gigantic arm projecting into the sea," wrote French geographer Elisée Reclus in 1855, shifting about and "spreading its fingers on the surface of the waters."[14]

Each avulsion (jump) of the channel, occurring roughly once per millennium for the past seven thousand years, created a new deltaic lobe, each with higher natural levees (ridges) of loamy soils abutting the main channel of the Mississippi. "You will observe that the land is of *peculiar* formation," wrote a Jesuit priest of lower Louisiana in 1750. "Throughout nearly the whole country, the bank of a river is the lowest spot; *here,* on the contrary, it is the *highest.*"[15] Lesser bankside ridges arose along secondary or abandoned distributaries, while in between formed low swamps, marshes, and *prairies tremblantes* (unrooted flotant), like webbing between the toes of an aquatic bird.[16]

Each of these deltaic geomorphologies served to intercept alluvium delivered by subsequent river overflows, such that past deposits begot future deposition. Le Page du Pratz, describing the process in 1758, wrote that sediments "are brought down and accumulated by means of the ooze which the Missisippi [*sic*] carries [during] its annual inundations. . . . Those oozy or muddy lands easily produce herbs and reeds; and when the Missisippi happens to overflow the following year, these herbs and reeds intercept a part of this ooze [and] the banks of the Missisippi became higher than the lands about it."[17] Even the most elevated land yielded by these processes might be better described as "muck," a term actually used by pedologists to describe local soils; words like "slippery," "wet," "rotten," and "decayed" are also apropos. "The upper surface is a marsh mud, extremely slippery as soon as wet, with a small mixture of sand," wrote architect Benjamin Latrobe in 1819, below which are "decayed vegetables, water at 3 feet, [and] large logs, [some] rotted. Such a soil [is] the result of the gradual accumulation of the deposition of the river, [and] logs & trees . . . descending the stream at every fresh."[18]

Water on the delta landscape came from five sources: from sixty to seventy inches of annual rainfall, particularly in the summer months; from high humidity, up to 85 percent on a winter day and 90 to 100 percent during summer; from lateral filtration into the groundwater from the channel of the Mississippi River and its distributaries; from springtime freshets (overflows) from those same riverine sources; and from daily tidal regimes and occasional storm surges from estuarine bays and Gulf waters.

"Oozy" deltas like that of the Mississippi, because of their prograding morphology and fluvial dependency, are among earth's most precarious land features. Geology that takes eons elsewhere happens in years here, even months. Fluidity is as intrinsic to these environs as it is antithetical to the hard continental environments in which humans evolved, and on which humans depend for their place-making. "Seemingly natural elemental categories such as earth and water lost their meaning in the delta," wrote geographer Adam Mandelman in his aptly titled book *The Place with No Edge*. What prevailed instead was a "watery in-betweenness [with] a sense of disorder and otherworldliness."[19]

To bring order here, rigid boundaries would have to be imposed on "the place with no edge," and in time they would take the form of levees, ditches, canals, walls, gates, and pumps. What they yielded was inhabitable space for sedentary living. But what got curtailed were those same oozy, muddy, fluid processes that had created the delta in the first place. Thus the paradox of New Orleans: that the very water-manipulating devices needed to make this city prosper would also render it precarious.

Adapt or Retreat

The first humans to inhabit the Mississippi River Deltaic Plain took a reactive tack on water. Indigenous peoples saw the hydric landscape as a condition requiring adaptation, rather than as a problem calling for a solution. Those living south of *Okwa-ta* ("wide water," today's Lake Pontchartrain), along *bayouks* (bayous, sluggish streams and rivers, or *ruisseau* in French) upon the deltaic plain of the *Michacépi* ("father of waters," the Mississippi) formed two differing cultural economies based on their hydrology.[1]

Those natives dwelling along the elevated riverbanks established small villages of circular palmetto dwellings surrounded by croplands of maize and millet. When freshets came, either by overtopping or crevasse, they moved to higher ground. Though they were farmers more than hunter-gatherers, and hardly nomadic, their place-making was portable, and they inhabited the delta provisionally.

Those natives of the saline marshes dwelled on interbasin ridges or relict beaches, and subsisted on fauna, fowl, finfish, and particularly "freshwater mussels (*unio*), brackish water clams (*rangia*), and saltwater species including oysters (*ostrea*)."[2] Shells were discarded in heaps (middens) which grew so high as to make living space—"an entirely new ecozone in the marsh," noted anthropologist Tristan Kidder.[3] "The shellfish is so abundant," marveled one colonial in 1803, "that the different tribes that inhabit these lakes make it their principal diet [and] pile them up [around] their villages, [creating] pyramidal forms that still grow."[4] Two middens on the western shore of Lake Salvador, Little Temple and Big (Grand) Temple, rose dozens of feet above the brackish bay, like calcium plateaus.

Adaptation ended where know-how began. Like humans anywhere, native peoples of the delta manipulated their environment to the extent of their technology. They burned fields, cleared forests, mass-captured fish, and overkilled game if the opportunity arose. They also dug ditches, diverted water, and shored up land to make for a better farm or village site.[5] But hydrological engineers they were not, at least not here. Lacking much natural topography and without a mechanized means for water-lifting, natives of lower Louisiana did not drain swamps and marshes.

They did, however, use the delta's labyrinthine waterways to trade far and wide, by pirogue and longboat, interacting with tribes from the coastal marshes to the south to the piney woods and loess bluffs to the north and west. Up to eight languages were spoken regionally, so many that the Choctaw called the area *Balbancha* or *Bulbancha,* "land of many tongues." A pidgin known as Mobilian Jargon developed for communication among the various *petites nations,* among which were the Mugulasha, Quinapisa, Oachas (Washa), Chaouachas (Chawasha), and Chitimacha in what is now greater New Orleans; the Annochy, Biloxi, and Mobile to the east; the Tangipahoa, Acolapissa, and Choctaw north across *Okwa-ta;* and the Bayougoula, Houma, Atakapan, Opelousas, Natchez, and Tunica to the west.[6] For all these nations, water underscored their transient nature—in more ways than one.

In stark contrast to the natives' adaptive culture, imperialists emphatically viewed water on the land as a problem to be solved, not a condition to be expected. Imperialism endeavors to possess and render into wealth, and French colonizers brought to bear their technical acumen to do exactly that.

Enter the engineer.

BIENVILLE

One August day in 1721, Adrien de Pauger, assistant engineer for the *Compagnie d'Occident* (Company of the West), stalked peevishly around a primitive village, map in his hand and mud on his boots. A "proud, proper, and religiously devout man," according to one historian, "one part idealist engineer and one part hot-tempered rogue," Pauger had much to steam about.[7] He didn't get along with his superior, Chief Engineer Louis-Pierre Le Blond de La Tour, who had his priorities elsewhere—namely New Biloxi, the recently designated company headquarters on the Gulf Coast. That meant Pauger got relegated to the disaster-in-the-making called *La Nouvelle-Orléans* on the lower Mississippi, where he would find his work cut out for him. And it put Le Blond de La Tour, perhaps wary of his ambitious underling, in a position to hobnob with *his* superior, Gov. Jean Baptiste Le Moyne, Sieur de Bienville, at the new coastal capital.[8]

Despite having fallen ill, Le Blond de La Tour proceeded to sketch a plat for New Biloxi for a site across the bay from the first French settlement in the Louisiana colony, Fort Maurepas (1699), because of its water problems. Being just north of what is now Oceans Springs, Mississippi, on the eastern shore of Biloxi Bay, "Old Biloxi" had been "situated in a place surrounded by marshes from which come out fogs that corrupt the air, which cause illnesses and deaths, which have carried off in the last six months nearly five hundred persons, [plus a] large number of sick. Furthermore the water that is drunk there is very bad and in the summer it is entirely lacking."[9] After a meeting in November 1720, the Council of Commerce decided to relocate the capital to a peninsula fronting the Gulf. Following French military engineer Sébastien Le Prestre, Seigneur de Vauban's guidelines for building forts, Le Blond de La Tour designed for what is now downtown Biloxi, Mississippi, an orthogonal street grid with a *place d'armes* and central church set within a star-shaped fortification.[10]

Governor Bienville joined Le Blond de La Tour on the New Biloxi project, but he was hardly the advocate local partisans might have expected. In fact, Bienville's heart was in New Orleans, which he had established in 1718, where he subsequently had granted himself vast land concessions, and whose site he favored as early as 1699, when he first saw it alongside his older brother, the late Pierre Le Moyne, Sieur d'Iberville.

Founder of the Louisiana colony, Iberville privately harbored dread as he first eyed the *aquatique* delta back in March of 1699. "All this land is a country of reeds and brambles and very tall grass," he confided in his journal, "nothing other than canes and bushes. The land becomes inundated to a depth of 4 feet during high water." To another crew member, it was "nothing more than two narrow strips of land, about a musket shot in width, having the sea on both sides of the river, which . . . frequently overflows."[11]

Iberville died in Cuba in 1706, probably of yellow fever, by which time his younger brother Bienville had risen in rank in Louisiana. In the decade ahead, the initial settlement at Fort Maurepas had expanded into a smattering of tenuous outposts from Mobile to Natchitoches, even as the whole Louisiana project labored under perilous conditions and scant resources.

King Louis XIV grew exasperated with the cumbersome colony, and in 1712 granted a commercial monopoly to financier Antoine Crozat to develop it pri-

Handwritten annotations across top:
1699 — 1712 — 1717 — 1719 — 1720 Louis
Bienville Louis XIV John Law new town Capital Stock
& NOLA Commercial 13 Stock floods Moved to
monopoly Crozat re development see p.14 Biloxi
of Louisiana ✱1 ✱2 worth
6 L

vately. The king died in 1715, and two years later Crozat gave up on Louisiana, his "three principal projects: discovery of mines of gold and silver, the establishment [of] plantations of tobacco, [and] commerce with Spain, [all] dissipated."[12] That same year, another financier, John Law, having learned of Crozat's renunciation, imagined the colony as a possible test bed for his theories on monetary policy. Law had posited that paper money could be backed by the promise of commercial wealth, rather than real wealth—gold, of which France had little. Louisiana could be the source of that wealth, Law argued, and France the beneficiary.

Law persuaded Prince Philippe II, Duc d'Orléans, acting as regent of France since the death of King Louis XIV, that stock sold on the commercial development of Louisiana could pay off national debts and enrich all investors. Philippe readily agreed, and on September 6, 1717, Law, now head of the newly formed *Compagnie d'Occident* (Company of the West), received a twenty-five-year monopoly charter on Louisiana, with the commitment to populate it with six thousand settlers recruited throughout Europe and three thousand Africans to be brought in chains from West Africa. Three days later, a scribe wrote in the company's register, "Resolved to establish, thirty leagues up the river, a burg which should be called New Orleans, where landing would be possible from either the river or Lake Pontchartrain."[13]

That resolution seemed to settle an ongoing debate about where, between Mobile and Natchitoches, a colonial capital and chief trading post should be located. Only after years of convincing himself that such a city had to be on the Mississippi, for reasons of defense, access, soil fertility, and commercial viability, did Governor Bienville come to accept that a perfect site did not exist; thus the best *available* site would have to suffice. To him this meant the elevated natural levee at a particular crescent on the right ascending (east) bank some thirty leagues up the Mississippi, a site he scouted out in February 1718. This sharp meander would expose enemy vessels to defensive firepower, while a convenient ridge and bayou formed a portage to Lake Pontchartrain and the sea, providing an alternative to the shoal-prone, tough-to-navigate Mississippi River—just as the 1717 resolution called for.

Sometime between mid-March and mid-April 1718, Bienville "arrived with six vessels, loaded with provisions and . . . thirty workmen, all convicts; six carpenters and four Canadians," according to colonist Jonathan Darby. "The whole locality was a dense canebrake, with only a small pathway [Bayou Road] leading from the Mississippi to the Bayou [St. John] communicating with Lake Pontchartrain." Darby recalled that "M. de Bienville cut the first cane," followed by "MM. Pradel and Dreux," after which workers cleared vegetation for provisional shelters, "made of standing boards and posts, with walls and chimneys of dirt and covered with cypress bark."[14] According to the memoir of geographer Chevalier de Beaurain, Bienville had "left fifty people . . . both carpenters and convicts" at the site, and assigned them "to dry up the land," after which they would "build some lodgings there."[15]

So began the dewatering of New Orleans, and watery it was. The outpost flooded on its first anniversary, April 1719, and remained wet for months. "The site is drowned under half a foot of water," Bienville groused, according to French historian Marc de Villiers du Terrage; "it may be difficult to maintain a town [here]." It was the worst deluge indigenous people had seen in years, and not only did it make a mess, it also added a new challenge: levee-building, a task that went to the young engineer Charles Franquet de Chaville, under the direction of assistant engineer Adrien de Pauger.[16]

Combined with reports of navigation difficulties on the Mississippi, the deluge convinced officials in Paris in late 1720 to relocate the colony capital from Mobile not to the troubled new company trading post of New Orleans, but rather to New Biloxi, which could serve as a coastal transshipment port for ocean-going vessels to transfer cargo to smaller longboats for interior navigation. It was at this point that a disappointed Governor Bienville dutifully departed with his chief engineer, Le Blond de La Tour, and a work crew to start building the new coastal capital on the Biloxi peninsula, while leaving other men behind to continue work at now-demoted New Orleans. Water problems had flummoxed progress at New Orleans, and while New Biloxi was hardly flood-safe, the shifted investment amounted to something of a conditional retreat from risk—not unlike the native strategy of delta living.

Then, in 1720, came news from Paris that financial chaos had befallen the Company of the West, as John Law's outlandish presumptions of commercial viability proved to be exactly that. Stockholders had caught wind of Louisiana's struggles, and by spring, company shares had plummeted in value. Investors raced to cash out, and riots erupted when banks ran out of coins. Bedlam prevailed within the company, distracting officials from their administrative duties. New Orleans, their conception from the halcyon days of 1717, thus became a third-rate priority of a second-rate company chartered by an indebted empire beleaguered by a cumbersome colony.

The turmoil showed in the streets of New Orleans—that is, if only it had streets. Driven more by expediency than vision, colonists had built huts in a desultory manner—"about a hundred forty barracks, disposed with no great regularity," wrote one contemporary, "a few inconsiderable houses, scattered up and down, without any order or regularity . . . common and ordinary buildings. . . . *New Orleans,* in 1720, made a very contemptible figure."[17] Out-of-plumb pathways yielded further disorder, as makeshift abodes on crooked lots became home to the struggling population of 327 free whites, 171 African slaves, and 21 Indian slaves, according to a 1721 census.[18]

This had not been Bienville's initial vision, of course. Possibly with the help of Mobile surveyor Jacques Barbizon de Pailloux, Bienville in 1718 had laid out a neat baseline angled at an azimuth of 37 degrees and set 700 feet back from the river. A map attributed to Le Blond de La Tour shows that line in red, labeled "Alignement Suiuant le projet de M. de Bienville des premieres maisons" (Alignment Following Mr. Bienville's projection for the first houses).[19] Apparently no one told the villages, and disarray soon prevailed—until that August day when Pauger arrived.

ADRIEN DE PAUGER

Assistant engineer Adrien de Pauger expected something better for New Orleans—planning, order, perhaps even grandeur. Having received a copy of Le Blond de La Tour's starlike plan for Biloxi, and having been similarly schooled in French military planning, Pauger adapted his boss's plat to the natural levee of the Mississippi. It would take some iterations, but eventually he delivered, on paper, a beautiful nine-by-six-block grid spanning a total of 620 *toises* by 360 *toises* in depth, its streets circumscribing an array of squares each measuring 50 by 50 *toises* (320 English feet). Each square contained two parallel sets of five lots separated by two slightly larger key lots, designed, Pauger explained, so that each parcel "may have the houses on the street front and may still have some land in the rear to have a garden, which here is half of life."[20] In the principal cell Pauger created a *place d'armes,* fronted by institutions of church and state, overlooking the *Flueve St. Louis* (Mississippi River). He split the blocks behind the church evenly with the grid's Y-axis, today's Orleans Street. The entire plat, angled to match Bienville's 37 degree baseline, neatly exploited the higher, better-drained natural levee.[21] Surrounding ramparts gave his better-situated, better-designed city protection from terrestrial assaults, while two corner bastions could confront approaching enemy ships from their riverside perches, the original reason for the 37 degree angle.

It was a rational, visionary, and grand plan, and Pauger was eager to survey it and start building New Orleans anew. Savvily, he sent a copy of his impressive map back to Paris, where Philippe II, the regent and Duc d'Orléans, had stepped in to administer Louisiana while the scandal-plagued Company of the West underwent restructuring. This intervention by the Crown may have made military defensibility more of a priority, for which New Orleans's strategic riverine position was critical—a site still worth holding on to, despite that Biloxi was now the capital. Additionally, Pauger, who had also been researching the hydrology of the Mississippi, came to the conclusion that the river's shoaling and strong currents were surmountable problems. He issued reports urging full navigational use of the Mississippi River, and boldly rebuked the "stubbornness" and "arrogance" of company managers who had forced "ships from France to be stopped at Biloxi, rather than enter the Mississippi . . . keystone of the country's establishment."[22]

The momentum growing for New Orleans got an additional boost from Pauger's beautiful map, with its neat streets and stout forts, emblazoned with a name heralding the Duke of Orleans. According to historian Marc de Villiers du Terrage, "the regent, god-father to the new capital, was necessarily flattered to see the project put into effect,"[23] really the first good news coming out of Louisiana in three years. "The year 1721 had been generally favourable to New Orleans," wrote Villiers du Terrage. "From a military post, a sales-counter, and a camping-ground for travellers, it had become, in November, a small town, and the number of its irreconcilable enemies began to decrease."[24]

The mood shift was hardly discernible on the ground, where Pauger steamed at the rogue villagers blocking his progress. Worse yet, they rebuked the lordly newcomer—no governor he!—as he confronted their transgressions in August 1721. Enraged at their insolence, Pauger huffed at one man who, he wrote incredulously, "wanted to build as he saw fit, without regularity and without plan." He nearly came to blows with an incensed housewife, and barely escaped a duel with her husband, when his projected straight street intersected with her crooked lot. In September, Pauger nearly got into fisticuffs with another man whose house he razed because it "was not in alignment of the street (having built it before the plan was proposed)." The man, amusingly named Traverse, sought indemnification from the Superior Council, but Pauger's authority prevailed, as did his vindictiveness. "Mr. Pauger sent to find him and, after having regaled him with a volley of blows with a stick, had him put in prison, with irons on his feet."[25]

This went on for a year—reports, complaints, the occasional demolition, additional violations, delays, and mounting criticism. "Pauger . . . has just shown me a plan of this own invention," wrote a dubious Father Charlevoix, "but it will not be so easy to put into execution, as it has been to draw [on] paper."[26] Besides, the priest grumbled, "the country [around] New Orleans, has nothing very remarkable; nor have I found the situation of this so very advantageous," as marshy soils' "depth continues to diminish all the way to the sea. . . . I have nothing to add about the present state of New Orleans."[27]

Good news was on its way. On December 23, 1721, superiors in Paris, partly distracted by the financial restructuring and partly swayed by Pauger's maps, decided to designate New Orleans as company headquarters and colonial capital. It would take some months for word to reach Bienville, and he rejoiced upon hearing it. "His Royal Highness having thought it advisable to make the principal establishment of the colony at New Orleans on the Mississippi River," Bienville beamed, "we have accordingly transported here all the goods that were at Biloxi," the previous capital. "It appears to me that a better decision could not have been made, in view of the good quality of the soil along the river [and the] considerable advantage for . . . the unloading of the vessels."[28]

And what of Pauger and the misaligned houses? Nature solved that problem—in the form of yet another threat. Around 9 a.m. on September 11, 1722,[29] "a great wind" swept the settlement, according to Pauger, "followed an hour later by the most terrible tempest and hurricane that could ever be seen."[30] "With this impetuous wind came such torrents of rain," recollected another colonist, "that you could not step out a moment without risk of being drowned[;] it rooted up the largest trees, and the birds, unable to keep up, fell in the streets. In one hour the wind had twice blown from every point of the compass."[31] The eye of the slow-moving vortex apparently struck directly over the city, and after it passed by dawn September 13, stunned colonists "set to work to repair the damage done."

New Orleans's first hurricane wiped out three years of building, having "overthrown at least two thirds of the houses here," wrote Pauger, "and those that remain are so badly damaged that it will be necessary to dismantle them. The church, the presbytère, the hospital and a small barracks building . . . are among [those] overthrown, without being, thanks to the Lord, a single person killed. . . . The river rose more than six feet and the waves were so great that it is a miracle that [all the boats] were not dashed to pieces."[32] Another witness fretted the loss of "beans, corn, and more than 8000 quarts of rice" as well as the destruction of "most of the houses in New Orleans with the exception of a warehouse built by M. Pauger," a commentary on the engineer's skill.[33]

While Pauger the colonist lamented the disaster, Pauger the engineer could hardly contain his glee. "All these buildings were old and provisionally built, and not a single one in the alignment of the new city and thus would have had to be demolished," he wrote. "Thus there would not have been any great misfortune in this disaster except that we must act to put all the people in shelter."[34]

Relishing the *tabula rasa,* Pauger set forth to lay out his plan. But first he made a major spatial adjustment, one he had mulled back in the spring of 1721. In a letter sent to Paris and copied to Le Blond de La Tour in New Biloxi, Pauger explained "the changes I have been obliged to make because of the situation of the terrain, which being higher on the riverbank, I have brought the town site . . . closer to it, so as to profit from the proximity of the landing place as well as to have more air from the breezes that come from it."[35] That is, the assistant engineer ignored the old Bienville-Pailloux baseline and, while retaining its 37 degree angle, shifted the city seven hundred feet closer to the river. It was a bold move for a subordinate, but his two superiors did not take issue. Learned in hydrology, Pauger realized the river current at this bend did not scour the bank; on the contrary, currents slowed here, dropping sediment and padding the bank with a sandy beach, or "batture." With precious topographic elevation so scarce and slight, why squander the crest of the natural levee? The realignment gave the city two extra feet of topographic elevation, enough to evade the backswamp deluges of 1816, 1849, and 1871 (and, under very different conditions, during Katrina in 2005). Had Pauger not made this change, today's Decatur Street would roughly align with Royal, and the rear fortifications behind today's North Rampart would be nearly at Marais Street. *Marais* means swamp—and for apt reason, as the backswamp in 1720 lay not too much further away.[36]

It was as if Pauger were putting a stick in the mud for New Orleans, figuratively speaking, and when he did so literally—to survey the streets—the era of adapting to or retreating from water came to an end, and the city's drainage history began.

2

The Ditch-and-Gravity Era,
1720s–1830s

ÎLES IN THE ÎLE D'ORLÉANS

Now working with Le Blond de La Tour, who had returned from his detail at New Biloxi, Pauger and his workers "cleared a pretty long and wide strip along the river" for today's Decatur Street, "to put in execution the plan [Pauger] had projected[,] traced on the ground the streets and quarters which were to form the new town, and notified all who wished building sites to present their petitions to the council."[1] Streets got names honoring French royalty, and, according to Jean-François-Benjamin Dumont de Montigny, writing in 1722, "New-Orleans began to assume the appearance of a city . . . and to increase in population."[2]

Now it could begin grappling with the *aquatique* problem, something so pervasive it affected the vernacular. In describing 1720s New Orleans, colonist Le Page du Pratz wrote that "the streets divide the town in to fifty-five isles[,] each of those isles [having] twelve emplacements . . . for lodging as many families."[3] To speak of people living on *îles* in a place called the *Île d'Orléans* is to limn the omnipresence of water in this *terre d'alluvion.* Early documents regularly refer to city blocks in New Orleans as isles, on account of their shored-up spoil barely rising above the dug-down streets. Even the dead occupied an isle, at the cemetery on Rue St. Pierre at Rue Burgundy, where barrow excavated from surrounding ditches was layered upon the square to make it slightly higher and less soggy for subterranean burial. Well into the 1900s, the local vernacular retained two old Creole words reflecting this urban microtopography: *islets,* meaning city squares, and *banquettes,* meaning either raised wooden walkways ("little benches") or the diminutive for the banks (*banques*) of a river.[4]

The century to come, 1720s to 1830s, might be described as the ditch-and-gravity era in the drainage history of New Orleans. The toil targeted not the standing waters of the backswamp—too vast, too distant, too deep—but rather the surface runoff and groundwater on the natural levee, where most people resided and most economic activity transpired.

Motivation for drainage was threefold. First, of course, was human habitability—to minimize nuisance, enable mobility, facilitate construction, and undergird all that entails a city. Next was economics: drainage on adjacent plantations aimed to regulate water levels depending on the crop; commodity agriculture would have been nearly impossible without it. Then there was basic human salubrity, as it was understood at this time. Medical thinking in this era, not yet enlightened to microscopic germs, bacteria, and viruses, relied instead on the sensorial to divine what caused malady. Drawing upon notions of natural philosophy scribed by Hippocrates, physicians perceived human health to reflect a balance of four bodily "humors"—phlegm, blood, yellow bile, and black bile. Insalubrity, then, came from imbalances, which, by extrapolating from the corporal to the environmental, they surmised has something to do with excessive heat, moisture, or decomposition. Swamps in "torrid" climes topped the list of feared environs, as their damp humus was thought to emit noxious vapors, or *miasmas,* from the Greek word for pollution or defilement. It didn't help that swamps were also dark, eerie, mysterious, impenetrable, and impeding of human endeavors. "The hot and humid climate of Louisiana and the colony's low-lying lands," wrote historian Marion Stange, "were exactly what contemporaries considered to be an unhealthy environment," and the proof was all around them. Thousands had perished since the Louisiana project had gotten underway, and by one 1725 estimate, only ten inhabitants out of a New Orleans population of nine hundred had not yet been afflicted by "fever."[5] Rampant sickness and high mortality stigmatized Louisiana and beleaguered every aspect of colonial progress, from the rallying of new investors in France, to the in-migration of colonists and slaves, to construction, resource extraction, and farming. All this cast excess water as not merely irritating but menacing, an "evil," as a later critic described the swamps near New Orleans—a "boiling fountain of death[,] dismal, low and horrid[,] belching up its poison and malaria[,] feeding the living mass of human beings [with] the dregs of the seven vials of wrath."[6] The conviction was as old as civilization: the Persians considered dirty water to be a sin; the Babylonians viewed urban uncleanliness as malevolence; and Christian writers used swamps as metaphors for perdition, such as the Slough of Despond (Swamp of Despair) in *The Pilgrim's Progress.*[7]

If swamp was evil, then drainage was pious, and engineers were saviors, agents of purification who could eradicate miasmas and end yellow fever, cholera, dengue, and malaria ("bad air"). But before New Orleans's backswamp could be eliminated, first New Orleans itself had to be dried, and that called for all hands, from the bottom to the very top of the French empire.

Draining New Orleans started concurrently with the surveying of today's French Quarter. Wrote Dumont de Montigny of the autumn of 1722, as Pauger laid out the streets and squares, "to each settler who [got] a plot," instructions were issued to "leave all around a strip at least three feet wide, at the foot of which a ditch was to be dug, to serve as a drain for the river water in time of inundation."[8] A few years later, having been informed by *ordonnateur* Edme Gatien Salmon of water troubles in this "flat and marshy country," none other than His Majesty, King Louis XV—the original drainage king, ruling from his throne in Paris—obliged his subjects in faraway New Orleans "to dig little ditches in front of their houses, one or two feet in width by a foot or a foot and a half in depth in order to drain off the water that seeps through the levee, when the river overflows, and the water from the rains which are frequent." As for the network of ankle-twisting "ditches cross the streets," the king called for "wooden bridges" to be built, "which must be repaired at least every year."[9]

In 1724, the *Conseil Supérieur* mandated each landholder would be responsible for digging and maintaining ditches fronting their properties, to the standards of the royal engineer, and if they failed, it would dispatch its own laborers and bill the proprietor. Subsequent decrees added a fine to the fee, suggesting that noncompliance was common. In 1726, the *Conseil Supérieur* appointed an *inspecteur de police* named Rossard and stipulated that one of his charges would be policing ditch compliance. These and other decrees make it clear that early streets got two curbside gutters with a crest in middle of the artery, rather than one centralized "French drain" and crests along the curbs. Only a few later-added narrow alleys got center drains, and some remain in place today, on Pirates Alley, Cabildo Alley, Pere Antoine Alley, and Exchange Alley.

In a fortified city, space was limited, and every inch devoted to drainage meant one less for urban development. Gov. Étienne de Périer, successor to Bienville as commandant general, explained to company superiors in 1728 his aim to have New Orleans "surrounded by a ditch which will drain off all the water which causes uncleanliness and diseases," but in a manner "not diminish[ing] the enclosed space of the town," possibly implying the creation of expulsion ponds to serve also as moats. The work would be done "in the winter," when rainfall was low, and by the labor of "twenty days given by each negro. These undertakings will cost the Company nothing or very little, only two men to direct the negroes who work."[10] The exchange implied that drainage, and by extension public health, were colonists' problems, not the company's, and that colonists' problems ultimately became the exertions of the enslaved. And so, from king to colonist to slave, the French set about draining New Orleans.

Fortification and moat construction became a priority following the Natchez Indian uprising at Fort Rosalie, which killed 250 colonists and caused fear colony-wide. The 1731 Gonichon Map shows channels under construction around the city perimeter which appear to have had both defensive and drainage purposes; they had been designed by the Royal Academy of Sciences naturalist and astronomer Pierre Baron, who served as engineer following the deaths of Le Blond de La Tour in October 1723 and Adrien de Pauger in June 1726.[11]

But as the shock of the uprising wore off, the defensive effort slackened, leaving some to ponder whether it had been launched "only to assure the citizens of this city who were extremely alarmed by the bold enterprises of the Natchez." The moat soon went from solution to problem. "In fact these ditches[,] the width of which was to be sixty feet and which were never more than two feet deep in some places[,] are at present filled up in such a way that almost no trace of them is visible." Their "wretched condition" was made worse by yet another blow—"the hurricane that ravaged the colony on the twenty-ninth of last August," 1732.[12]

Disasters were tough on early New Orleans, and early New Orleans was tough on its engineers. "At the end of 1726," wrote historian Gilles-Antoine Langlois, "not a single one of the original Louisiana engineers, who in such a brief amount of time helped to shape the city's development, remained in the colony," having either been reassigned or sent to an early grave by insalubrious conditions.[13] Nevertheless, by the early 1730s, a rudimentary urban drainage system came to be, looking "like a manmade bog with intersections," as historian Lawrence Powell put it.[14]

First, ditches would be dug around each square, and shaped into tapered gutters lined with bricks if available. Second, the barrow would be mounded upon the square and leveled off, plateau-like, creating the "isle" where houses would be built. Gravity would thus send rainwater off the squares, while the flanks of the ditches would filtrate groundwater and dry the soils. Compaction and overburden (the weight of buildings) would squeeze out more soil water and send it into the ditches. Third, the backslope of the natural levee, which drops by about three to four inches per hundred feet, would steer the ditchwater backward, street by street, toward the backswamp. There, the slope weakened to "less than one inch on the hundred feet"—nearly flat, which is why a swamp formed.[15]

How to keep runoff flowing smoothly as it moved across river-parallel streets? One proposal, sketched by Ignace-François Broutin in 1732, suggested steering the *ecoulement des eause* (flow of water) through *fossez* (ditches) fronting the *banquette au tour des maisons pour les gens de pied* (pedestrian sidewalks fronting the houses) into *rigollés* (gutters) dug through each intersection in the form of an X. This cruciform interconnection allowed water levels in any of the four feeder ditches to equalize within subterranean brick chambers and eventually discharge beyond the rear of the city. Broutin also sketched plans for similarly sturdy brick pedestrian bridges over the ditches, each with little Roman arches under which the runoff would pass.[16] A bit ahead of their time, the X ditches and crested brick chambers were probably never installed, possibly because bricks were costly to import and not yet locally produced with sufficient quality and quantity. Rather, open ditches along the river-perpendicular streets were simply extended across the intersecting streets and covered with wooden bridges to allow street traffic to pass, a strategy visible in the 1734 map *Partie du Plan de la Nouvelle Orleans Pour Faire.*[17]

Once the runoff made it off the cityscape, it discharged into the moat beyond the rear fortification (today's North Rampart Street), else dispersed through the forests, accumulated in the swamp, and trickled out Bayou St. John into Lake Pontchartrain. One 1730 map proposed routing the outflow into a circular basin directly behind the city with an outflow to Bayou St. John, an idea that would not come to fruition until the 1794 excavation of the Carondelet Canal (today's Lafitte Greenway).[18]

Ditch and gravity: that's how engineers drained New Orleans for its first century. It worked in theory, but execution all depended on quality, declivity, capacity, precipitation, and maintenance. There were no pumps, no supplemental sources of energy; the slightest perturbation in slope and clearance hindered the meager gravitation force, which caused blockages and overflows—and any problem downslope propagated upslope. Too weak a head, and the water stopped; too strong, and the water scoured and overflowed. Worse yet, colony policy outsourced the ditching and maintenance to completely unqualified property owners, and the supervisory engineers really had no quantitative metrics for system design. Not until the late nineteenth century did researchers work out the mathematical relationships among land area, precipitation, declination, soil type, and system capacity, laying the groundwork for the field of stormwater management. Engineers in 1720s New Orleans were left to improvise.[19]

What resulted was a mess, even a century later. "The soil of the Crescent City," one observer wrote, "is filled with humidity, [even] in the driest time of the seasons . . . to its utmost capacity, at two feet from the surface." So saturated, the soils absorbed very little runoff, leaving the rest to overwhelm the drainage ditches. Engineers tended to dig those ditches with the same width and depth everywhere, instead of expanding their capacity as they ran farther back and drained a larger catchment, all of which led to more overflows. And that's all if the ditches were clean and clear of debris. They weren't. "Shunning the river, the choking gutters send their burdens swamp-ward, littering the angles of pavement with clumps of cotton and wool, heads of barrels, hogsheads sometimes; broken paper-boxes, bits of pasteboard, twine and bagging rope; all of which the ever-thirsty swamp licked, in course of time, into its capacious maw."[20] In effect, New Orleans's early drainage system, like that of the Romans of old, had taken on both stormwater and wastewater removal roles, neither adequately.

Add to this the threat of fluvial flooding via the Mississippi, and the forbidding delta environment even got to Bienville. The city founder and largest *concessionaire* in the region usually struck a tone of stoicism in his communiques, but he betrayed misgivings in a March 1734 letter to the Count of Maurepas: "The river has been very high for three months and has overflowed . . . so that more than half of the lands of the inhabitants are submerged and they cannot do any ploughing. . . . This country is subject to such great vicissitude[;] now there is too much drought, now too much rain. Besides the winds are so violent."[21]

Drainage ditches were like hydrological barometers of colonists' ability to wrest New Orleans from nature. The devices intended to eliminate flooding often ended up causing exactly that—which forestalled efforts to make them better. "The reasons for the settlers' negligence were not only laziness or indifference," wrote historian Marion Stange. "Some colonists viewed the drainage ditches as additional sources of infection rather than as a means of protection from disease and epidemics."[22]

In truth, urban drainage, like levee-building, cannot be outsourced to amateurs and enforced by policing and penalizing. Deceptively simple, these hydrological interventions must be done by experts, top-down, through good governance, funded by a regular revenue stream and maintained by skilled full-time workers. Such misapprehensions were not exclusive to New Orleans; Paris, London, and other cities in this era also tried to send the task of drainage and sewerage downstream to the masses, only to see the problems back up and overflow.[23]

ADAPTATIONS, INTERVENTIONS, AND CONCESSIONS

If nature proves too much for human engineering, people's initial perception of a problem transforms to a grudging acceptance of a condition, and the expectation of a solution gives way to pragmatic adaptation. During New Orleans's initial decades, that adaptation was architectural. The city's earliest buildings largely resembled their counterparts in France, raised on brick piers only a few inches above the mud, their cross-timber walls of *bousillage* unprotected from heavy rainfall, and their steep roofs adjoining walls with hardly any overhang. Architectural historian Samuel Wilson Jr. described these early French colonial buildings as looking almost "medieval," and they hardly made sense for this subtropical climate and delta geography.[24] Builders built what they knew, and they were all new here.

But over the next few years, builders gradually began to adapt to Louisiana conditions, while retaining fundamental French design sensibilities. What resulted was a second generation of "creolized" buildings, embodying experimental tweaks and improvements based on learned lessons. Most importantly, they were raised higher on brick piers, not by inches but many feet, as if in expectation of occasional flooding. Cross-timber walls were covered with clapboards and later stucco, and roofs became oversized, pavilion-like, which made space for an airy gallery or verandah, while their steep double-pitch hipped shape and overhanging eves deflected rainfall away from the walls. To wit, the new manager's house at the King's Domain (formerly the Company Plantation, now Algiers Point) looked like what was swiftly becoming "a typical . . . Louisiana plantation house," wrote Wilson, "illustrat[ing] the evolution of a regional style required by the exigencies of climate."[25] Other modifications included plenty of apertures for ventilation, center chimneys, and an outdoor staircase to save indoor space while capitalizing on the spacious gallery.

The adaptations caught on, and Creole builders replicated them throughout French Louisiana. The "plantation houses . . . in this part of the world would appear to have been built, all of them, after the same model," wrote a visitor a full century later, sketching alongside his journal entry an archetypal example with a wrap-around gallery, hipped roof with double pitch and center chimneys, all raised on piers. "The climate requires [that] the building is surrounded with galleries[,] for shade," he noted, and "there are no cellars, for two feet digging brings you to water."[26] Well into the twentieth century, New Orleans architecture, regardless of style or typology, reflected design responses to wet soil, possible flooding, heavy precipitation, and high temperatures. Only when local society gained confidence that these once-tolerated conditions had become solved problems did this responses slacken—and the results were disastrous.

While adaptive design kept water away from structures, it did nothing for feet on the street. Drainage remained an imperative, but who would pay for it? Not the Crown, nor the Company of the Indies (the new name for the Company of the West after its financial restructuring), as both institutions viewed drainage as a colony concern. So colonists themselves would bear the cost, and pass the toil to their slaves, either by directing them personally or providing their labor to colony administrators.

In one case, the aforementioned Salmon, the colony's financial officer and paymaster, proposed to construct brick-lined drainage ditches and brick bridges with revenue yielded by a tax of "five livres per head of negroes." (King Louis XV said no.)[27] In most other cases, white inhabitants were required to lend "each negro" they owned to the company for "thirty days of statute-labor for the public works," including drainage. The French called it *corvée*—that is, owned bodies put to public chores, like a levy payable in humanity. "Several inhabitants have begun to furnish them," reported Governor Périer, "to cut down the trees at the two ends of the town as far as Bayou St. John in order to clear this ground and give air to the city and to the mill," a reference to the *moulin a vent et a cheval* (wind- and horse-mill, probably for rice) that Pauger had designed and erected at what is now the Canal Place mall.[28] "We are expecting to have the canal begun in the direction of the mill which will go to the said bayou and which will drain off all the water."[29] That canal may have referred to a vision, circulated since 1725, for "a canal communicating between the river and Lake Pontchartrain"[30] for both drainage and navigation purposes. That river-to-lake waterway, proposed repeatedly for a century to come, was never actually dug, but the plan later gave Canal Street its name, and it was enslaved Africans who first cleared its space. The larger concept of the river-to-lake canal finally came to fruition fully two centuries later, with the excavation of the Inner Harbor Navigation (Industrial) Canal across the Ninth Ward during 1918 to 1923.

While slaves working under *corvée* dug miles of ditches to drain New Orleans, their brethren dug similar channels on that "tongue of land," the natural levee, "bounded in front by the river and in the rear by . . . very low cypress swamps" which were flooded "three-fourths of the year."[31] There, on that higher ground, administrators dispatched *arpenteurs* to survey so-called French long lots, elongated parcels typically 40 to 80 arpents deep (an arpent measuring 192 English feet, thus 1.5 to 3.0 miles) to be granted to empowered colonials (*concessionaries*) to establish plantations. "By a peculiarity perhaps unique, the highest places in all these lands are the banks of the Mississippi," explained an observer later in the colonial era. "This high ground provides the only means they have of establishing plantations; and it generally consists of good soil, rarely with too much clay[,] often loamy with an adequate mixture of sand. . . . [Thus], concessions of land are granted in arpents measured fronting on the waterways and going back as far as possible to the muddy lands at the rear."[32]

The French long-lot cadastral system was essential to colonialism itself—that is, to the seizing of land to create wealth and enlarge the empire. *Concessionaries* were the selected stakeholders to initiate this process; concessions were where it would happen; slavery would provide the workforce; and delta hydrogeology was what imbued these particular concessions with prospective value. The "great advantage" of the river, wrote Gov. Étienne de Périer in 1731, "is that each inhabitant enjoys the same convenience as his neighbor," in terms of transportation access to fertile land, and the long-lot surveying method sliced up fertile land evenly and conveniently.[33] The original valuation of all future wealth production rested squarely on that initial titling—be it for extracting timber, growing crops, raising livestock, leasing land, or using it as collateral to raise capital. *Concessionaries* became the first landed gentry in Louisiana, and the forefathers of the future property-owning class. Thus, the long-lot concessions, the paperwork backing them up, and the surveying logic behind them, all figured heavily in creating the Louisiana society, economy, and landscape we have today.

But the lands themselves, even the highest, were of little value until their water content was lowered. Without drainage, lands in deltaic Louisiana could be used to hunt game, harvest timber, gather firewood and moss, and if the land was cleared, maybe graze livestock and plant rice—that's about all. But *with* surface drainage, land could be used to build communities, erect facilities, grow vegetables, cultivate export crops, raise animals, create industries, and sustain larger populations—in sum, to host a society. In this environment, *drainage was nearly as intrinsic to the concept of "original value" as landownership itself.* Likewise, slavery was as critical to building those drainage systems as drainage was to raising land values. Governor Périer said as much in 1731, when he explained to his superior in Paris that "from sixty leagues above New Orleans to the lower part of the river"—that is, the deltaic plain—"the lands can be drained and freed from water only by those who have negroes, since the work on levees and drainage is difficult and hard. Even though a man were not sick, no matter how good a settler he may be, in an entire year he would not put one *arpent* of land in condition to be planted." How much more valuable would slave-drained land become? "Infinitely more than those of the upper part of the colony," wrote Governor Périer.[34] That assessment remained the same two decades later. "The settlers [in Louisiana] are all asking for negroes and really cannot succeed without that," wrote a marine official to the secretary of the navy in 1752. "[S]end here good peasants, farmers, and decent people, and a supply of negroes.... You know, my lord, that one must sow in order to reap."[35]

The goal of agrarian drainage was to modulate water content and organic matter in loamy soils in accordance with the intended use of the land. Colonists were new to this environment, and every year brought different conditions. "We are all pupils who are learning every year what the climate and the land require," wrote Governor Périer in 1730, acknowledging that "the Company is impatient to see the success of this colony, [and] with a little patience and time it will ... begin to develop."[36] Water content first had to be reduced just to enable the breaking of sod and furrowing of rows. Afterwards, agrarian drainage depended on the crop: more water was needed for rice, less for sugar cane; more water for tobacco, less for indigo. Drainage systems could also help during droughts, by serving as irrigation systems by means of gated sluice flumes cut through the river levee. This is how rice fields were flooded—by reversing drainage canals, and inundating the water-tolerant crop so as to kill water-intolerant weeds. But a rice planter's solution became a sugar planter's problem: "*water* is the greatest enemy of the cane," wrote a later advocate of Louisiana sugar; "you must have sufficiency of ditches to carry off rain water at once. No money spend in ditching is thrown away, the more ditches you have the better."[37]

Agrarian drainage worked like a rural version of urban street gutters: ditches were dug parallel to the river to intercept runoff, draw out groundwater, and steer it into wider lateral ditches perpendicular to the river. There, it would flow down the backslope of the natural levee, from a crest (*écore,* or high riverbank) of ten to fifteen feet to the backswamp, a foot or so above the brackish tidal lagoons of "Lakes Maurepas and Pontchartrain on one side and on the other by those of the Washas," today's Lake Cataouatche and Lake Salvador on the West Bank.[38] The drainage networks iterated the geometry of the long lots, with main outflow ditches coinciding with property boundaries, and together they established a spatial framework within which most human life would play out. To this day, the humanized landscape along the lower Mississippi fits perfectly, almost eerily, into the sleeve of the centuries-old forty-arpent and eighty-arpent lines, and the former plantations' boundaries therein are still typically scored by drainage ditches.[39]

Land concessions were privileges, but they came with responsibilities, and drainage topped the list, preceding levee building and following only land clearing. In a typical arrangement, Governor Bienville wrote in 1726 of a "good settler" named Rivart who offered to clear out Bayou St. John in exchange for title to bayou-side land, "which he would drain by ditches in order to make pastures." Drained land, productive pasture, a navigable bayou: the governor deemed this offer to be so "advantageous" and "so inexpensive for the Company, and of such great assistance to New Orleans, that I do not think it ought to be rejected."[40] Governor Bienville later wrote to the Count of Maurepas in Paris about the "idea of the distribution of the lands along the river . . . to forty arpents as far as the cypress swamps," adding hopefully that "some of these cypress swamps . . . could be drained in the course of time and be made into meadows."[41]

Given the difficulty of digging, planters were inclined to get multiple uses out of their drainage ditches. If widened, ditches could be used to float cypress logs cut from the swamp up toward the river to be milled. A little wider, and they could become a channel to float a raft loaded with backswamp clay to make bricks, or shells from an old Indian midden to make lime for mortar. A pirogue could be paddled down for hunting and fur trapping. And if widened even more and paralleled with a towpath, a drainage canal could become a navigation canal; conversely, a navigation canal could double as a drainage canal, if properly sloped and gated to prevent backflow into feeder drains. The towpath could become a toll road—another source of income—and the topography it created could support fruit trees or other profitable plants. Hoping to launch a silkworm and silk-making industry, colonial administrators once encouraged *concession-*

aries to "plant mulberry trees [along] the drainage ditches that each one will have dug on his land."[42] Today's greater New Orleans landscape retains imprints of numerous such water-management projects of centuries ago, among them the Marigny, Carondelet, Gormley, New Basin, Company, Harvey, and Algiers canals—all of which were both drainage and navigation canals at some point in their histories.

CLAUDE JOSEPH VILLARS DUBREUIL

An early maestro of the multipurpose drainage canal was landholder, ship-builder, and building contractor Claude Joseph Villars Dubreuil Sr. In the late 1730s, Dubreuil directed his slaves to dig a twenty-five-foot-wide channel across his Barataria plantation forty arpents back to the densely wooded swamp. First of its size on the West Bank, Dubreuil's canal drained his croplands while enabling the extraction of cypress logs to build twenty vessels to fight the Chickasaw War.[43] Later in the 1700s, timber extracted from this and other multipurpose canals enabled an industry in ship lumber, masts, planks, stave wood, box boards, and other wooden products for export to Cuba and the United States.[44] Years later, Dubreuil's canal came into the possession of Antoine Foucher and François Gardère, by which time it was known as the Gardère Canal. As a landscape feature, the old channel is gone today, but hydrologically it lives on in the form of the Gardere Drainage Canal in modern Gretna and Harvey.[45]

Dubreuil owned another plantation on the East Bank, where his endeavors are also imprinted in the modern cityscape. In the early 1740s, he gained title to a valuable parcel immediately below New Orleans. Needing lumber for construction projects in the adjacent city, Dubreuil floated cypress logs from his West Bank plantation over to this East Bank holding. There, sometime before 1753, he had his slaves excavate a canal to power a *moulin à planches* (sawmill) by diverting high river water to turn a waterwheel about two hundred feet inland.[46] Risky as they seem, such mill races were not uncommon. An English spy who surveilled Louisiana in 1770 noted, "many of the planters have saw mills, which are worked by the waters of the Mississippi in the time of floods, and then they are kept going night and day till the waters fall."[47] A decade earlier, the Company Plantation at present-day Algiers Point, which had a depot, forge, brickworks, croplands, slave encampment, and hospital, also featured a prominent rice mill and sluice flume, through which high river water was diverted to turn a millstone to separate bran and grain.[48]

Dubreuil's mill race bisected his wedge-shaped plantation, which was aptly positioned at a sharp river bend where current velocity and flow direction were optimal.[49] As the water flowed back toward Bayou St. John, the race became a drainage outlet for the crops Dubreuil's slaves raised elsewhere on his plantation. It's possible that lumber hewed at Dubreuil's mill remains inside the Old Ursuline Convent, built by Dubreuil in 1752 and still standing at 1111 Chartres Street.

In 1805, engineers used the trajectory of Dubreuil's mill race to become the central artery and drainage canal of the newly created subdivision of Faubourg Marigny. In 1831, that trajectory, today's Elysian Fields Avenue, was extended to Lake Pontchartrain for the track bed of the Pontchartrain Railroad, and in the 1900s, it became the axis to which many modern Gentilly subdivisions would be oriented. In this manner, water-management decisions made in the 1740s became inscribed into the built environment of the 2000s, influencing the lives of New Orleanians in nearly every movement of their everyday lives.

Dubreuil's work is but one such example. Nearly every metro-area neighborhood has comparable hydrological backstories, be they traceable to the 1900s, 1800s, or 1700s, and credited to skilled engineers or subjugated laborers. According to historian Gwendolyn Midlo Hall, Dubreuil owed some success to his "African slaves' technological knowledge—how to dam and control the waters of the rivers and bayous." That may also be said of many early drainage systems, and the development that followed. If we call it a "built environment," due credit must go to the builders.

Since 1754, a series of battles that became known as the French and Indian War had raged between French and British colonists, each side allied with various native nations. After seven years of conflict, King Louis XV of France realized a British victory was imminent, and in late 1761 endeavored to make the best of the aftermath. Early the next year, in one of the more consequential letters ever addressed to Louisiana, the king informed Governor Kerlèrec of his recent secret treaty made at Fontainebleau, in which the monarch "ceded a part of the province of Louisiana to the King of England," meaning the eastern side of the Mississippi River. Then came the shocking news: "I have decided to give the other to my cousin the King of Spain." Subsequent communiqués indicated no one in the colonies, French or Spanish, completely believed the news. But true it was, and on November 13, 1762, King Carlos III of Spain wrote back to his "very dear and dearly beloved cousin," accepting "the country known under the name of

Louisiana as well as New Orleans and the island on which this city is situated," a reference to the Bayou Manchac distributary, which, as depicted on maps, made the East Bank seem enough like an isle to justify not ceding the city to the hated British. Once again, water dramatically affected the destiny of New Orleans.[50]

The goodwill between France and Spain did not extend much beyond the two Bourbon cousins. It would take another seven years, along with acrimony and some bloodshed, before Spain fully took control of Louisiana in 1769.

In the three decades to come, Spain administered *Luisiana* in an ambivalent manner, seemingly distracted by other priorities but at times acting with intentionality and vision. The new regime got off to a violent start, as Gen. Alejandro "Bloody" O'Reilly swiftly suppressed an insurrection among French loyalists with executions and property confiscations. Yet it readily hired French Creole engineers and bureaucrats, and its own administrators did not hesitate to marry into the local francophone aristocracy. In regard to race relations, Spanish leadership allowed for *coartación* (manumission through self-purchase) among slaves, giving the hope of freedom to bondsmen and helping create an influential caste of free people of color. Yet it also revived the African slave trade, this time drawing from the Congo region (as opposed to the Senegambia region, where the French had drawn from primarily in the 1720s), and cracked down mercilessly on escaped slaves (*cimarróns,* or maroons) seeking refuge in the swamps and marshes.

If the Spanish *dons* had one overriding goal in *Luisiana,* it was to maintain the colony as a *barrera* (barrier) between the British foe to the east and the prized *Nueva España* to the west. That required peopling the vast expanse, by issuing generous land grants "free of charge," wrote geographer Adam Mandelman, "so long as they adequately leveed and drained the parcel within a three-year period."[51] Among the recipients were *les Acadiens* ("Cajuns") exiled from French Canada, British Tories fleeing the revolt in the Thirteen Colonies, *Granadinos* and *Málagueños* from the Spanish peninsular, and *Isleños* from the Canary Islands. The drainage and leveeing they oversaw on their Spanish land grants laid the groundwork for much of south Louisiana agriculture and the communities where their descendants live today.

THE VERY ILLUSTRIOUS CABILDO

In New Orleans proper, the Spanish found something of a French Caribbean village—Gallic in tongue, Creole in culture, bucolic in ambience, low in density,

Spanish changes

mostly wooden in its edifices, dusty when dry, and muddy when wet, which was most of the time. But changes were afoot, for one of Spain's lasting legacies in New Orleans would be its emphasis on urbanization.

Whereas the French oversaw Louisiana remotely through a Superior Council—thus all those letters from Bienville to Paris—the Spanish, namely Governor O'Reilly, formed a local advisory council of *regidores* (councilors) *1799* known as the Very Illustrious New Orleans Cabildo. The building where they met effectively became "city hall," and the circa-1799 Cabildo, later the New Orleans City Hall and now a museum, still stands fronting Jackson Square.

Because city problems affected them personally, *regidores* were inclined to engage them, and they became a sort of *de facto* city council and planning commission (minus the democracy). Their deliberations in the *Actas de Cabildo* are a fascinating and sometimes amusing registry of familiar controversies and complaints: about loud cabarets, code violations, steep fines, unpopular land-use rules, proto-zoning variances, malodorous fishmongers, hunters peddling spoiled meat, inadequate nighttime illumination, and potholed streets. Because *regidores* could only advise the governor, not all they decreed and recorded got actualized, making the *Actas* more of a sounding board than a leger of executed projects.[52]

Muddy streets were a common complaint, and in many ways, the Spanish Cabildo picked up where the French Superior Council left off, devising ad hoc drainage solutions and berating citizens for not embracing them. *Regidores* spoke at length of *puentes* (bridges), wooden planks extending the width of the *banquettes* such that they bridged over the deep ankle-breaking gutters, "thus avoiding the odor of corrupted and stagnant waters." *Puentes* were especially critical where gutters crossed the river-parallel streets, as they allowed traffic to pass over smoothly without damaging the ditches. As in French times, the Cabildo tried requiring property-owners to construct the *puentes* nearest their houses; in fact, in the early years of the Spanish regime, Governor O'Reilly "specifically forbade the Cabildo to maintain the gutter system with public funds."[53] What resulted was an inadequate and fragmented system, starved of official commitment. Only later did the widespread noncompliance force the Cabildo to levy a tax on the owners and in 1785 propose a tax on carts (two *reales* per month), so that publicans (contractors) could be hired to do the specialized work. "The financial burden for maintaining the city's gutters was considerable," wrote historian Gilbert C. Din, costing "slightly more than five pesos to repair each existing gutter and about thirty pesos to construct each new gutter."

[Handwritten margin notes: "1774", "6 Hurricanes", "2 fires", "1782", "1792–94", "3 tropical storms"]

Ditching projects came in spurts, with up to sixty gutters constructed at once, costing 1,750 pesos.[54]

Street gutters were not the only excavations needed. As the levees grew more effective in constraining the Mississippi, water pressure increased on their soil mass, leading to percolation and lagoon-like accumulations. In 1799, the attorney general called for a ditch to be dug along the foot of the levee, to be tied into the street gutters for a backward flow beyond city limits. This phenomenon, of river water making it into the drainage system, persists today: at any given moment, "constant duty" pumps remove a certain amount of water which had never fallen from the sky or flowed across the land.[55]

Publicans also regraded streets, by using pickets and a special level to attain the proper inclination. Some residents took it upon themselves to fill their street's own annoying potholes, as residents are wont to do today, but because they dug soils nearby, they caused "pools [to] be formed and the water [to] become stagnant which is detrimental to the public health." The Cabildo thus decreed they "bring [earth] from the outskirts of the city, instead of . . . the central section . . . for repairing the sidewalks."[56] Some folks obeyed and carted home batture sand to make "their own ditch and sidewalk," but they often did so "without paying attention to grading it to the proper level of the city," which brought matters back the original problem—and called for more publicans to the rescue.

The Cabildo also created the city's first street-sanitation service, providing wagons, fodder for beasts of burden, and funds to be paid to slaveholders to hire out their enslaved workers to collect refuse. But the horses and mules also fertilized the streets, as did the occasional cow and pig, which, along with puddles, potholes, overflowed ditches, and splintery *banquettes,* made a walk across town hazardous to one's health, never mind the miasmas.[57]

So went the war on mud in late-1700s New Orleans, made all the worse by six hurricanes between 1776 and 1782, followed by two catastrophic fires (March 1788 and December 1794) which together destroyed over a thousand buildings and clogged the drainage system with ash and debris. The Cabildo authorized funds to rebuild the gutters in June 1788, after the first fire, but left that task to property owners after the 1794 blaze.[58] Once again, hyper-local outsourcing failed, and officials later noted the "lack of proper sewerage and the amount of stagnant water in the streets due to the fact that after the fire each individual made his own ditch and sidewalk without attention to the proper level."[59] To make matters worse, three additional tropical storms struck during 1792 to 1794.

CARLOS TRUDEAU

Amid the disasters, opportunities arose. Within days of the 1788 blaze, Madame Maria Josefa Delondes Gravier and her husband, Bertran Gravier, decided to subdivide their plantation immediately above the commons by the upper fortifications. These twelve arpents had once been part of the former Jesuit tract, and previously Bienville's own land grant, described by a visitor in the 1750s as "the finest plantation in the colony."[60] Like all such lands in this era, the tract had an agrarian drainage system in place by the time Madame Gravier inherited it in 1785.

The task of designing and laying out the subdivision went to Spanish city surveyor Carlos Laveau Trudeau, who labeled the subdivision the *Suburbio* or *Arrabal* (suburb) *Santa Maria,* or Faubourg Ste. Marie to the French. Occupying the same natural levee as the original city and subject to the same soil-water conditions, New Orleans's first suburb, now the Central Business District, would get the same ditch-and-gravity approach to drainage, from Common Street up to roughly Howard Avenue. Some lower sections may have drained into the moat along the fortification, but most would have discharged into the woods behind *Calle de la Barona* (Baronne Street) and either filtrated into the soil, ponded in the backswamp, or drained out Bayou St. John.[61]

The 1788 move by Gravier and Trudeau, which capitalized on a disaster by creating new space for rebuilding, initiated a process for urbanization that would, in time, utterly transform the metropolitan map. In the century ahead, urban expansion in New Orleans would fit within the preexisting spatial framework of long-lot plantations, usually a dozen or so arpents in frontage and forty to eighty arpents in depth, a geometry that was iterated by agrarian drainage systems. One by one, plantations became neighborhoods, and the hydraulics needed for farming shifted to those needed for development. In effect, the future urban geography of New Orleans was preconfigured by the needs of enslaved commodity agriculture as devised for a fluvial delta.

A few months later, the Cabildo deliberated another land-use problem: the old French burying ground on St. Peter Street, by now a fetid nightmare. "[Such] a great number of people [are] buried in the cemetery," read the *Actas,* "that there is no room for any more; and at the time of digging the graves, the remains of other deceased are found, which not only cause annoyance but [also] bad odor, which due to the proximity to the City may be the cause for infection [and] epi-

demics of disease. [Thus] it was deemed necessary to establish another cemetery, located further from the City."[62] Importantly, the *regidores* did not explicitly
identify groundwater as a main problem. So long as it's on the natural levee, and
the standard ditching-and-mounding isle-making treatment was performed, it
was possible to bury below-grade in New Orleans; the French had done so for
years. Rather, the *regidores* cited overcrowding, malodor, and public health as
the problems.

Their solution—the city's now-famous above-ground tombs and mausole-
ums—is often upheld as Exhibit A for how New Orleans historically responded
to a high water table. But while above-ground tombs did indeed reduce the need
to ditch and drain, their original primary motivation was to increase capacity
(you can store many more bodies in a multi-story tomb than a grave of the same
footprint) and to free up scarce urban space amid the wreckage of the 1788 fire.
Serendipitously, the Spanish necrotic custom of above-ground entombment offered the perfect solution: culturally appropriate, geographically rational, hydrologically convenient, and architecturally appealing. In 1789, Spanish authorities
created a new cemetery outside the fortifications, pushing this unwanted land
use backward, away from the city, onto slightly lower and wetter ground. The
cemetery was not alone in this regard; backward went anything purged from the
city proper, even the leper colony, to "a quantity of land in the rear of the city . . .
appropriated for . . . that loathsome malady."[63]

In the new cemetery, masons erected brick tombs and arranged them along
walkways, much like buildings along streets, and a "city of the dead" arose, superior in every way to the old graveyard on St. Peter Street. Later, St. Louis Street
would be extended adjacently to this new cemetery, by which time a second
space would be created (1822) four squares back. They would become known
as St. Louis Cemetery No. 1 and No. 2, both owned by the Catholic Archdiocese
and still active today. The above-ground cemeteries have fascinated visitors ever
since, and most come away with the overly reductionist understanding that soil
water was their sole motivator, whereas it was actually one of three, and not the
first. Among them was Scottish James Edward Alexander, who visited in 1833
and wrote:

> New Orleans is called the "Wet Grave," because, in digging "the narrow house,"
> water rises within eighteen inches of the surface. Coffins are therefore sunk
> three or four feet, by having holes bored in them, and two black men stand on
> them till they fill with water, and reach the bottom of the moist tomb. Some

people . . . dislike this immersion after death; and, therefore, those who can afford it have a sort of brick oven built on the surface of the ground, at one end of
which, the coffin is introduced, and the door hermetically closed, but the heat of
the southern sun on this "whited sepulchre," must bake the body inside, so that
there is but a choice of disagreeables after all.[64]

HÉCTOR DE CARONDELET

Eyeing potential enemies in the form of the British and later the Americans, the
Spanish strove to complete the city's long-neglected defensives, building *reduc-
tos* (redoubts, or bastions) named Borgoña, San Fernando, San Juan, San Louis,
and San Carlos into the ramparts surrounding the city. The earthen berms were
fortified with timber *estacados* (stakes) and fronted by moats thirty feet wide and
four feet deep, which were fed by the ditch-and-gravity drainage system of the
city therein.[65]

The man pushing the upgrade was Don Francisco Luis Héctor, Baron de
Carondelet, a capable Walloon in service of Spain who had the diligence of Bienville, the perspicacity of Pauger, and the popularity that neither of his predecessors enjoyed. Transferred from San Salvador in the Province of Guatemala to
govern Louisiana and Spanish West Florida (the latter attained after the British
lost their colonies in the American Revolution), Governor Carondelet arrived
in January 1792 and assessed New Orleans's panoply of problems. Over the next
five years, he intervened effectively on ward administration, street lighting, policing (night watchmen), sanitation, and other city services, while easing rules
on slavery to prohibit work before dawn or after dark, and to permit Sunday
off days.[66] He also helped launched the city's first newspaper, *Le Moniteur de la
Louisiane.*

Drainage became a priority for Governor Carondelet. His administration
hired two professional engineers to grade the gutters, Gilberto Guillemard (who
later designed the Cabildo) and Nicolas de Finiels (who later laid out the Faubourg Marigny). It also adopted a procurement system for ditch maintenance
and construction to encourage bidders, and hired the American contractor Ro-
berto Jones (who built Madame John's Legacy in 1788, still standing on Dumaine
Street) and Bartolomé Lafond (Barthélemy Lafon) to do the work.[67]

Operational problems persisted, and to the extent that they were tactical in
nature, Lafon proposed to line the gutters with locally made bricks or imported
stone, rather than wood planks. Governor Carondelet sensed that the inade-

quacy was more systemic, as evidenced by the narrow curving outlet by which urban runoff flowed to Bayou St. John, if it made it out of the city at all.[68] In contemplating this and another water-related problem, Governor Carondelet devised a solution that would address both issues within the same space.

That other problem was the two-mile-long Bayou Road portage to Bayou St. John and Lake Pontchartrain, whose watershed abounded in natural resources such as timber, tar, pitch, shells for mortar, clay for bricks, game, finfish, and shellfish. Why not provide an alternative to that old muddy Indian path by digging a drainage canal to discharge urban runoff into the bayou, and when finances allowed, widen and deepen it into a navigation canal? Schooners and tow barges could then bring those products directly to New Orleans, while gravity could "rid it of the stagnating waters which contribute in a great degree to its insalubrity and the vast quantities of musquitoes [sic] which render it unpleasant in summer." Such dual-purpose infrastructure "brings to view the future greatness of the city" in terms of "encreasing commerce, and presses the necessity of opening a communication with the sea thro' the lakes."[69]

An announcement of the project appeared in Le Moniteur on May 24, 1794, and probably reflected the sentiments of the governor himself, because, as a later court document pointed out, that newspaper was "printed under the eye of the Baron de Carondelet."[70] In June, after communicating with his superiors in Spain, Governor Carondelet moved forward with his drainage and navigation plan, both of which promised to be the largest to date locally.[71] His superiors approved the project but did not fund it, so the governor resorted to the labor of one hundred presidios (convicts) and sixty slaves, thanks to a request to "the inhabitants of the city such numbers of negroes as they could spare to cut down trees."[72]

Phase one of the project began in late 1794, as workers cleared vegetation and dug a 1.6-mile-long channel from today's Basin Street by St. Peter Street, and heading straight back to Bayou St. John, six feet in width, nearly as deep, and with a declivity of "less than one inch on the hundred feet."[73] (It was during this time that the second great fire, on December 8, charred the heart of the city a few blocks away.) Phase two, in 1795, involved 150 enslaved men who widened the channel to fifteen feet, shored up the banks into guide levees with towpaths, and erected a drawbridge over Bayou St. John so schooners could pass. Plans called for "banquettes [to be] planted with rows of trees [to] afford an agreeable promenade," indicating that authorities hoped to get multiple benefits out of this infrastructure. Indeed, as early as summer 1795, they noted a "marked diminution of mortality which prevailed [last] September and October," thanks to "the disgorgement of waters which stagnated behind the city."[74]

The canal, even at this early stage, was almost too effective as a drain. "In heavy rains," one witness said, runoff "brought so much dirt that the canal would soon be filled up." Governor Carondelet himself intervened, instructing workers to install long wooden boxlike gutters on either side of the channel, into which surface waters would flow for temporary storage. After the sediment fell out, a worker would open a valve to discharge the clear water into the canal, and clean out the residue with a shovel. This specially appointed "keeper" ranks among one of New Orleans's first drainage-system employees.[75]

Later came the canal's expansion for navigation purposes, for which a turning basin (thus Basin Street) was built nestled into the corner of the *Fuerte San Fernando* bastion and *glacis* (rampart, now North Rampart Street), an aperture known as the San Fernando Gate. This task called for another donation of enslaved labor from "planters and citizens," to the tune of "one negro each for three days—a service of little moment," authorities assured, because it "will rid them totally of the stagnating waters and consequently of the sickness common in the fall—while it would allow the completion of the port for schooners . . . to come up to the city." The city's first mayor later testified that "the inhabitants furnished their negroes cheerfully," creating a workforce of "160 to 175" on average. "The negroes dug and the convicts carried away the dirt."[76] Maps strongly suggest the turning basin and moat were one and the same body of water, and served as an outflow reservoir for the urban drainage system—all this just a block from the new Spanish cemetery.

In August 1796, the Illustrious Cabildo dedicated the "Canal Carondelet," with a marble marker placed at the gate to hail New Orleans's latest drainage king, and the new dual-use infrastructure went into action.[77]

But the next year, Spanish superiors decided to transfer Governor Carondelet to Quito, Ecuador. Having lost its champion, the Carondelet Canal fell into disrepair, and within a few years it became an "unwholesome morass," its "depth, which once reached seven to eight feet, was [by 1802 only] two to three feet."[78] (The Carondelet Canal would later get a second life as a navigation asset, only to return to a quagmire a century later, by which time it was known as the Old Basin Canal. The city had it filled in the 1920s, and in the 2010s, advocates had the grassy swath landscaped into the Lafitte Greenway linear bikeway. It still serves a drainage function, hosting an important concrete culvert connected with Pumping Station No. 2 to relay runoff out the London Avenue Outflow Canal. The slight indentation of the original Carondelet Canal can still be discerned near its junction with Bayou St. John.)

Monumental geopolitical changes brewed in the waning years of the eighteenth century. A revolution among discontented British colonists had launched a new nation on the Eastern Seaboard, the United States of America, and enabled New Spain to encircle the Gulf of Mexico by shifting West Florida from British to Spanish control. Americans would soon press westward into their new territories, over the Alleghenies, down the Ohio, and into the Mississippi Valley. They fought and displaced Indians, cleared forests, planted crops, raised livestock, established a frontier-exchange economy, and eventually developed a surplus of produce and a deficit of currency. What they needed was an export market, a source of revenue and imported goods, and for that they needed a transshipment port. They needed New Orleans—mud, floods and all.

Across the Atlantic, discontented French subjects toppled the Bourbon monarchy and eventually set Napoleon Bonaparte on a pathway to power, while in the Caribbean, a caste war erupted in the lucrative French sugar colony of Saint-Domingue and developed into a full-blown slave insurrection. Napoleon had dreamed of reconstituting New France, starting with Louisiana and ending with Saint-Domingue, to which he dispatched troops to quash the rebellion. To that end, he successfully pressured Spain, overextended and fighting its own insurgencies, to secretly retrocede Louisiana to France in 1800. But in the effort of regaining control of Saint-Domingue, French troops fell to yellow fever, and the insurgents went on to launch the independent Republic of Haiti. His best-laid plans subverted, Napoleon lost interest in Louisiana and found his empire in debt, even as his eyes turned to Great Britain and the European theater, where he would soon wage the Napoleonic Wars.

All this made Napoleon a prospective seller on the colonial real estate market, particularly upon learning of the Americans' keen interest in acquiring New Orleans, that once-orphaned entrepôt now so vital to their commercial interests. While negotiating with American diplomats in Paris, Napoleon stunned his guests by offering them not just New Orleans but all of Louisiana. What better way to keep Louisiana out of British hands than by selling it to the Americans, who would surely fight their former colonizers, leaving Napoleon to battle Great Britain itself?

The Americans pounced at the astounding offer, and over months, negotiated the details of the Louisiana Purchase, culminating with a final transfer cer-

Louisiana Purchase Dec 20 1803

emony at the *Place d'Armes* in New Orleans at midday December 20, 1803. "The French colors were lowered and the American flag was raised," recalled Louisiana's last French colonial prefect, Pierre Clément de Laussat, as he "handed over the keys to the city."[79] The interregnum of 1800–1803 was over; this was American mud now.

It would take years to effect, but that transaction, starting that sunny noon in late 1803, would irrevocably shift the vectors of cultural influence upon New Orleans. In colonial times, the city operated as an apogee in the Franco-Hispanic/Afro-Caribbean world of the south Atlantic, a far-away cog in an Old World imperial machine that was monarchical, mercantilist, and enslaving. Now New Orleans would become a vital entrepôt in an expanding New World republic, one that was selectively democratic, unabashedly capitalistic, astonishingly innovative, dashingly confident—and also enslaving.

Nearly every aspect of local life began to change. American administrators assumed authority. Anglo-Americans began to migrate in. Ministers brought Protestantism to Catholic parishes. Jurors introduced English common law into local Roman civil jurisprudence. The English language made inroads in discourse once dominated by French. American surveyors laid out orderly township-and-range grids all around serpentine French long lots. American architects began to introduce Neoclassicism to Creole design idioms. And American engineers joined their local counterparts in the renewed imperative of dewatering.

Eyeing profits, landowners around New Orleans readied themselves for the forthcoming urban expansion, and drainage topped the list of needed improvements. Bernard Marigny was the first to pull the development trigger after Americanization, when in 1805 he hired a designer and surveyor to create a new faubourg. That man was Barthélemy Lafon, who would become New Orleans's next drainage king, responsible for creating roughly half the city's expanding footprint, while finding innovative ways get water off its sole.

BARTHÉLEMY LAFON

Barthélemy Lafon hailed from a family of minor noble stock in Villepinte, a French town near the Spanish border and adjacent to the *Canal Royal en Languedoc.* Perhaps inspired by this engineering marvel linking the Atlantic and Mediterranean, the multilingual Lafon trained in surveying and engineering while studying Classical Greek and Roman civilization. An illustrious career serving the Bourbon monarchy awaited the budding professional—until the outbreak of the French Revolution in 1789, whereupon the twenty-year-old fled his home-land and found safe harbor in the Francophile society of Spanish New Orleans.[80]

Equipped with the right technical and language skills, Lafon found plenty of work in the needy colony. Cabildo records show him contracting for work on levees, drainage ditches, bridges, and levees, and building the fish market and the jail, while advertisements in *Le Moniteur* have him selling cypress planks and soliciting a manager for his lands in the Chef Menteur region to the east.[81] But while prosaic tasks earned Lafon a living, grander schemes occupied his mind, and they tended to enmesh his two driving passions: water-related design and Classical urban planning. He sketched plans for a "magnificent public bath in neoclassical style," proposed a public theater, and, citing examples from ancient Rome, waxed eloquent on the role of the performing arts in a society, as well as "cafes, concert halls, [and] public balls; in one word, one could assemble there the useful together with the pleasurable."[82] After the 1794 fire, which destroyed 220 structures and prompted new construction codes, Lafon designed and built a number of Spanish-style structures, some of which still stand. He also became a landowner, businessman, and land speculator. Yet in retrospect, it seems as though Lafon was biding his time, waiting for history to catch up with his lofty aspirations.

That moment came with the Louisiana Purchase. Working under the new American governor, William Charles Cole Claiborne, Lafon became the acting deputy surveyor of the Territory of Orleans, an area that extended up to the thirty-third parallel, now the Louisiana-Arkansas border. "Much of Lafon's job as the chief surveyor consisted not only of verifying land grants, land purchases, and establishing exact borderlines [of] French long lots," wrote historian Ina Fandrich and anthropologist Jay Edwards; "it also included carefully measuring the borderlines between the urban lots in the city of New Orleans" and evaluating which lands "were . . . suitable for agriculture or for urban development, and which parts were swampy wastelands."[83] In effect, Lafon had assumed the role Carlos Laveau Trudeau occupied in Spanish times, a man to whom Lafon had answered as a contractor in the 1790s, and with whom he shared many professional talents. It was Trudeau who had laid out New Orleans's first suburb, Santa Maria (Ste. Marie) in 1788, and now with expansion imminent, Lafon was positioned to design—and drain—the next wave of development in New Orleans.

In an age before disciplinary specialization, Lafon was the rare jack-of-all-trades who mastered many, starting with *ingénieur géographe* (engineer-

Lafon's skills

geographer), surveyor, city planner, hydrologist, cartographer, architect, builder, and landowner, and ending with sailor, speculator, impresario, privateer, and reluctant spy. No aloof theoretician he, Lafon did not hesitate to get hands dirty and boots wet—literally. "Frequently, we encounter him in Southern Louisiana's ubiquitous cypress swamps," wrote Fandrich and Edwards in reviewing his survey records, "commenting with apparent frustration '*and I had to end my operation because of the waters.*'"[84]

Over the next decade, Barthélemy Lafon comes across as omnipresent. He's *all he did!* mapping the Mississippi Valley, laying out a town (Donaldsonville), planning neighborhoods, designing buildings, sketching forts, surveying properties, issuing *procès verbal,* buying and selling land, serving in the military, running for office, compiling a city directory, fathering a mixed-race family (while also owning slaves), and all the while "engag[ing] actively in the city's social, cultural, and political life."[85] His productivity came from expertise and vivacity as well as opportunistic collaboration. "Lafon not only drew on Spanish charts when creating his maps," wrote Cameron B. Strang in a critical history; "he relied heavily on a Caribbean community of experts, particularly geographer refugees who had migrated to New Orleans from Saint Domingue. . . . Lafon both incorporated the skills of these geographers and used them as envoys to help forge stronger ties between New Orleans and eastern centers of power."[86] As one contemporary said of the man, "Lafon says that anything at N. Orleans may be done by intrigue."[87] So inclined, Lafon's various pursuits eventually entangled him with the occasional rogue—or two, as it turned out.

Speaking of entanglements, Lafon's dual careers as a government official and land investor marked a moment when the public-private line began to blur in regard to drainage. Whereas in times prior, most of the urban ditching and draining had been government directives backed by compulsory private support (labor, taxes, or "donated" slave labor), now drainage activity would be increasingly spearheaded by private agents in the local real estate market, backed by government support. In much of the urbanization to come, drainage and development would take the form of ad-hoc partnerships between private developers and public interests, the former eyeing profits; the latter, municipal coffers. Lafon was the first with a boot in both puddles, and he understood that money could be made from dewatering them.

So did other powerful men, right up to the president of the United States. In 1808, President Thomas Jefferson wrote a letter to local landholder Armand

Jefferson drainage plan

Duplantier, apprising him of the counsel he had given the famed Marquis de Lafayette, to whom Congress had granted some land in what is now the Tremé neighborhood for his support of the American Revolution. "Much of the lands adjoining [New Orleans] were covered with water, but very shallow," recounted Jefferson from his desk at Monticello. "I recommended . . . an extension of the sales of lots there to raise money for surrounding the submerged part with a ditch & dike sufficient to reclaim it. . . . I conjectured . . . that a common ditch of 3 feet deep & wide would draw off the water, & with the earth which came out of it, form a dyke which would keep out the surrounding water." With those words, the president of the United States foresaw the basic drainage strategy for a century to come in New Orleans, and understood its relationship to real estate value.[88]

Powerful local men understood as well. Lafon's first neighborhood-making experience came when Bernard Xavier Philippe de Marigny de Mandeville, whose family had acquired the former Dubreuil plantation below the city's east gate,[89] hired him to survey a new subdivision. French-born engineer Nicholás de Finiels did the design work in 1805, in which he devised a clever street network to reconcile orthogonal arteries with a sharp bend of the Mississippi, while straddling the *Canal del Molino de Don Pedro de Marigny*—that is, Dubreuil's circa-1740s mill race, now known as the Marigny Canal.[90] "The old sawmill canal determined the direction of the new streets," wrote architectural historian Samuel Wilson Jr., "the canal itself becoming the center of the principal street to which was given the name *Champs Elisées*."[91] With a *boussole* (compass) in hand and a *graphomètre* (tripod-mounted sight) in front of him, Lafon in late 1806 took Finiels's plan to the land and "set boundary markers at the corners of all the squares and to lay out all the lots," creating the Faubourg Marigny, New Orleans's second suburb and first to extend downriver.[92] For this Lafon was paid $700, indicating that he had been wearing his private-contractor hat and not his deputy survey hat when executing this project. His client, Bernard Marigny, deemed the work had "been executed by the said Sieur B. Lafon to his entire satisfaction."[93]

With the streets demarcated, next came drainage. As in the city proper and in Ste. Marie, gravity-fed ditches would be dug to reduce soil moisture and drain runoff. But unlike those other areas, the Faubourg Marigny had the benefit of the Marigny Canal to speed outflow. Water draining from a *poisson*-like array of feeder gutters poured into this centralized channel and flowed down the natural levee, elbowing westward onto a rectified natural tributary, passing through a culvert beneath the Gentilly Road, conflowing with Bayou St. John and finally discharging out Lake Pontchartrain. Not coincidentally, that elbowed junction aligned with the forty-arpent rear lines of downriver plantations, in what is now Bywater and the Upper Ninth Ward, Holy Cross and the Lower Ninth Ward, Arabi, and Chalmette. As these parcels later urbanized, their ditch-and-gravity systems sent their runoff into the Bayou Bienvenue swamp opposite the forty-arpent line, which discharged eastward into Lake Borgne.[94]

Perhaps Lafon's greatest contribution was his creation of what is now known as the Lower Garden District, the closest thing New Orleans has to a Grand Manner or Baroque urban plan. It came from a rare opportunity: four large contiguous plantations, on a broad natural levee immediately upriver from New Orleans, coming onto the market at once.

Originally claimed by Bienville, this land had become the upper half of the impressive Jesuit plantation until their expulsion from the colony in 1763, after which it was sliced into six parcels and auctioned off. The two parcels closest to the city became the Gravier plantation and the Faubourg Ste. Marie in 1788. The next one, along today's Howard Avenue, came into the hands of Madame Silvestre Delord-Sarpy, who in 1806 decided to have it subdivided. To devise a street plan, she hired Lafon around the same time he worked on the Faubourg Marigny, putting him in the unique position of expanding New Orleans at both ends simultaneously.

Lafon's initial plat for "Faubourg Delord" featured a broadening of the existing Ste. Marie street grid to span this wider natural levee. But just as Lafon completed the sketch, Madame Delord-Sarpy sold her parcel to Armand Duplantier, who in 1807 rehired Lafon to update his work. This was the same Duplantier who in 1808 heard from President Jefferson about drainage strategies in New Orleans.

Around this time, Lafon had either encouraged, gleaned, or convinced himself that the families owning the next three plantations adjacent to Duplantier's—the Saulet (Solet), Robin, and Livaudais clans—would soon also decide to divest of agriculture and opt to urbanize. Perhaps with their passive consent or collaboration, Lafon went ahead and sketched a full-blown plan spanning all four plantations and, importantly, ignoring the property divisions therein. Such arbitrary hindrances would not vex this visionary![95]

Drawing from his Classical education, Lafon turned the faubourgs Duplantier, Saulet, Lacourse (Robin), and Annunciation (Livaudais) into something of a

Grand Manner Plan, with diagonals (present-day Annunciation/Tchoupitoulas, Prytania/Camp, and Camp/Coliseum) imparting drama while drawing the eye to prominent nodes and plazas. He created *Place du Tivoli* as a rotary (later Lee Circle, now Harmony Circle) unifying four radials, the park itself intended for fountains and amusements like those of the villa Tivoli near Rome. He created a traditional plaza (*Place de Annunciation*) fronted by space for an *eglise,* matching parks in the city's three other neighborhoods. He created a riverfront emporium at *Place du Marche* (hence Market Street) and another one farther back, on *Cours de Dryades* straddling *Rue Melpomene,* which would later become the Dryades Market. He made space for a Classical college, or *Pritanee* (thus Prytania Street) and an outdoor *collesée* for public gatherings, thus Coliseum Square. His gazetteer drew liberally from Greek and Roman mythology, as he named streets after the gods Hercules, Appollon (Apollo), Bacchus, and Triton; the nine Greek muses; and the nymphs of water (Nayades or Nyades, now St. Charles Avenue) and forests (Driades or Dryades, now Oretha Castle Haley Boulevard).[96]

Lafon's "best talents and imagination," according to Samul Wilson Jr. (who coined the term "Lower Garden District" for this area), went into making utilitarian drainage needs into beautiful landscape features. In the ninety prior years, the words "beauty" and "drainage" were rarely said in the same sentence, be it in French, Spanish, or English. But consider how Barthélemy Lafon handled drainage: runoff would flow into a scenic Amsterdam-style open canal on what is now Camp Street and connect with a semicircular water-storage *bassin,* which would then gravitate through the neighborhood via a tree-lined open canal on Rue Melpomene and eventually out Bayou des Cannes, a tributary of Bayou St. John. A similarly landscaped canal would direct water to "flow gracefully around [Tivoli] Circle, much like a well-planned stream in a formal French garden," out a grand canal in the middle of *Cours des Tritons* (Tritons Walk, today's Howard Avenue, thus its great width), and disperse in the woods by the present-day Superdome.[97] Combined with green spaces and colorful nomenclature, Lafon produced, at least on paper, "an environment of classical beauty and formality in one of the earliest expressions of the Greek Revival to appear in New Orleans, [of which] water was an essential element."[98]

Lafon's street network came to fruition by 1811, and many of his Classical toponyms still grace the map. But most of his features did not. Tivoli eventually got a monument, but not amusements; Annunciation Place never got its church; and neither the market on Market, the amphitheater at Coliseum, nor the prytaneum on Prytania were ever built. Evidence suggests his canals eventually fell in line with the squalor New Orleanians had come to associate with drainage.

Nevertheless, his urban vision made for an interesting and livable neighborhood, particularly Coliseum Square—"planted with luxuriant trees," wrote resident Thomas K. Wharton in the 1850s, "surrounded with beautiful houses, and gardens filled with the choicest flowers," enjoyed by "all of the children in the neighborhood [who] gather on the Square in the evenings."[99] As for Lafon's Camp Street catch basin, it had an erratic history, including a time in 1881 when the city installed underground brick chambers to store runoff. Sidewalk repairs in the mid-2010s broke open the tops of these vaults, revealing their interior brickwork and invoking the imaginative mind of Barthélemy Lafon, the drainage king of the 1810s.

JEAN GRAVIER, EDWARD LIVINGSTON, AND THE ST. MARY BATTURE

In New Orleans's water tussles, pitting fluidity against rigidity, rarely did nature toss humanity a gift, much less a doozy at the right place and time. But that's exactly what delighted eyes saw in those sandbars forming along the Mississippi just off Tchoupitoulas Street starting in the early 1800s.[100]

What happened was that the ever-shifting river channel had reached a point by which its current slowed measurably as it made its way around the East Bank crescent, allowing sediment to fall out of the water column. Anglophones called such depositions "beaches" for their sandy composition, sand being the coarsest and heaviest alluvial particle. Francophones called them "battures," probably from *battre,* meaning "to beat against" and denoting the space between the levee and the main channel. Battures are dry when the river is low, inundated when high, and "beaten against" by currents in between. They form all along the point-bar sides of meandering rivers (as opposed to the cutbank side, where stronger currents scour bankside soils).

Colonials had understood battures since the early 1700s, and according to the colony's Roman Law–inspired civil code, such ephemeral lands were considered a public resource. Whenever river stage dropped and battures appeared, citizens could promenade upon them for leisure, fish or gather driftwood, or cart away their sand for use as fill. Nature would replenish it.

But what happened starting around 1803 was different. This was not just any point-bar; it was the Faubourg Ste. Marie riverfront, on the fast-growing upper side of town. And it was no ordinary time; there were rumors of an American acquisition of the colony, which would surely bring a boom in river commerce. Americans brought with them the sensibilities of English common law, which, unlike Roman civil law, viewed such alluvions not as public space but the private property of adjacent proprietors.

This boded well for the Gravier, Duplantier, Solet, Robin, and Livaudais families as well as the Ursuline Nuns, all of whom owned riverfront land in this vicinity. One particular proprietor, Jean Gravier, saw economic opportunity in the batture, and in 1803 hired workers to erect a small levee around the *terre d'alluvion* forming off his parcel on Tchoupitoulas Street between Julia and St. Joseph. In effect, Gravier self-privatized what most New Orleanians considered to be public space, and in doing so, unwittingly initiated New Orleans's largest reclamation project of the nineteenth century. But unlike most reclamations, nature would do most of the sediment delivery, while levee-builders and lawyers did the rest. The case would become known as the St. Mary Batture.

Most citizens were irked by Gravier's usurpation and pushed officials to act. The *Conseil de Ville* (City Council) responded in 1804 by reiterating that the batture was in fact in the public domain and could not be fenced off as private property. But as the batture grew, so did the economic pressure.

That same year arrived a lawyer named Edward Livingston, former mayor of New York City and brother of the famed Robert Livingston, who had negotiated the Louisiana Purchase and would later help develop the steamboat. According to historian Ari Kelman, Edward Livingston "brought with him to New Orleans an American perspective on property rights, a New Yorker's eye for the value of riparian land, a debtor's nose for easy money, and one of the keenest legal minds in the nation."[101] Livingston took on Gravier's case in exchange for ownership of the heart of the St. Mary Batture, riverside of Tchoupitoulas from Common to St. Joseph streets. In October 1805, he sued to secure his client's right to the disputed beach.

This time, the forum that would hear the case was not the Creole-dominated *Conseil de Ville* but a new American territorial court. In May 1807, the Justices of the Territorial Court ruled in Gravier's favor, and an elated Livingston set out to shore up his booty. He sent laborers to the St. Mary Batture to start building a levee and dock for what would essentially be a private mini-port.

But when his laborers arrived, angry citizens, described as a "mob" by Livingston, gathered and chased them away. The workers returned another day, only to be confronted again by larger crowds, backed by city support. Protests and counterprotests intensified throughout the summer of 1807.[102]

The stakes were high, because what courts decided in New Orleans could form a legal precedent throughout the Louisiana Territory. On one side were advocates for the *Conseil de Ville*'s call for the continued public ownership of the batture. On the other were those who supported the Territorial Court's decision implying that the batture could be privatized. Both parties appealed to the territorial governor, William C. C. Claiborne, for resolution, who in turn sought the advice of President Thomas Jefferson, warning him that blood may spill in New Orleans if the matter were not settled.

Even though his Anglo-Saxon heritage might have aligned him more with the common-law partiality for privatization, President Jefferson sided resolutely with the public use camp favored by most Creoles, declaring that the batture belonged to the United States. His reasoning might have been tinged by personal animus toward Livingston, but Jefferson probably wanted to keep the peace in postcolonial New Orleans, whose port was critical to his dream of an agrarian nation. Jefferson also sought to ensure free and open river access by farmers throughout the hinterland, and legally upholding the public ownership of riverbanks was one way to do that. All this made the St. Mary Batture a national issue.

Standing in the way was Edward Livingston and his levee-ringed shoal off Tchoupitoulas Street. In early 1808, President Jefferson dispatched US Marshals to evict the outspoken esquire. Unvanquished, Livingston took his case to Washington and then to the people, publishing polemical pamphlets on the plight of the private-property owner. During 1810 to 1813, he again sued, targeting President Jefferson (by this time out of office) in a court in Richmond, as well as the US Marshal who had evicted him, in a court in New Orleans.[103]

Livingston lost the Richmond lawsuit but won the New Orleans case, giving his argument new life. He traded legal volleys with the city for years to come, and the St. Mary Batture became a never-ending news story throughout the 1810s, as the reclaimed land grew to 3,400 feet lengthwise and on average 470 feet in width.[104]

Finally, in September 1820, Livingston and the city negotiated a compromise. "All the soil between the present Levée and the river shall in future be held by the city for the purposes of navigation," explained John Adems Paxton in 1822, "but that no buildings whatsoever shall be erected thereon." In other words, land riverside of the new levee would have a public maritime servitude, as it does today. As for the deposition on the interior of the new levee, Paxton wrote, "This piece of property, which has made so great a figure in the history of litigation, is now divided among a great number of proprietors . . . and a liberal arrangement with the corporation of the city has put it in a situation in which it may be improved and made useful to the public, as well as a source of profit to the owners." He concluded, "All [our] commerce centers on the Batture, and it would be difficult to select in any city in the world a spot in which more extensive business is done in the same space. The property then must soon become invaluable."[105] Though litigation would continue for decades, the 1820 compromise paved the way for the St. Mary Batture to be incorporated into the cityscape.

This was natural reclamation; the river dumped the sand and took its waters elsewhere, padding the East Bank with new land. One observer noted how a cotton press "built in 1832 . . . upon the edge of the river" was by 1834 "now distant at least three hundred feet."[106] Workers helped secure the new land by realigning the levee farther outwardly, then carting or pumping in artificial fill to equalize its elevation with the crest of the natural levee along Tchoupitoulas Street. Similar reclamations occurred throughout what is now the Uptown New Orleans riverfront; for example, the former batture that is now home to Children's Hospital and the former Marine Hospital had first been leveed off from the river in the late 1700s, and in the 1850s was raised "to the height of the levee, at a cost of over $70,000" using nearby river sand dredged and pumped in by steam engines.[107] Likewise, the Audubon Riverfront Park popularly known as "the Fly" was once batture, and after various iterations of reclamation and landfill, is now one of the highest areas in town.

But nature's gift to the East Bank came at a cost to the West Bank. There, powerful currents leaned into the shore, eroding sediments and threatening levees. Straight across from the St. Mary Batture, the riverfronts of upper Algiers, McDonoghville, and Gretna lost nearly as much land as the East Bank had gained. So too did parts of the East Bank farther downriver, where the dynamics reversed due to a sharp meander. The installation of revetments by the US Army Corps of Engineers in the late 1800s and early 1900s stabilized bank erosion, and no longer are cutbanks as prone to major scours. But the effort is a constant one, as humans and physics had become one and the same agent in the water battles of New Orleans.

Paxton was right in predicting the value of the St. Mary Batture. The reclaimed land hosted lucrative steamboat and flatboat wharves in the 1800s, and an industrial and warehousing district by the early 1900s. Its fortunes declined later in the century, only to see revitalization following the 1984 Louisiana World Exposition (its theme: "the World of Rivers"), when the area was rebranded as the Warehouse District. The old St. Mary Batture is now home to condominiums, restaurants, art galleries, the Riverwalk, the Convention Center—and some of the priciest real estate in town.

Clues divulge the area's unusual provenance. Blocks of the St. Mary Batture are shaped like parallelograms, rather than the squares found inland, because of the geometric effect of extending old interior streets beyond the natural curves of the original riverbank. Its street names, meanwhile, are decidedly American (Commerce, Peters, Fulton, Front, Delta, Water), sans the euphonious syllabus of the older interior names (Gravier, Notre Dame, Julia, Tchoupitoulas). Front Street is a particularly good barometer of batture reclamation, because it got extended as soon as the new levee had been built; in fact, its original name was "*Rue de la Levee Neuve.*" The Tanesse Map first put *Levee Neuve* on the map in 1815; the S. Pinistri Map of 1841 shows Front Levee Street running up to Race Street; the W. Walter Map of 1855 has it extending to Joseph Street, and the Robinson Map of 1883 shows Front Street fully extended from the upper French Quarter to State Street and the old Marine Hospital.[108] And if one looks closely at the foot of Henry Clay Avenue by Children's Hospital, one can still see the land surface decline slightly as it crosses Tchoupitoulas Street, which three hundred years ago had been the crest of the natural levee.

The apotheosis of riverfront reclamation came in the early 1900s. Following the launch of the river-accessing Public Belt Railroad in 1904 and a rigorous modernization campaign, officials with the Port of New Orleans aimed to replace the antiquated cotton presses on the original batture with a huge modern facility accessible to trains and ships. No such space existed, so the Port created it—by, once again, pushing out the levee along the "belly" of the crescent, from Louisiana Avenue to State Street, and infilling the gap with sediment. Upon these hundreds of newly reclaimed acres was erected the gargantuan Cotton Warehouse and Terminal (1915), largest in the world, capable of storing two million cotton bales.[109]

In all, between the early 1800s and the mid-1900s, upwards of seven hundred acres had been reclaimed along the East Bank of the Mississippi within Orleans Parish. That's about one-third the size of the city's biggest and best-known reclamation project, the Lakefront Project of 1926–34. Unlike that concerted effort, the riverfront reclamation happened piecemeal, with no "drainage king," no one agency in charge, and little public notice. It was more a process than a project, started by delta physics and aided by engineering, and it changed the shape of the urban core.[110]

The reclamation of the batture also benefited Barthélemy Lafon's residential subdivision of 1806–10, today's Lower Garden District, in that it created new space for riverfront commerce while buffering interior blocks from riverfront nuisances. One can only imagine what the Classicist could have done with the new space. But Lafon by this time had gotten distracted by other dalliances, and his imprudence ended up destroying his illustrious career.

His bad decisions came at a time when they were tempting to make. The interregnum of the early 1800s was a confusing time in Louisiana, and that meant opportunities on the margins of the law. Because each administration (Spanish, French, American) insisted on regulating trade through taxation, economic pressure mounted to meet marketplace demand by rerouting supply chains around legal authorities. "When it came to obeying the rules of mercantilism," wrote historian Lawrence Powell, "Louisiana ranked among the New World's worst scofflaws. The entire economy was steeped in smuggling[;] it thrived on contraband trade."[111] Citizens outmaneuvered bureaucracy with such ease and reward that pretty much everyone, from merchants to mothers, partook of the black market. It was the moral equivalent of working off-the-books today—nothing to shout about, but hardly scandalous, and in some circles, a mark of savvy.

But for a black market to exist, someone had to do the dirty work, and that meant sneaking through backswamp bayous, muskets at the ready, evading authorities, contraband in the hold. Most notorious was the Barataria region to the south, "arguably the largest smuggling depot and irregular naval base in North America," according to historian Robert C. Vogel, "and the presence of so many armed adventurers on Louisiana soil had become a national scandal."[112] Lording over the Baratarians were the brothers Jean and Pierre Lafitte, and it was through his association with these infamous rogues that Barthélemy Lafon went awry. Crossing paths possibly through their mutual membership in the local Masonic Order, Lafon got personally involved in the Lafittes' Barataria racket "at different times," according to anthropologist Jay D. Edwards, "command[ing] at least four armed raiders whose names have been recorded."[113] The biggest money was on slave smuggling, and that was the Lafitte brothers' main contraband, and Lafon's too. It is difficult to reconcile the Classical designer pondering his Greek muses by day, and the profiteering pirate trafficking in human beings by night.

Lafon probably would have gotten away with it but for the War of 1812. As British military planners scanned their maps in 1814, the Barataria region became a potential theater of conflict, and the Lafittes' privateers could become combatants, should the Brits decide to buy their friendship. So the Americans sent Master Commandant Daniel Todd Patterson on a preemptive raid of their lair at Grande Terre in Barataria Bay. The surprise incursion netted a number of "hellish banditti," as Maj. Gen. Andrew Jackson called the Baratarians. It must have come as a shock to the Americans to find among them the former acting deputy surveyor of the American Territory of Orleans, Barthélemy Lafon.

Jackson reluctantly arranged with Jean Lafitte for his Baratarians to side with the Americans, in exchange for clemency. Their participation in the Battle of New Orleans played a role in the overwhelming American victory at Chalmette on January 8, 1815, whereupon the pirates became heroes. Lafon, too, had made himself useful to the American cause, making maps, building fortifications, and organizing defenses, even as Jackson looked askance at him.

But after the hoopla, Lafon found himself estranged from American authorities, and his work dried up. He decamped with the Lafittes to Galveston, then still part of New Spain, and entangled himself in a dangerous web of espionage and filibustering, at times paid by the Spanish Crown.[114]

Lafon returned to New Orleans in 1818 hoping to restore his reputation, but now tainted as a foreign spy as well as a pirate, he failed to restart his career, struggled with his assets, despaired over the subjugation of his mixed-race family, and hoped to return to France. Instead, he succumbed to yellow fever in September 1820, at age fifty-one, and was entombed in St. Louis Cemetery No. 1.[115]

Today, Barthélemy Lafon, who made many lasting contributions to New Orleans and also committed his share of crimes, is little remembered by locals, whereas Jean Lafitte, who lived a life of crime and redeemed himself only once, enjoys folk-hero status. Yet even in death, Lafon's influence on New Orleans persisted, through his tutelage of the next generation of drainage kings, namely Jacques Tanesse and Joseph Pilié.[116]

JACQUES TANESSE

A native of Saint-Domingue, Jacques Tanesse gained cartographic experience working under Lafon, and upon becoming city surveyor applied his skills to create a number of high-quality maps and new subdivisions for New Orleans. One was the Faubourg Daunoy, the first below the Faubourg Marigny, which, with

Lafon's surveying, served as a template for development down to today's Lower Ninth Ward.[117] Now forming part of the Bywater neighborhood, Tanesse's 1810 plan foretold the conversion of these plantations' forty-arpent line into what is now Florida Avenue, putting it in an eventual position to become selected, in 1895, as the city's main outfall canal to Bayou Bienvenue and Lake Borgne.

Tanesse did this at roughly the same time that the city acquired land behind the French Quarter from property owner Claude Tremé, and, during 1810 to 1812, had Tanesse subdivide it as the Faubourg Tremé. This site straddled Bayou Road (today's Gov. Nicholls Street) on the slight upland known as the Esplanade Ridge, such that runoff here flowed to either side of the ridge, rather than backwards. This had long been the case for the farms along other narrow ridgetop arteries such as the Metairie and Gentilly roads, but it was a new condition for an urbanized neighborhood. Tanesse arranged for ditches to direct runoff laterally; that which flowed west of the ridge went to a drainage canal on Orleans Avenue, while the east side went to a canal on St. Bernard Avenue. From there, both channels flowed to the backswamp and out Bayou St. John.[118]

Also in 1810, Tanesse took on that great cleave in the urban fabric, the commons between the present-day French Quarter and the Central Business District, which since the 1720s had been reserved as a firing line for the upper fortifications. No longer needed for military defense, the weedy interstice became an eyesore, impeding "the communication between the town and the suburb St. Mary" and creating "receptacles of stagnant water and of all manner of filth which engender disease."[119] Technically federal land, the newly formed city council in 1805 petitioned Congress for the space, and in 1807 Congress recognized that claim, provided that the city reserve space "to continue the canal of Carondelet from the present basin to the Mississippi" and leave undeveloped "any lot within sixty feet of the space . . . which shall forever remain open as a public highway."[120] It was Jacques Tanesse who took that directive to paper, in 1810, as he designed "Canal Street" to span fully 171 feet, with a 60-foot central right-of-way, while deftly conflating Adrien Pauger's 1722 city grid with Carlos Trudeau's 1788 Ste. Marie grid. These are the blocks and lots spanning from today's Iberville Street to the aptly named Common Street.

As it happened, Barthélemy Lafon had secured rights to some land in this swath, granted to him back in 1798. But he was forbidden to erect buildings on it, and his envisioned foundry never came to fruition. Neither did the navigation canal—thankfully, as such a waterway could have rived the city in two if its lock at the river had failed. But the name stuck, and we call it Canal Street to this day. Never did schooners and towboats sail up and down the "Great Wide Way," but Canal Street did once have an open drainage canal, of the ditch-and-gravity variety, and today has subterranean culverts.

Jacques Tanesse's *Plan of the City and Suburbs of New Orleans from an Actual Survey* (1815), among the best city maps of the era and redolent of Lafon's influences, captures the drainage system in this key period of metropolitan growth. There is the still-envisioned channel on Canal Street; Lafon's Camp-Coliseum-Melpomene drainage system as well as "Canal du Tivoli" on Delord (now Howard Avenue); the Canal Gravier as the main outlet for Faubourg Ste. Marie runoff; the Canal Girod on Orleans Avenue; the Marigny Canal; and an elbowed drainage outlet on what is now St. Bernard Avenue.[121] All steered their outflow to what Tanesse labeled as *terres vacantes,* and illustrated with watery hachures and shrubby tufts. The cartographic orderliness betrayed the hydrological reality of the city, which the visiting architect Benjamin Henry Boneval Latrobe put more bluntly. "Mud, mud, mud," he despaired in his 1818–20 journal; "this is a floating city, floating below the surface of the water on a bed of mud."[122]

JOSEPH PILIÉ

Like Tanesse, Gilbert Joseph Pilié was also a refugee of Saint-Domingue who started his New Orleans career contracting under Lafon. As his mentor had surveyed the Faubourg Marigny in 1806, Pilié laid out an extension in 1809 known as Faubourg Nouvelle Marigny, which made use of a new drainage channel called the St. Bernard Canal. In 1818, during Lafon's twilight years, Pilié followed Tanesse in the position of city surveyor, while also working as an artist and architect of numerous Creole-style buildings. Pilié later served as chief surveyor of the Second Municipality, and both his son and grandson followed in his footsteps as surveyors. The family legacy continued into the 1910s, having started under Lafon's wing a century earlier.[123]

Drainage during the Pilié years had become extensive in size but still perfunctory in operation. There were four polders, each with its own subsystems: Lafon's Coliseum/Melpomene arrangement for the upper faubourgs (today's Lower Garden District); the Poydras, Gravier, and Canal Street canals for the Faubourg Ste. Marie (Central Business District); the Carondelet Canal as well as a new channel on Orleans Street for the French Quarter and part of the Fau-

bourg Tremé; and the Marigny Canal and St. Bernard Canal for the lower faubourgs of Marigny, Nouvelle Marigny, and Daunoy. All runoff discharged either
into tributaries or the main channel of Bayou St. John, else dispersed into the
soils or woods.[124]

What changed in Pilié's era was a new source of water wetting the cityscape.
There had always been rainfall, of course, as well as river filtration, saturation
from the backswamp, and the occasional levee overtopping or breaching. In the
1810s, private companies added more moisture to the mix by drawing river water into raised tanks and piping it gravitationally to paying subscribers for domestic use. In the early 1820s, famed architect Benjamin H. B. Latrobe mounted
a steam pump in a three-story pump house to draw water from the Mississippi
by the French Market, store it in raised cast-iron reservoirs, and distribute it to
nearby basins and through cypress pipes to customers' residences. Latrobe's waterworks were completed three years after the architect's death (to yellow fever),
and served the city during 1823 to 1836. His son, John H. B. Latrobe, described
the operation in 1834: "The water works erected by my father are in operation[;]
I saw this morning the water bubbling up from the pipes into the large cast iron
box around them, and running off in a rapid stream through the gutters. At every corner were crowds of negro women filling their buckets and water carts."[125]
Other companies installed similar systems in adjacent neighborhoods in the
1830s. A rare map from this period shows scores of "fire plugs and pipes" from
private waterworks, all clustered in the wealthier "front of town"—an example
of how urban improvements were denied to the marginalized populations in
the rear precincts. Surplus water delivered by these systems flushed through the
gutters, cleaning them but also pushing the detritus to the back-of-town, while
burdening the drainage system with additional volume to remove, lest it puddle
and flood.[126]

By 1834, comparable water and drainage networks had also been developed
for the newly subdivided Jefferson Parish communities of what is now Uptown:
in Carrollton (subdivided 1833), with a main drain on Canal Avenue (now South
Carrollton Avenue); in Hurstville (1834) and the Faubourg Bouligny (1834, with
a main drain on Napoleon Avenue); and in the faubourgs Plaisance, Delassize,
Livaudais, Lafayette, and Nuns in what is now the Irish Channel, Garden District, and Central City.[127] Part of this latter area, which became the City of Lafayette in 1833, drained out the Gormley Canal, a mile-long channel dug in 1828 by
William Gormley. A big slaveholder who spent years in court to gain legal title
to this right-of-way, Gormley likely used his slaves to dig the four-foot-deep,

eighty-foot-wide channel (now roughly St. Andrew Street) and build a harbor and quay on what is now the block bounded by O. C. Haley Boulevard and Felicity, South Rampart, and St. Andrew streets. Evidence suggests that this was a resource-extraction channel as well as a runoff-removal canal, with Bayou des Cannes and the rear swamps serving as outfall.[128]

The 1834 *Topographic Map of New-Orleans and Its Vicinity,* by Charles Zimpel, himself a multiskilled surveyor-engineer-cartographer, shows one particularly interesting drainage feature. On "Florida Walk," the early name for what is now Florida Avenue following the old forty-arpent line, two "basins" are shown, pond-like and ornamentally shaped, connecting with the rear drainage canal fed by the streets of the Faubourg Marigny and Nouvelle Marigny. Both have parks around them: Commerce Place at the Elysian Fields Avenue (Marigny Canal) intersection, and Constitution Place at Franklin Walk, now Franklin Avenue. The generic toponym "walk" in this era implied a recreational intention; for example, there was also a Carondelet Walk developed along the canal of the same name, and it became a popular pleasure garden. There was certainly public demand for water recreation; as early as 1795, according to one historian, the Carondelet Canal basin had become "a popular swimming place for society ladies who . . . dived 'head and all' into whatever water was there until summer droughts turned it into a mudhole." An 1828 ordinance expressed forbade "bathing naked, in the waters of the Basin and Canal Carondelet, and in all other ditches or basins."[129]

It's unclear if Florida Walk ever came to fruition; only a disheveled lane called Constitution Place and four city-owned lots stand there today, straddling the Florida Avenue Drainage Canal. Nevertheless, the twin polygonal basins on Zimpel's 1834 map represent early examples of landscaped holding ponds on public space, wherein runoff would be stored and beautified rather than just left to mud. The Zimpel map also shows an articulated drainage system in McDonogh (McDonoghville), the first of its type on the West Bank—though agrarian systems and mill races had long been on plantations, while a basic ditch-and-gravity system operated in what is now Algiers Point.

"DANG, ORDURE OR FILTH WHATEVER"

By 1835, according to a retrospective published fifty years later, street gutters in New Orleans had been "gradually extended into the swamp, and a few draining canals had been made, viz.: The Melpomene, from St. Charles to Willow streets, the Canal Gravier, on Poydras from Baronne street to a branch of Bayou St. John,

Canal street from Claiborne street to a branch of Bayou St. John, and Orleans street from Claiborne street to Bayou St. John; St. Bernard from St. Claude street to Bayou St. John, and the old Marigny canal from Elysian Fields street, via Marigny avenue, to the Bayou St. John; in Claiborne, from Canal Carondelet to Ursulines street."[130]

Yet hardly did these various polder networks comprise a genuine citywide system. There was no drainage "czar," nor city drainage department, thus no one could be held accountable when drainage failed. Hierarchical oversight and engineering standards were limited, and funding and maintenance came mostly from beleaguered homeowners subjected to tasks, taxes, fines, and noncomplying neighbors.

The ad-hoc nature of drainage was built into the laws that purported to govern it. A public health ordinance passed in 1816 made it "the duty of all owners, tenants or keepers of houses, court-yards, or other lots or grounds . . . to sweep [and] properly clean [with] several buckets of clean water . . . the gutter of the banquette that lies before their respective premises." This had to be done every morning before 10 a.m., and in a manner that did not "convey any mud or filth [to] any neighbour's premise," else face a fine of one to three dollars—two days' pay at the time. Those owning corner houses were also responsible to clean out the cross bridges, "made of plank two inches thick, and at least twenty feet long . . . nailed . . . on strong joists [and] laid on a level with the road."[131] Apparently there was much to clean, because a later article provided a laundry list of items forbidden in ditches and canals: "oyster-shells, hay, straw, dang, kitchenstuff, broken-glass, . . . leather, paper, or cloth, any shavings or chips, or other ordure or filth whatever." The ordinance also forbade the emptying of any "vessel containing feces . . . in the gutter of any banquette in any ditch, basin or canal," stipulating instead that it "be emptied into the current of the river"—which was the city's main drinking water source.[132] Another ordinance passed in late 1817 upped the fine for anyone "obstructing the natural draining [or] stopping up . . . any ditch or canal . . . necessary for said drainage" to fifty dollars plus the cost of the cleanup, reimposed for every fortnight the problem persisted.[133]

SLAVES, DITCHERS, AND SYNDICS

Given the bone-wearying travail of excavation and futility of homeowner maintenance, drainage labor in the early 1800s often ended up in the hands of the enslaved, as it had in French and Spanish times. People in bondage were "hired out" by individuals, by firms such as improvement banks or general contractors, or by the Louisiana Board of Public Works, which provided state-owned slaves as subsidies to support internal improvements. Prisoners were also availed to drainage projects, be they free white, free Black, or enslaved Black, in exchange for payments made to the wardens.[134] This began to change in the 1820s, when steam dredges could be floated in on barges to dig backswamp canals. But urban ditching and cleaning remained manual labor.

Enslaved teams toiling on drainage systems were an everyday sight in antebellum New Orleans. They occupied the opposite end of the social caste system from the drainage kings, but were just as critical to the effort. "The cleaning of the streets," wrote an 1830 visitor, "is performed . . . by slaves . . . even females . . . chained together, and with hardly any clothes on their backs, sent [by] their masters, as a punishment for some delinquency, [for] about one shilling Sterling per day."[135] Imprisoned Blacks—including those unable to show freedom papers or account for their owner—were by law "put to the chain [and] employed in the works of the city," else whipped. City-controlled chain gangs deployed on drainage and other public works were guarded by two white overseers, who, six days a week, marched them to the work site at dawn and worked them until sunset, save for a two-hour noon break. Enslaved women, meanwhile, cleaned gutters, streets, and *banquettes.*[136]

Council proceedings abound with official deployments of enslaved chain gangs for every conceivable municipal project: digging canals, building levees, repairing wharves, paving streets, excavating graves, fighting fires, and constructing public buildings like Charity Hospital. Firms behind the region's biggest internal improvements of the 1830s, including the New Orleans Drainage Company, Barataria and Lafourche Canal Company, New Orleans Canal and Banking Company, and Pontchartrain Railroad Company, all counted the enslaved among their corporate assets, and/or depended on enslaved workers for their business.[137]

The labor market changed in the 1830s, when Irish immigration increased the pool of workers for companies to hire as contractors. Low-paid hands were completely expendable, whereas enslaved workers embodied fiscal value and entailed medical liability. Dollar-a-day Irish "ditchers" became the standard labor source for many drainage and navigation canal projects during the middle decades of the nineteenth century, ranging from the New Basin Canal on the East

Bank to the Destrehan (Harvey) Canal on the West Bank, and numerous smaller projects in between.

Ditching was brutal work, and a tough sell to any young man with other options. Toiling with shovels and wheelbarrows, workers endured a blazing subtropical sun, with fetid water stewing below, and insects and stultifying humidity all around. Deep slippery mud sucked off footgear, so many workers worked barefoot, and swinging implements often struck limbs or faces. Many perished of arbovirus diseases, chiefly yellow fever.

Drainage management, meanwhile, remained a localized responsibility. The closest thing New Orleans had to a citywide drainage authority came in the form of the city surveyor's involvement in the layout of new subdivisions, and in the provision of an untitled "syndic" (government official) to each ward or subdivision. Syndics were responsible for making annual inspections of drainage ditches, levees, bridges, and roads and to prorate the costs according to the width of the adjacent parcels. "The works necessary for keeping in repair the . . . drainage ditches," stated an 1817 ordinance, "shall continue to be executed at the expense of the owners of the adjoining lots or grounds, each being answerable for the repairs of the portion lying in front of their respective property."[138]

Syndics were the unloved drainage kings of their respective wards, the first generation of regular government workers whose job description was drainage. Unfortunately, they did so mostly by harassing residents to do the work, or sticking them with the contractor bill. Giquel, syndic of the Second Ward in what is now the Irish Channel and Garden District, issued a press release in 1827 instructing parcel owners in these new faubourgs of their responsibilities. Calling them out by name, he tasked them to build three-foot-wide cypress-plank bridges, sidewalks eight feet wide (twelve feet on the wider arteries of Jackson and Nayades, now St. Charles Avenue), and gutters two feet wide and one and a half feet deep, and up to three feet deep as they ran back to Bayou des Cannes. "*[Build] a bridge jointly*"; "*[dig] their gutters*"; "*filling up the street*"; "*dig a gutter all along their front, their great drainage ditch and their boundary ditch, as far as Mr. L. Foucher,*" read Giquel's tart scolds.[139] Such mandates yielded varied responses: some complied; others let the syndic charge them to hire a contractor; and still others were no-shows or absent owners. Even if everything got done on time, rarely was it all up to par. As a result, New Orleans's drainage system in this era was an improvised patchwork of independent, idiosyncratic ditch-and-gravity subsystems installed by private developers and their contract engineers, inspected by syndics, and maintained by browbeaten property owners. It was a water-management scene that Adrien de Pauger would have recognized in its design, but found to be perfectly appalling in its administration.

Citizens protested, and sensing the city was deaf to their complaints, they took their case to the state legislature. "The undersigned inhabitants of that part of New-Orleans, which lies above Canal-street," went one such grievance in 1826, "respectfully represent . . . that by neglecting to abate nuisances, to drain or to fill up stagnant ponds, [and] remove offensive and deleterious substances from the streets[,] the city continues to be unhealthy, the premature death of many citizens is annually caused thereby, and emigration to the state is prevented. . . . By a judicious administration, the principal sources of disease might, in a few years, be removed from New-Orleans, and the city be made a healthy place."[140]

LA MARE À BORÉ

Not all ponds were deemed "offensive and deleterious." Curiously, people's stigmas about wetlands diminished as the "*wet-*" dominated over the "*-lands.*" That is, in the few morasses where the forest canopy opened up and bayous widened into ponds, that sense of foreboding gave way to appreciation—of natural resources, of healthful recreation, even of aesthetic beauty. Perhaps the best example of this open-water preference was *La Mare à Boré* (Boré's Pond), in the backswamp of present-day Uptown.

The name inferred Jean Étienne de Boré, the planter who in 1795 documented procedures to granulate locally grown sugar juice, triggering a massive shift of regional agriculture toward cane cultivation. Those experiments took place on the plantation De Boré had acquired fourteen years earlier immediately downriver from present-day Audubon Park, with a frontage along Tchoupitoulas Street. In 1785, he attained a Spanish land grant to extend his parcel to a depth of eighty arpents, terminating in the backswamp where Earhart Boulevard now intersects State Street.[141]

According to De Boré's grandson Charles Gayarré, who was born in 1805 and raised on the plantation, that bog was neither completely wooded nor perfectly flat. "On the Boré plantation," he wrote in his elder years, "midway between the river bank and the cypress swamp, there was a depression in the land, where, in consequence of it, a large pond of standing water had been formed." Rimming that micro-basin was what Gayarré described as "La Terre Haute," not the natural levee per se, but probably the ridge-and-swale topography sometimes seen on its backslope, the *terre haute* being the ridge and the "large pond" having formed in the swale.

"This pond, known far and wide," Gayarré wrote, "was called La Mare à Boré, [and] all around this pond, the soil was of a marshy nature, full of tall weeds, sheltering a multitude of wild game, such as snipes, water-hens, rails, etc." As the only open water body in the vicinity, La Mare à Boré became a favored stopover for migratory birds. "During the winter it was the resort of innumerable flocks of ducks, that successively came to it in the evening until it was completely dark." To locals, this made La Mare à Boré a happy hunting ground. "As they passed over their expected shelter," Gayarré wrote of the ducks, "the ambuscaded hunters rose from their concealment and emptied their guns. Hence this was called La Passée. . . . This pond and marshy ground was a famous shooting spot at that epoch," the late 1810s and 1820s.[142]

Late Saturday afternoons, he recalled, were when "the élite of New Orleans—lawyers, physicians, commission merchants, brokers, bankers" came out for some shooting at the pond. "On such occasions we could hear from our dwelling-house a lively rattle of gun-firing, as if a skirmish was going on. Some even camped there, to be ready for the sport early on the next morning." For it to have become this renowned suggests that La Mare à Boré was a salient feature, the most accessible "sportsman's paradise" closest to the urban population. "Fires were lighted, tents erected, and the comforts and wants of the human body attended to with proper care," wrote Gayarré, implying that libations flowed liberally at the popular getaway. "Jokes were cracked, tales related by the blazing piles, pranks perpetrated, and to speak the unpleasant truth, there ensued, although rarely, quarrels that led to duels." Local slaves, he recalled, "connected that spot with hobgoblins and apparitions, among others the ghost of a colossal raccoon."[143]

Gayarré made an interesting comment regarding pond access. "In any other country this sporting ground would have been jealously guarded, but in Louisiana this would have been looked upon with extreme disfavor. Hence this pond . . . was treated as public property, without any interference from the owner." Perhaps Gayarré, a learned esquire and proud Creole, was invoking the spirit of Roman civil law, which tended to view riparian lands and battures as public domain, or at least publicly accessible. American notions of English common law, on the other hand, tended to give the upper hand to private ownership. But Gayarré also made clear that this pond did have a rightful owner, his grandfather, and when the elder died in 1820, Gayarré inherited part of the property. Subsequent sales put the various subsections of Jean Étienne de Boré's once-vast colonial-era holding on track toward their later drainage and urbanization.

Where precisely was La Mare à Boré? An inset in Francis P. Ogden's *City of New Orleans* map (1829) shows a dendritic network of now-gone bayous draining the Uptown watershed, of which one tributary, Bayou des Cannes, drained the present-day Central Business District and Lower Garden District. This was the outflow to which Barthélemy Lafon connected his drainage canals, after which the water merged with other bayous and eventually discharged through Bayou St. John. Bayou des Cannes's largest Uptown tributary is shown on Ogden's map as originating namelessly in present-day Carrollton and flowing through Fontainebleau and Broadmoor, where it widened to seven hundred feet for a distance of about two thousand feet, in perfect alignment with the rear of the former Boré plantation. This, clearly, was La Mare à Boré.

It would take decades of drainage efforts to draw down and dry out this lovely oasis, though its memory lives on to this day, as neighbors recount rumors of some sort of "lake" lasting in Broadmoor into the 1920s. Overlaid on a modern-map, Ogden's depiction puts the lost paradise around Nashville Avenue's intersections with McKenna and York streets, from Calhoun Street to the Fontainebleau Drive–South Salcedo intersection.[144]

La Mare à Boré would become collateral damage in the upcoming war against the backswamp. What motivated the effort was fear of malady and desire for expansion, and what it obliterated was natural beauty as well as ecological services, in the natural storage of excess water and the recharging of the groundwater. But it was the loss of the pond's cultural meaning that most affected Charles Gayarré as he reminisced about his childhood. "Page after page could be written about the many occurrences which in those days contributed to the fame of La Mare à Boré," wrote the octogenarian in an 1887 essay. Now, the world of his youth all drained away, Gayarré saw himself "standing alone in the arid and parched wilderness of the past, forgotten, but trying in vain to forget and to close my eyes to the shapeless shadows that beckon me away. But enough."[145]

LA SOCIETÉ MÉDICALE AND THE PHYSICO-MEDICAL SOCIETY

While miasmas were a fallacy, the association of filth with malaise was spot-on, even if the causative relationship was not understood. Death rates in New

Orleans ranged around 7 percent in the late 1700s and 4.3 percent throughout the 1800s, and approached 10 percent in epidemic years, compared to 0.8 to 0.9 percent in modern America. The worst maladies included dysentery, typhoid, malaria, dengue, cholera, and most of all, yellow fever.

Yellow fever first struck New Orleans in 1796, killing over 600 people at a time when the city had a population of 8,756. Another seven outbreaks occurred during 1799 to 1812. It would take a century for medical researchers to understand that the vector was *Aedes aegypti,* a mosquito that likely arrived here as eggs on vessels from the Caribbean or directly from its native West Africa. Able to spawn in droplets, the insect, known as *Culex* in the nineteenth century, was well-suited for warm, humid urban environs like New Orleans, where droplets abounded, and human blood meals could be found adjacently. From 1796 to 1905, over 100,000 Louisianians including roughly 40,000 New Orleanians lost their lives to this viral disease.[146]

Not all of those afflicted succumbed. Those born locally tended to develop youthful immunity and were said to be "acclimated," and those of African ancestry were thought to be more resistant as well, perhaps for similar reasons. Most at risk were newcomers: transients, sailors, visiting businessmen, and most of all, immigrants, because they tended to live and work in crowded neighborhoods replete with nuisances. For this reason, people called yellow fever "the strangers' disease," else "yellow jack" or "the saffron scourge," for the jaundice it induced. That was the least frightful of the symptoms; one Louisianan in 1801 described the effects like a scene out of Bosch's *Last Judgment.* First came "lassitude, and a violent pain in the kidneys," then "fever," after which "the mouth becomes parched, and the respiration difficult, the tongue thickens, and [turns] black. . . . Delirium succeeds, and the patient is violent agitated, and would destroy himself if not prevented." The culminating phase brought "spitting of thick and black blood" and a terrifying "inflammation of the eyes [with] rupture of the vessels," after which "the patient sinks into a stupor and dies." Victims went from healthy to agony in forty-eight hours, and over the next two days would either perish or recover.[147]

Reports like these rarely ran in the local press. They would be bad for commerce, and commerce was booming. The port bustled; fortunes were made; population grew, and the city expanded. New Orleans earned contradictory sobriquets—"Queen of the South" and "Necropolis of the South"; "Crescent City" and "the Great Southern Babylon." The ironies gave cause for pundits to wax cynical. "New Orleans is of course exposed to greater varieties of human misery, vice, disease, and want, than any other American town," wrote Timothy Flint in 1826. "Here misery and disease find a home, clean apartments, faithful nursing, and excellent medical attendance."[148]

In fact, medical attendants strove tirelessly to solve the mystery of the plagues, and collaborated across cultural lines. Francophone Creole and "foreign French" doctors formed la Societé Médicale de la Nouvelle Orléans and tended to advise gentle, sparing interventions, while anglophones (Americans) organized the Physico-Medical Society and were predisposed to "heroic medicine," extreme measures such as bloodletting and purging. Each had their hands full: yellow fever epidemics erupted in the 1820s, followed by terrifying cholera epidemics in 1832–33, which wiped out roughly six thousand New Orleanians in three weeks, including five hundred in a single day. Blinded by its own acknowledged ignorance, the city fought back with desperate and sometimes dangerous measures: fumigating with chlorine and sulfurous acid gas, dumping carbolic acid, sprinkling lime, burning tar, and on one occasion, firing cannon. What was really needed was basic enlightenment, and more researchers seeking it.[149]

In 1834, seven young doctors resolved to create that workforce by establishing a physician-training program. Their effort won the support of the state legislature, which in April 1835 passed "an Act to incorporate the Faculty of the Medical College of Louisiana . . . in the city of New Orleans [to] advance the cause of science [and] preservation of health[,] trade and commerce[,] agriculture[, and] arts and sciences." Using a ward offered by Charity Hospital as a provisional classroom, faculty and students got to work, and after the first year, eleven candidates earned medical degrees at the first commencement, on April 5, 1836. By 1840, the growing enrollment studied in larger rented space described as "the New Hall of the College on Canal Street."[150]

During 1837 to 1843, another fifty-five hundred New Orleanians perished to yellow fever, and the Medical College of Louisiana rose to meet the challenge. In 1843, faculty members leased a parcel on what is now 900 Common Street and commissioned architect James Dakin to build an academy for $15,000. According to a contemporary medical journal, the new facility, "adorned with two very rich Corinthian columns[,] contains on the ground floor a large and well arranged lecture room, [for] two hundred students, [and a] chemical laboratory[,] library and reading room." On the second and third floors were an amphitheater, a dissecting room, and a museum.

Impressed, local leaders took up the cause of the medical college at the Louisiana Constitutional Convention in 1844, aiming to expand it into a university. After much wrangling, the legislature in 1847 transformed the private Medical College of Louisiana into the public University of Louisiana.[151]

Later that year, the faculty commissioned a monumental $40,000 expansion. Opened by 1848, the symmetrical trio of Greek temple–like edifices formed a stately visage befitting a scholarly institution in a great city—a city that had just lost another twenty-three hundred citizens during the previous summer's scourge. In these halls, in what would later become Tulane University of Louisiana, medical researchers plotted their data and posited their theories—with the backswamp on their one side, the populace on the other, putrid puddles at their feet, and *Aedes aegypti* buzzing in the air. So began the next era in the drainage history of New Orleans.

3

The Polder-and-Paddle Era,
1830s–1850s

INTERNAL IMPROVEMENTS

We tend to think of infrastructure today as a government charge. But in the America of two hundred years ago, federal authority lacked the resources to keep pace with the expanding nation. It did what it could for what was known at the time as "internal improvements" and relied on states and private interests to do the rest, offering whatever fiscal subsidies and legal accommodation it could muster.

In Louisiana, state government did its part, knowing well that navigation, levees, and drainage undergirded practically the entire economy. In 1831, Gov. Andrew Bienvenu Roman, a Creole from Opelousas, got the state legislature to create a board of public works to spearhead navigation improvements, a vital interest to the region from which he hailed. In possession of newfangled steam shovels and dredges as well as 150 slaves, the board dispatched its assets on various projects, including clearing the Atchafalaya River of logjams, and subsidizing the Barataria and Lafourche Company to dig a canal from present-day Westwego to Thibodaux.[1]

The private sector got involved through the "improvement bank," a financial institution designed to build infrastructure (as opposed to a commercial bank for merchants, or a property bank for mortgages) like canals, locks, and roads, and earn back their investment by running them as a business. Eyeing the profitable trade of Lake Pontchartrain basin resources, a group of Anglo-American financiers won a state charter on March 5, 1831, to form the New Orleans Canal and Banking Company. This improvement bank proposed to build a six-mile-long navigation channel through the backswamp, with a turning basin at the city end and a harbor at the lake. Importantly, from the standpoint of hydrology, the six-foot-deep, sixty-foot-wide waterway would be lined with guide levees and tow roads. The company would then charge tolls for passage and fees for every other conceivable use. With that promise of profit, the N.O.C.&B.C. began offering stock in April, and by early June had raised an astonishing $4 million (over $100 million today) through offices in Boston, New York, Philadelphia, and London—global capitalism, 1830s-style.[2] The three-hundred-foot-wide canal right-of-way ran in three segments through the rears of plantations owned by Macarty and other families in what would soon become Carrollton. Hundreds of Irish ditchers were recruited to start digging the channel in 1832.

One year into the work, disaster struck. "Pelicans have been seen flying over the city," reported one account on September 6, 1833, "which according to popular belief are sure presages of stormy weather at sea."[3] The visiting Scottish Capt. James Edward Alexander told what happened next:

> In the middle of the night I was awoke by the noise of the doors and windows violently agitated by the wind; it increased to the hurricane roar, lulled, and rose again, and blew with appalling force from the opposite point of the compass, rain at the same time deluging the city. Thus it continued all next day: the sea rushed into Lake Pontchartrain; behind the town it burst its banks, and the city was under water, the Levee only being dry. . . . Many houses were unroofed, and almost all damaged[;] many lives were lost [and] the unburied dead were laid in their coffins in the grave-yard, and floated about till the waters subsided. . . . [T]he stench was horrible.[4]

In an era when New Orleanians feared fluvial flooding far more than sea surges, the September 1833 hurricane was an exception. In time, such exceptions would become the rule, and Captain Alexander seemed to sense as much, predicting "one day this city, rapidly increasing as it is in wealth and consequence, will be swept into the Gulf of Mexico."[5]

Few others had Alexander's premonition, but many others savored the "wealth and consequence" Alexander alluded to, including the N.O.C.&B.C., which got right back to digging its channel to the very lake that had just flooded the city. Ditchers excavated a series of sections and piled the barrow to form the guide levees. Once completed, the walls between the sections were removed so that lake water incrementally filled the channel. The first half of the canal, from Mobile Landing (the turning basin at what is now Howard Avenue at Loyola

Avenue) to the Metairie Road (now City Park Avenue) was completed in August 1834; the section to the lake, at what is now West End, was finished in 1835.[6]

Because most urbanites knew the canal for its turning basin, they called it the "Basin Canal"—specifically the "new" one, to distinguish it from the older Carondelet Canal, which became the "Old Basin Canal." The New Basin Canal earned $405,563 in its first year, delivering freight such as sand, gravel, shells, bricks, lumber, firewood, charcoal, fruits, vegetables, finfish, shellfish, game, and livestock, while sending city goods and imports to Mandeville, Madisonville, Covington, and other communities across the lake.

While everyone recognized the economic significance of the New Basin Canal, a few also perceived its hydrological consequences. Its major impact entailed not so much the water it carried as much the topography it created—that is, the twin guide levees along its flanks, topped with tow roads (causeways). The levees kept navigable water within the confines of the channel, and the roads allowed mules or oxen to tow barges along the waterway, as well as carriages paying a toll to use the "turnpike." What was enormously consequential about the guide levees was that they curtailed the Uptown backswamp's ability to drain out Bayou St. John, and thus marked the beginning of the end of using gravity to drain New Orleans. At least one city surveyor came to view that moment regretfully, writing in 1888 that both the New and Old Basin Canal, while "important as arteries of commerce, were so located as to interfere seriously with the natural drainage of the City, [creating] apparently permanent obstructions."[7]

Others viewed the same circumstance as an opportunity: the guide levees effectively broke the watershed into two smaller subbasins, like Dutch polders. Whereas the original swamp had been too unwieldy to dewater, the partitioning by the canals' guide levees had turned it into a defined and manageable problem. Once the swamp was divided, it could be conquered (drained), and once it was drained, it would become valuable.

Land speculators soon scooped up cheap swamplands, confident that drainage would one day boost their value. In an 1839 editorial titled "Speculation—Swamp," the *Daily Picayune* noted the recent "sale at public auction much property in the Faubourg Treme, then inundated with water," and disabused readers from thinking such transactions preposterous. "Inundated with water!—So what if it were? Is not a great part of Egypt . . . ? Was not Holland . . . ? The entire State of Louisiana? . . . Thus our speculators may be supposed to have reasoned when they bid with such avidity for the *land and water* question."[8]

THE NEW ORLEANS DRAINAGE COMPANY

That "speculation mania," as the editorialist called it, caught the attention of financiers who had access to engineering know-how. In 1835, local investors formed the New Orleans Drainage Company, aiming, according to a later report written by City Surveyor Louis Surgi, "to drain and reclaim, by means of canals and ditches, the land comprised between the upper limits of Suburb Livaudais, the line of the New [Basin] Canal to Lake Pontchartrain, along the shore of said lake to Bayou Cochon, in a straight line to Fisherman's Canal down thence to the Mississippi River."[9] The Draining Company, as the firm was often called, earned a recommendation from the board of public works for the state legislature to grant it a twenty-year charter on March 19, 1835.[10] The company raised $360,000 on the first day, and substantial investments from both the city and the state (which had issued the bonds) made it akin to a public corporation—another example of the cross-sector approach to building infrastructure in nineteenth-century America.[11]

Capitalized at $640,000 and equipped with two steam engines, the Draining Company set forth to "drain the swamps between the city and Lake Pontchartrain on the same plan that is adopted by Holland, by hydraulic machines. The profits are derived from the increased value of the lands drained."[12] If all went well, the gains could be lucrative: "The Legislature guaranteed to the company a claim on any lands they may improve, equal to one half of the benefits derived from such improvement."[13]

For the first time in New Orleans, commercial interests got into the business of reclamation and drainage—not just to dry out a new faubourg on the natural levee, but to remove "several feet of water" from "about 35 square miles" of the lowest bottomlands, to allow for "excellent cultivation with cane."[14]

Also new was the technique. The alluded "Holland plan" meant polderization, which entailed erecting a back levee across the backswamp (during low-water conditions) to reduce the expanse to a manageable subbasin (polder), which can then be dewatered via an outlet (either natural or manmade) connecting to an outfall, or discharge area. In this case, the polder would be formed with the help of two interior levees, both along manmade waterways—the New Basin Canal and Fisherman's Canal—along with the natural levees of the Mississippi to the south. The area targeted for reclamation spanned from today's Harmony Street in Central City, to the I-10 corridor roughly from the Smoothie

King Center to the cemeteries, up along Pontchartrain Boulevard to the lake, and back down across Old Metairie to today's Hollygrove and Carrollton/River Bend neighborhoods. The outlet would be Bayou St. John, and the outfall would be Lake Pontchartrain.

And then there was the new energy source. Steam power was nothing new to Louisiana; by one account, it began in 1803, when two entrepreneurs tried to fit a primitive steam engine to a vessel. By 1807, a steam-powered sawmill operated at Manchac, and soon, sugar mills and other operations were adopting the technology. Famously, in 1812, the *New Orleans* docked at its namesake city, propelled from Pittsburgh by a thirty-four-inch-cylinder steam engine. By the early 1830s, over 180 steamboats plied western rivers, each capable of countercurrent navigation, all helping enrich the Queen of the South. In the city proper, steam power provided, among other things, water for domestic use, by pumping from the river into raised reservoirs for delivery to subscribers.[15]

So while locals in 1835 were familiar with steam engines, the Draining Company's proposal nevertheless represented a bold new application. The technology could liberate engineers from the frustration of weak hydraulic head by poldering inundated spaces and using steam pressure to push, paddle, and pump the water upwardly and outwardly. "The Draining Company have entered into a contract with Messrs. Merick and Harper to construct the necessary machinery for thoroughly draining the first division of the swamp," reported the *New Orleans Bee* in 1836, alluding to the aforementioned polder.[16] Operated by company engineer Mr. Commani, the three-horsepower steam engine sat on a five-by-fifteen-foot-long chamber-like "floating bridge," with one end sucking up swamp water and the other end expelling up to a million gallons per hour into Bayou St. John via a "mammoth" cast-iron wheel with two dozen wooden paddles.[17] Three such "draining machines" were installed at various junctures with Bayou St. John, the first in 1837, another in 1840, and a third in 1845, each paddling out their own polders.[18] They looked a bit like the wooden wheels on a country grist mill, only with no millhouse, with steam power instead of a brook, and with the movement of water as discharge rather than input. "Not only will a large quantity of valuable land be brought into use," wrote the *Bee* of the expected result, "but a great source of disease will be removed."[19]

Sanguine as it was in November 1835, the *Bee* expressed an interesting caution in June 1836: "By the present plan of the draining company, the swamps may certainly be drained," the paper allowed. "But as certainly, proprietors of

[said] lands will have to elevate the surface of their lots by thick coating . . . or refuse of some kind, if intended for building. *The drainage cannot give substance to the spongy soil.*"[20] In that passing comment, the unnamed journalist correctly surmised that drainage would cause subsidence. Had he carried the concept to its logical conclusion, he might have predicted that, someday, draining New Orleans would sink it below sea level. He never realized he had touched upon the premier geographical story of the city's next century.

GEORGE T. DUNBAR'S PLAN

As part of the public subsidy, the Louisiana Board of Public Works had the state engineer, George T. Dunbar, aid the private drainage effort by conducting a survey of the backswamp. Dunbar's *Report on the Draining of the Back Lands Beyond Claiborne Street,* according to the US Army Corps of Engineers, "was the first drainage plan for the city that was based on New Orleans's topographic and environmental conditions," although his work was actually a collaboration with the Drainage Company's project.[21] Dunbar's findings, communicated to the company's president, Felix Garcia, in February 1840, recommended the use of storm drains (sewers) feeding underground pipes (rather than open ditches) to rid the streets of water, which would flow by gravity to lift pumps (rather than push paddles) to draw the liquid from below-grade basins and eject it into higher receptacles. Neither the storm drains nor the underground pipes materialized, probably because both were ahead of their time—by fifty-five years, to be precise.[22]

The New Orleans Drainage Company persevered amidst persistent problems. While their steam paddlewheels gushed impressively, that million-gallon-per-hour figure equated to only thirty-seven cubic feet per second. Even in triplicate, with everything running smoothly, that was not enough to offset the incoming runoff and river infiltration, much less draw down standing water. The Panic of 1837, meanwhile, quashed land speculation and vaporized capital, after which the company waded in murky financial and legal waters. State stakeholders sought to divest in the early 1840s, and inundations from the lake and river in 1846 and 1849 swamped the whole project.[23]

The company managed to put the engines up and running again, but they made more noise than progress. Deeply in debt, and never having paid dividends to shareholders, the company sold off its "31 Negroes," enslaved to drain, and came to the end of its twenty-year charter in 1855.[24] Its assets reverted to the city, which continued operating the three paddlewheels into the 1860s.

While the New Orleans Drainage Company failed to reclaim the Uptown backswamp, it succeeded in expelling water far faster than gravity would have, and managed to dry out those blocks along the polder's uppermost perimeter. The company also left behind extensive drainage infrastructure, including the engines and wheels along Bayou St. John and mile-long feeder ditches on Canal Street, Broad Street, Hagan Avenue, Carrollton Avenue, Orleans Street, and Carondelet Street, into the heart of the Faubourg St. Mary.[25] Their effects in (somewhat) improving street conditions caught the attention of a sardonic *Daily Picayune* reporter, who wrote in 1844 how "two large viaducts, well paved over, carry off the superabundant water of . . . 'Lake Poydras,'" a notorious mere known to form along Poydras Street from St. Charles to Magazine—"perhaps one of the greatest achievements of civil engineering of which the present age can boast."[26]

DRAINAGE AND FLOODS

Drainage begets floods. It's one of the great paradoxes of human geography—that the very effort to dewater lures humans to settle in hydric places, thus subjecting them to deluge if and when those dewatering devices fail. "Floods are 'acts of God,'" geographer Gilbert F. White allowed, "but flood losses are largely acts of man."[27]

Louisiana is also a paradox, and here the opposite axiom holds true as well: *Floods beget drainage.* That is, the occasion of a deluge triggers calls to prevent such an "act of God" in the future. The ensuing effort to build levees historically dovetailed with new policies to drain the land behind the levees, creating a source of revenue—through agriculture, pasture, resource extraction, or urbanization—to fund more levees, which in turn brought about more drainage and development.

The New Orleans flood that helped bring about the nation's most far-reaching drainage law began in the rainy spring of 1849. Waters swollen by regional runoff and upcountry snowmelt rose on the lower Mississippi, evoking all-too-familiar fears of a *crevasse*—"a fissure or breaking of the Levée," as one local explained, "occasioned [firstly by] the yielding of the Levée; and secondly, the sinking of the bank of the river."[28] If a crevasse is severe, wrote another observer, "the waters rush [through] with indescribable impetuosity, with a noise like the roaring of a cataract, boiling and foaming, and . . . excit[ing] universal consternation."[29]

Levee crevasses were the premier cause of flooding in and near nineteenth-century New Orleans, such that locals worried far more about river deluges than those induced by hurricanes. But not all crevasses were serious, and some brought benefits. "As the water passes the breach and begins to spread over the lower plains, its velocity is diminished, and its earth matter . . . is deposited," explained the *Picayune* during the gigantic Bell Crevasse on the West Bank in 1858. "Thus the cultivatable lands along the margin of the river become greatly widened by every Crevasse, and the subsequent increase in the fertility of plantations is a measureable compensation for the disadvantage of an overflow."[30]

Another example affected the Macarty plantation in present-day Carrollton. A breach in Macarty's levee in May 1816 flooded the backswamp clear up to the rear flanks of New Orleans proper, such that "one could travel in a skiff from the corner of Chartres and Canal . . . throughout the rear suburbs."[31] The floodwaters damaged that which had been developed, but benefited that which had not, namely the backswamp. "The receding water," noted one historian, "filled the low terrain with alluvial deposits, enriching the soil as well as elevating the swamp sections."[32] Not coincidentally, the year 1816 proved to be unusually healthy for New Orleans—only 651 deaths, compared to 1,252 in 1815 and 1,772 in 1817—probably on account of the massive spring cleaning of detritus and stagnation.[33]

Such cases beg the question of what exactly constitutes a disaster. But for those in harm's way, a crevasse meant destruction, and that's what seemed to be brewing in the spring of 1849. The Fortier plantation in present-day Waggaman was the first to go, when its levee breached on April 17 and unleashed a "rush of water [of the] most awful destructive appearance [with a] noise . . . heard from a long distance." Two miles upriver was the sugarcane plantation of Pierre Sauvé, positioned on the erosive cutbank side of the river, where historically a distributary had bisected the backswamp with the Metairie-Gentilly Ridge.

On May 3, 1849, Sauvé's levee ruptured and quickly widened into a 150-foot-long, 6-foot-deep crevasse. A torrent of river water roared across present-day River Ridge, along what is now Sauvé Road, and accumulated in the swampy lowlands known as Hoey's Basin, named for the plantation of John Hoey. Once that bowl filled, the deluge, bounded by the Metairie Ridge (Metairie Road) to the north and the natural levee of the Mississippi to the south, spread eastward, subsuming the New Basin Canal on May 8, and creeping uphill into the rear of today's Uptown. The track bed of the present-day St. Charles Streetcar Line blocked the water in some areas, but in others the inundation surpassed St. Charles Avenue (named Nyades at the time—Lafon's water nymphs), reaching nearly to Magazine Street. Downtown, floodwaters reached as far as Bourbon Street, though they went no farther east than the Old Basin (Carondelet) Canal, whose guide levees, unlike those of the New Basin Canal, withstood the pressure.

Now bounded, the floodwaters swelled to heights that no one had seen before. A journalist climbed to the 185-foot-high cupola of the St. Charles Hotel on June 4 and described the view: "Far away [to] Carrollton . . . to the lands in the vicinity of the Sauvé crevasse, the surface of the country on the left [East] bank of the Mississippi is one sheet of water, dotted in innumerable spots with houses. . . . The streets in the Second Municipality are now so many vast water courses [issued] from the bosom of the swamp. . . . New Orleans [looks like] the city of Venice."

Among the damages: the New Orleans Drainage Company's prized steam engines and paddle wheels, the very devices devised to dewater, instead were left at "the extreme verge of humanity . . . their engine houses wide apart and isolated . . . as so many arks of civilization."[34] Additional flooding sprang from another crevasse at English Turn, from the still-open Fortier breach, from a break in the Kenner levee, and from weak spots at Bonnet Carré which would later rupture. Sauvé's Crevasse was a disaster by anyone's definition.

Conditions improved as the various breaches helped lower the stage of the river, and rainfall and snowmelt diminished. Volunteers on June 20 finally succeeded in plugging the main Sauvé Crevasse, but not before 220 city blocks, two thousand structures, and twelve thousand residents suffered damages. The floodwaters receded through evaporation, soil percolation, and outflow through the New Basin Canal and Bayou St. John. Citizens cleaned their disheveled abodes, and the city passed a special tax to fund repairs to damaged infrastructure. "May Heaven avert from us such another catastrophe!" wrote the *Picayune* journalist. "May our citizens, in their foresight and their intelligence, devise some means of raising an insuperable barrier to another inundation from [the Mississippi River]!"[35]

Sauvé's Crevasse became a rallying cry for better levees, and its high-water line would serve as a local benchmark for subsequent levels of protection. An

1858 state law, for example, mandated that the levees along the New Basin Canal "shall be raised to the level of the high water from Sauvé Crevasse of eighteen hundred and forty nine, so as to protect the city from inundation from any future crevasse."[36]

Nationally, Sauvé's Crevasse added momentum to calls for federal intervention in flood control and land reclamation ongoing since 1844, when the last major freshet roiled down western rivers. Advocates made their case for levees and reclamation by pegging it to trade and transportation. "Arguing that the destructive waters originated in remote parts of the Union," wrote two historians, "proponents urged Congress to appropriate funds for building levees just as it provided aid for oceanic commerce."[37]

Congress had no such revenue for a commitment of this magnitude, but it did have an abundance of land—including swamplands "unfit for cultivation" titled to the federal government, courtesy of the Louisiana Purchase. Why not let Louisiana have a go at draining them? It would be tough work, for sure, but the state would gain title to those lands and could keep the revenue from their future improvement or sale. The bill went before Congress, and on March 2, 1849, *the Act to Aid the State of Louisiana in Draining the Swamp Lands* became law. "Be it enacted," read the one-page decree, "that, to aid the State of Louisiana in constructing the necessary levees and drains to reclaim the swamp and overflowed lands therein, the whole of those swamp[s] shall be . . . granted to that State."[38]

Two months later came Sauvé's Crevasse, which gave further cause for states to push for federal swamps-for-drainage swaps. Congress responded by expanding the Swamp Lands Law in 1850 and again in 1860, further encouraging states to drain wetlands and boost their values through economic uses—which would generate more revenue for more levees, more drainage, and more wetlands reclamation.

In this manner, floods begot drainage: the deluges of 1844–49 led to the so-called "swamp buster acts" of 1849–60, which eventually accounted for the transfer of 64 million acres from the federal domain to fifteen states, of which most were in the South, and of which 9.3 million acres (15 percent) were in Louisiana, second only to Florida.[39]

Those engineering achievements produced abundant resources and great wealth, and helped make Louisiana what it is today. But they also incurred costs, as they lured people to settle in flood plains and set them up for the sort of losses that geographer Gilbert F. White called "acts of man."[40] And in that manner, drainage begot floods.

DRAINAGE AND POLITICS

The "swamp buster acts" applied to those wetlands that had become federal following the Louisiana Purchase. Closer to New Orleans, most wetlands were privately owned, and their hydrological fate lay in the hands of market forces within competing political jurisdictions whose boundaries rarely coincided with watersheds.

The City of New Orleans had its share of rivalrous factions, and each had a geography and therefore a hydrology. Its society was structured as a tripartite racial caste system, backed by violence and fractured by nationality, language, and religion, with francophone Creoles on one side, anglophone Americans on another, and immigrants caught in the middle. Rancor between Creole and American elements mounted throughout the 1810s and 1820s. The soul of the city was at stake, and so was its day-to-day management, because if different peoples in different spaces didn't work well together, neither would their drainage systems.

They didn't. On March 8, 1836, the state legislature rived New Orleans "into three separate sections, each with distinct municipal powers,"[41] yoked together in an unholy trinity. Ethnic geography drove the new political map: the First and Third municipalities, below Canal and Esplanade respectively, would be mostly Creole, the latter also heavily immigrant. The Second Municipality, above Canal, would be mostly Anglo-American, with plenty of Irish and German immigrants. Each municipality would have its own Council of Aldermen, which together would form a General Council serving under a single mayor. Everything else was in triplicate, from police and fire service to finances, schools, wharfingers, sanitation—and drainage.

The municipality system, which lasted from 1836 to 1852, was confusing, redundant, wasteful, inefficient, and divisive. It especially beleaguered drainage, as water disregards political divisions. And there weren't just three: upriver of New Orleans, in 1833, residents of the Jefferson Parish faubourgs of Nuns, Lafayette, Livaudais, and Delassize petitioned the state to incorporate themselves as the City of Lafayette, comprising today's Irish Channel, Garden District, and Central City. That same year, farther upriver, developers laid out the subdivision of Carrollton on the former Macarty plantation, which incorporated as a town in 1845. One year later, residents living between at Lafayette and Carrollton formed the Borough of Freeport, which in 1850 became the City of Jefferson, from Toledano Street up to Eleonore and later to Lowerline Street.

Each of these six jurisdictions (that is, three semiautonomous municipalities within the City of New Orleans in the Parish of Orleans, and three separate towns or cities in the adjacent Parish of Jefferson) had it its own set of ordinances, syndics, taxes, and policies on their ditches and culverts, and variations in quality and enforcement spanned the gamut. Yet all their runoff flowed into the same hydrological basin, and each entity viewed that backswamp as equally abominable. Maps of these times show a veritable glossary of terms used to describe it, including "sea marsh," "overflowed prairie," "trembling prairie," and "little woods" (*petit bois*) for the tide-affected areas by the lake, and *ciprière,* "woods," "undergrowth," "willow swamp," and "reed jungle" for the forested swamps of the interior. The only thing everyone agreed upon was getting rid of the whole loathsome morass.

Leaders understood this task required some bureaucratic order, and began supplementing their syndic systems with formal drainage commissions. One of the first was the Second Municipality, which in July 1840 appointed three "commissioners to estimate and assess the damage and expense of opening and draining the streets" of Canal, Common, Palmyra, Gravier, Perdido, Poydras, Hevia (now Lafayette), and Cypress, which at the time was an extension of Girod Street. These were all river-perpendicular streets, and they gathered the runoff from the present-day Central Business District, Warehouse District, and Lower Garden District. The ditches steered the surface flow into broader channels on Claiborne and Johnson, after which natural topography, in the form of the Bayou des Cannes tributary, sent the runoff to the "draining machine" on Bayou St. John.[42] All three commissioners of the Second Municipality had Anglo names—Charles Diamond, Robert Douglas, and George Y. Bright—befitting this Anglo-dominant jurisdiction. The other two Creole-dominant municipalities, as well as the three outlying cities, formed comparable drainage entities, be they formal commissions, informal committees, or initiatives within the comptroller's office.[43]

DRAINAGE AND MEDICINE

The most vocal advocates for drainage were not drainage experts; they were health experts. Despite the high death rate, no public health commission existed in New Orleans until the summer of 1837, when civic boosters called for "organizing a board of health, [as] the season has now arrived, when yellow fever usually . . . visits as an epidemic. . . . A board of health is a desideratum long wanted in this city, and it is to be hoped the gentlemen of the medical faculty in this city will unite, and form one without further delay."[44] In September, the city's popular new *Picayune* newspaper asked, "Why can we *not* have a Board of Health in New Orleans?"[45]

The call was heeded, but because of the municipality system, some leaders spoke of forming three separate boards of health. Level heads realized this was a ridiculous idea, and by 1841, the Board of Health of New Orleans was speaking with one unified voice.[46] The decade that followed marked a heroic era in this city's tumultuous public health history: the board of health did its due diligence in recording and reporting statistics; the Drainage Company deployed its steam wheels toward what it understood to be a salubrious mission; the various political jurisdictions launched their drainage commissions; and the Medical College of Louisiana expanded into the University of Louisiana and became the premier pipeline of expertise to the board of health. As a result, New Orleans, despite the political schisms, reached mid-century far more seriously committed to improving conditions than ever before. The 1840s saw drainage operations start to evolve from a bottom-up reactive task of contract engineers, beleaguered citizens, and hired-out slaves, to a top-down political engagement of commissions, boards, and engineering companies.

The emerging civic commitment fueled heady predictions. "[By] the end of the nineteenth century, this metropolis will in all probability extend back to Lake Pontchartrain, and to Carrollton on the course of the river," wrote one influential publisher. "The swamps, that now echo only to the course bellow of the alligator, will then be densely built upon, and rendered cheerful by the gay voices of . . . at least *a million of human beings.*"[47] The drainage commitment certainly had the blessing of the elite class, which had one eye on its own well-being and the other on its pocketbook. In 1850 an informed planter named Thomas Affleck, who had been keeping abreast of the developments, issued an entreaty for authorities to heed. "Undue moisture and filth," he implored, are "the leading causes for the insalubrity of the city. . . . The absolute saturation of the ground . . . with many gases offensive and injurious to health . . . is literally at the foundation of the evil." And as to pocketbooks: "The value of property is lessened and the expense of building greatly increased. Witness the settling of most of the large buildings [and the] greatly enhanced [expense] of a gas or water pipe." The solution, Affleck beseeched: "thorough drainage," which would "yield a large revenue to the city, instead of being an expense. . . . The yield would be incalculable."[48]

4

Plagues and Progress,
1850s–1860s

On May 9, 1853, the square-rigger *Northhampton* arrived from Liverpool and docked along the Fourth District riverfront to discharge four hundred passengers, mostly Irish immigrants in steerage. While cleaning the ship the next day, deckhands discovered a troubling sight: a pool of black vomit in the sick ward. One soon fell sick, by which time it had come to light that "several persons [had] died on the voyage, and [another] man, a steerage passenger, whilst coming up the river." Another, James McGuigan, took ill upon settling into an Irish Channel boardinghouse, where his condition only worsened. He checked into Charity Hospital, and within two days, McGuigan convulsed "with black vomit" and expired. This was Patient One.

Two days later, on May 30, a sailor from Germany named Gerhardt Woette, infirm for five days, arrived at the same hospital. He was dead by that evening, of "yellow fever with black vomit," *Patient Two.* In subsequent days, in similar ways, perished "a seaman born in Scotland," a "laborer, born in England," "a laborer, born in Ireland," a woman "born in Germany," and "a man named Kein," also German.[1] More followed, and still more.

The "stranger's disease" had arrived at New Orleans, apparently on the *Northhampton.* Or had it been here all along? That fellow Kein, after all, had "worked in the swamp[,] living on Gormley's Canal [which] drains from certain streets." And those two siblings from the Irish Channel who caught the contagion? "The Dr. does not think these cases were caused by importation of fever from abroad, but by the locality[,] the neighborhood being wet, without drainage, containing many small tenements, crowded with destitute emigrants."[2] Whatever the origin, the pestilence was spreading fast—and it was not yet summer, much less "the sickly season" of August through October, when epidemics were worst.[3]

So began the deadliest season New Orleans would ever see, and it stultified all aspects of urban life. Streets emptied. Vessels departed. Commerce ceased. Those with means fled—"either to the North or to the pine woods," or to coastal retreats like Biloxi or Grand Isle.[4] Quite literally the liveliest places in town were the cemeteries: whereas under ordinary times, the area's twelve "cities of the dead" accepted around 500 corpses per month, in 1853 those interments numbered around 670 in May and in June, then rose to 2,132 in July and 6,198 in August—two hundred per day.[5] Lafayette Cemetery alone, located in the heart of the affluent Garden District, took in 1,177 new bodies within its one-block space in August. The citywide flow of cadavers into catacombs declined to 1,621 in September and leveled off in the mid-700s through December, still 50 percent above average, and a bellwether of subsequent scourges in 1854 (2,316 deaths) and 1855 (2,615 deaths).[6] Most of the excess mortalities, of course, were attributable to yellow fever.

In all, of the city's population of 154,133, fully 29,120 people contracted yellow fever during 1853, or 1 in 5, and of them, 1 in 4 perished—at least 7,048 documented deaths (probably an undercount, in light of the cemetery data and those missed altogether). Some accounts put the toll at 10,000 or more, and as geographer Craig Colten has pointed out, "that number is made more dramatic by the fact that about half the city fled at the epidemic's outbreak." The two uptown districts above Canal Street suffered the highest infection rates, on account of "strangers"—that is, their large foreign and domestic migrant populations, as well as transient seamen and rivermen. The two downtown districts had markedly lower rates, possibly for their predomination of Creoles (those born locally), who were understood to be "acclimated" to the disease through childhood exposure.[7]

EDWARD HALL BARTON AND THE SANITARY COMMISSION

If the 1840s saw the beginnings of professionalization in public health and water management, the calamity of 1853 brought about a local age of medical enlightenment. It began soon enough, in September, when conditions improved but danger still lurked. The board of health at that time had created a sanitary commission and staffed it with five of the city's best doctors, putting Dr. Edward Hall Barton in charge and Mayor A. D. Crossman as chair. The board charged the commission to determine the epidemic's origin, means of transmission, and the efficacy of quarantine in controlling it. To the question of urban conditions, the board wanted answers on "the subject of sewerage and common drains . . . and their influence on health," and sought "a thorough examination into the sanitary condition of the city."[8]

In their research, Barton and his commission aimed to learn everything they could about yellow fever, surveying communities throughout Louisiana and adjacent states, plus Mexico and the Caribbean and as far away as Ecuador and Brazil. Their methods included testimonies, historical data, and geographical analyses, including maps and graphs overlaying qualitative and quantitative data through time.

The scholarship reflected Barton's academic rigor. Born in Virginia in 1796, Barton studied medicine at the University of Pennsylvania and settled in St. Francisville in 1820, where he became involved in the Louisiana medical profession and developed a healthy skepticism of "heroic medicine," meaning draconian treatments such as bloodletting. He moved to New Orleans in 1833 and specialized in what today would be called biostatistics, in which he integrated medical data with social, geophysical, and climatic information to tease out determinants of public health. A teetotaler in a city that was anything but, Barton once calculated that New Orleanians spent $7,449,989 annually guzzling 1,824,471 gallons of alcohol, at great cost to health and virtue. He was among the first faculty members of the Medical College of Louisiana, serving as its dean from 1836 to 1840, and joining both the Anglo-dominant Physico-Medical Society and the Creoles' *Societé Médicale.* As one of the few Louisiana doctors to be a member of the American Medical Association, Barton in 1849 presented a paper on sanitary conditions in New Orleans, a topic that would dominate the rest of his life. His statistical analyses led him to courageously repudiate his own earlier public plaudits on health policy, and he became notoriously frank in contradicting those medics who whistled past the graveyard in the interest of sustaining commerce.[9]

So Barton had done his homework when, at a meeting of the New Orleans Academy of Sciences where he had presented his data, he was asked whether an epidemic might strike that summer. "Judging from the past, if the facts exhibited by the chart were not mere coincidences," he replied, "the present year would be marked by a great augmentation of disease."[10] The date of that exchange was June 6, 1853, the same week of the first deaths of the Great Yellow Fever Epidemic of 1853.

That sort of fact-backed intuition is what got Barton selected to lead the sanitary commission, and his scientific data pervades nearly every chapter of the commission's final report, released in 1854. While the document does allude to miasmas, evil, vice ("insalubrity and immorality have a similar paternity"), and the "hand of God" ("Providence permits no evils, without there being corresponding remedies"), Barton devoted many more of the six hundred pages to documenting geophysical problems and discussing what today would be called urban water management. He even mentioned mosquitoes, four times, albeit passingly.[11]

Barton was *onto something.* He had collected and analyzed massive amounts of raw data, and strove mightily to connect dots and tease out relationships. He meticulously plotted annual mortality rates from 1787 to 1853, graphing a line as jagged as lion's teeth, and underscoring each posting with corresponding water and soil conditions. To wit: *1785 Crevasses affecting the city; 1796 Canal Carondelet dug . . . Trenches dug around the City & Swamp Exposed . . . First yellow Fever; 1811 Hurricane damaging the city much; 1820 Wooden sidewalks removed; 1824–1828 Gormley's basin and Canal prepared to drain upper & back part of the City; 1832 Extensive digging of the basin of the Bank [New Basin] Canal; 1836 Drained the rear of 2nd District; 1848 Immense excavations . . . for foundation of Customhouse; 1852 Cleaning out Canal Claiborne in August; 1853 Cleaning out Carondelet Canal, digging its new basin . . . Widening canal . . . Digging for Railroads . . . Digging in streets for water & gas . . . Heavy rains . . .*[12] Every watery disturbance flagged in the report also appeared on Barton's high-quality *Sanitary Map of the City of New Orleans,* the best documentation of the drainage system of this era.

The commission's major finding, which members found to be "so unequivocal and so constant," got italicized emphasis. "*No epidemic* has occurred that has not been preceded and accompanied by a great disturbance of the original soil [such as] digging and clearing out canals, basins, &c. . . . The numerous undrained, unfilled lots and squares dotting the surface of the city, becom[e] muddy pools . . . for filth and garbage, [making] the numerous low, crowded and filthy tenements [into] 'fever nests.'"[13]

Barton and the commissioners were spot-on in associating stagnant water with yellow fever, even if they did not understand that the vector was the virally infected *Aedes aegypti* mosquito breeding thereupon. They were also right to recommend a campaign of urban sanitation to abate these nuisances.

But we now know the commission strayed in its conflation of urban nuisances with *all* waters, including those naturally ponding in rural backswamp, and in overplaying the sanitation card, regarding it as a causative agent of the disease. Viewed through the lens of our modern knowledge, this is where the commission erred. One of its culminating recommendations was "the drainage, by machines, in the rear of the city [which] should be so effectual that no water should exist within two to three feet of the surface. . . . The swamps . . . must be effectively drained . . . *at first, thorough and complete* . . . and that hot-bed of pestilence removed."[14] The commission next called for "*a perfect pavement* [which] should consist of materials that would neither admit of absorption nor evaporation. . . . The best protection that exists [is] a *pavement* that will neither absorb or retain water or anything else."[15]

As if to further mortify modern urban-water advocates, the commission also cast an accusatory eye toward forests, viewing them as passive vectors for disease. "An extensive, dense forest growth not only invites moisture, (that is rains), but retains it. Its removal, in clearing the country, is known by experience, to dry up springs, and actually lessens precipitation. . . . Clearing the low country then, and thoroughly draining it, dries it, and as it has been shown, greatly tends to improve its sanitary condition, is urgently demanded here."[16] To be fair, the commissioners did see some value in foliage, sensing that heat played a role in yellow fever, and recommending "planting trees in the public squares and broadest streets, furnishing shade and pure air during the day, and absorbing the noxious gases during the night."[17] They just couldn't see the forest for the trees.

Once the 1853–55 plagues passed and the economy reinvigorated, Barton faded from the public eye. He died in 1857 in North Carolina, of causes unknown, perhaps feeling his work had been for naught in the city he loved.[18]

In fact, Barton's report helped establish an ethos that would deeply inform local sensibilities: that water was the enemy; that drainage was an absolute good; and that dewatering ought to be thorough—removing swamp water, runoff, and soil water, every drop. If Barton wasn't a drainage king, he was the court's chief physician, and his word mattered.

His work, and the trauma behind it, garnered momentum for drainage progress. In 1854, the State of Louisiana enacted a statute for New Orleans to borrow up to half a million dollars ($17 million today) by raising thousand-dollar bonds payable at 6 percent interest over forty years, "to reclaim and drain the swamp and overflowed lands within the corporate limits." Using existing polder perimeters such as canal levees, railroad berms, and natural ridges, the statute broke the city into four "sections." First priority to "drain and reclaim" was section one (roughly today's Lakeview and City Park) and section two (today's Broadmoor, Fontainebleau, greater Carrollton, Hollygrove, and Lakewood up to West End), parts of which were in Jefferson Parish. Third in priority was Section 3, today's Gentilly, and last was Section 4, contiguous with today's Eighth and Ninth wards, the least populated area within municipal limits.

In addition to the bonds, the statute permitted the city to impose a drainage tax based on the size of the real estate holding, as most swamplands near the metropolis were privately owned. This is the origin of the modern funding mechanism for drainage, whereby revenue comes from a tax millage on real estate, rather than the user's fees in place for drinking water and wastewater—something today deemed problematic. The statute also did not give the city power of expropriation, nor control over land use or other aspects of watershed management. And it tended to defer to the New Orleans Draining Company, that 1835 private-public partnership that still splashed away in Bayou St. John with its steam-driven paddle wheels.[19]

Shortfalls and all, the 1854 statute legally sanctioned New Orleans's post-plague war on water and put money on the table to move forward. It also officially codified, for the first time, the city's polders, using the term "sections" for what would later be called "drainage districts."

LOUIS PILIÉ'S REPORT AND THE 1858 BREAKTHROUGH

Momentum for drainage started to grow within city hall, too. In January 1855, the city took the advice of its Committee on Health, a subsidiary of the sanitary commission, and passed an ordinance to establish a *bona fide* Health Department. This was not a temporary board or committee but a permanent city department, with staff and budget, and it continues under that same name today.[20]

Later in 1855, Mayor John L. Lewis nominated "commissioners of swamp drainage" to each of the four "municipal districts," the new nomenclature for the three former municipalities of 1836–52, plus the now-annexed former City of Lafayette. What the commissioners needed was good topographical data on the backswamp, where only a few dozen inches can spell a world of difference. In 1856, the state legislature appropriated five thousand dollars for this purpose, but City Surveyor Louis H. Pilié, son of Barthelemy Lafon's protégé Joseph Pilié, saw New Orleans as his domain, and went ahead to produce his own plan.

Pilié's 1857 *Report on Drainage* focused on the area lakeside of South Claiborne Avenue, which was roughly the rural fringe at that time, and implied that there would be outfall canals dug directly to Lake Pontchartrain, rather than natural flow through Bayou St. John.[21] He also recommended building a levee along the lake, thus polderizing the entire expanse to its south. "As his report was not accompanied by maps, the proposed location of the pumping machinery necessary to lift the drainage water is not clear," wrote a historian in 1922, "but probably it would have been placed along the lake shore."[22] If so, Pilié's report foresaw a number of elements that eventually came to fruition: open outfall canals with ejection into Lake Pontchartrain, lakefront levees, and lift pumps as opposed to push paddles. It also had one element that *should* have come to fruition: lake-shore pump locations. An official city publication from 1895 described Pilié's 1857 report as "the earliest report on the subject of drainage," though that could also be said of the 1840 plan of State Engineer George Dunbar, in affiliation with the New Orleans Draining Company.[23]

Pilié's work spurred the state legislature to pass Act 165 in 1858, which, according to historian John Smith Kendall, for the "first time took definite steps towards the solution of the drainage problem in New Orleans."[24] This law reworked the four municipal-district-based jurisdictions into three "drainage districts," a term that would soon become ubiquitous throughout Louisiana. The First Drainage District covered the urban core, between the Old and New Basin canals, and outflowed at Bayou St. John; the Second spanned Uptown and flowed out a tributary of the bayou; and the Third reached downriver and discharged via "the Draining Engine at the intersection of London avenue with Gentilly road"—the origins of today's London Avenue Outfall Canal.[25] Each drainage district would have an appointed commissioner (president) and a board with powers of assessment and taxing based on land area, the revenue to be used to construct the elements that Pilié had described. The law also augmented these bodies' legal "right, at all times, of entering on the lands . . . and of placing thereon their engines and machinery . . . and of digging all necessary canals and drains."[26]

The 1858 state law gave New Orleans and Jefferson Parish communities the beginnings of a drainage bureaucracy, with each district getting its own office, president, commissioners, secretary, treasurer, attorney, and engineer. Motivated by yet another major yellow fever epidemic in 1858—the second-worst in New Orleans history, claiming 4,855 lives—legislators passed two additional drainage acts in 1859 and 1860. Among other things, these laws gave district boards the ability to issue thirty-year bonds for $350,000 per district, by which time Maj. P. G. T. Beauregard had gotten involved as an advising engineer. The bureaucracy had support from other sectors of the economy: the private sector provided the contractors; the mostly immigrant proletariat class and the enslaved caste did the hard labor; and local industries built the heavy machinery for the dewatering. Leeds & Co., for example, manufactured "steam and horse-power draining machines," and John Armstrong's Foundry and Boiler Manufactory made "draining machines."[27]

Was the new momentum equal to the task? The 1856 state law allowing real estate taxes for drainage got applied sporadically, leaving some property owners aggrieved that theirs were "the only lands in the city which have been especially taxed for drainage, all others having been drained at public expense."[28] The 1858 state law deploying Pilié's plan allocated only $81,000 to do the huge project, to be divided equally among the districts—whose commissioners would draw no compensation, never a good way to run a serious operation.[29]

With revenues disputed, allocations limited, and staff not professionally committed, the effort equivocated. By decade's end, only four steam-driven drainage machines churned away (at the rear of Dublin Street into Carrollton's outfall canal on Seventeenth Street, on the Melpomene Canal at Claiborne, on Bayou St. John, and on Bayou Gentilly), their 28.5- to 34-foot-diameter wheels and four-to-seven-square-foot paddles lifting the water by three to five feet—all still "completely insufficient for the drainage requirements of the city."[30]

RAYMOND THOMASSY'S COLMATES PLAN

From the post-plague "age of enlightenment" emerged one of the most innovative ideas of the era—an argument not to drain the backswamp but to fill it. Behind the concept was a Renaissance man by the name of Raymond Thomassy, who, according to one local researcher, gleaned it while studying the Vatican's *Galleria Delle Carte Geografiche,* the famed topographical paintings by Ignazio Danti commissioned by Pope Gregory XIII in 1580.[31] Along with observations of the Pontine Marshes near Rome, a sound understanding of hydrological engineering, and an ahead-of-his-time willingness to meet nature halfway, Thomassy proposed a system of what he called "colmates"—that is, "artificial aqueducts," as a medical journal explained in 1859, "for transporting the water itself to any spot where its sediment is needed. [Colmates are] contrivances to put water under the will of man . . . for the benefit of the health, agriculture and commerce."[32]

A colmate (from the past participle of the French verb *colmater,* meaning to patch, plug, or fill) was an integrated sediment-delivery system whereby gravity would be used to divert turbid water from the Mississippi through sluice gates and onto aqueducts to the backswamp, where it would fill a checkboard of small polders, that is, pond-like impoundments lined with manmade berms. Once the polders were topped off with muddy water, the sluice gates would be closed, and the standing water would evaporate or infiltrate into the soil, leaving behind a layer of alluvium. Repeated every few days or weeks, the elevation of the polder would gradually rise to match the height of its rims—theoretically raising the backswamp nearly to the height of the natural levee.

If successful, Thomassy's colmates could revolutionize the whole concept of water management in New Orleans. By treating water as an asset instead of a liability, and by rewatering instead of dewatering, colmates could turn the topography of the deltaic plain from a slope to a plateau, shoring up the lowest areas to protect the city from river floods and storm surges. Additionally, the entire soil body would remain wet, as deltas need to be, lest they subside. It was an informed and enlightened solution, and it worked well on paper.

But colmates presupposed many things. For one, much of the backswamp would have to be expropriated, cleared of vegetation, and topographically reconfigured to form the checkerboard of polders. Aqueducts would have to be built through neighborhoods, and god forbid the sluice gates might fail and cause a crevasse (after all, they could only be opened when the river was high, fast, and powerful). All declinations would have to be perfectly executed; all polder rims strong; and the supply of suspended sediment at the top of the river's water column would have to be consistently sufficient to do the land-building at the right pace and timing. Even if everything worked perfectly, the resulting filled land could only be used for agriculture, so that it might be periodically rewatering and replenished. Urbanization would preclude rewatering—lest it flood people's houses—and the deprivation of new water and alluvium would only allow the new soils to subside back into the shape of a bowl, returning us to the original problem.

Yet Thomassy's colmates were worth an airing, and on February 1, 1858, Lewis G. DeRussy submitted to the Louisiana Board of Commissioners a *Special Report Relative to the Cost of Draining the Swamp Lands Bordering Lake Pontchartrain.* In the report, DeRussy explained the "filling" concept, and priced out the needed network of levee-rimmed polders (which he called "sections" and proposed five) and aqueducts (which he designed as canals, about a dozen per section). DeRussy did incorporate some steam engines in the project, but only as auxiliary sources of surplus water mobilization; mostly, gravity would do the trick. Total cost: $1,021,426.[33]

Colmates never got past the envisioning stage. Not until the late 1900s would the concept return to the public discourse, in function if not in name, in the form of river diversions and sediment-siphons to reverse saltwater intrusion and restore coastal wetlands. The fact that they made it to the state's attention in 1858 speaks to the innovative thinking at work in the years following the 1853 plague.

Unfortunately, the drip-drip-drip of actual drainage progress during those years fell short of yielding an effective city drainage department (as had been done for health, with paid professionals and adequate resources). "Large expenditures have been made [on the] questions of levees, drainage and paving," wrote a political scientist in 1888, looking back on the prior half-century. "The trouble in these matters, as in most of our American cities, has been to secure some system that should be *continuous, consistent,* and *rightly administered,* [whereby] work on the levees, the drainage and the paving of the city shall be *harmonized,* and continued without interruption until it shall be *completed in a manner worthy of the place.*"[34]

All that remained elusive at the dawn of the 1860s. Instead, an ad hoc assembly of overcommitted quasi-volunteers worked with private contractors to bootstrap a patchwork of subsystems. Together, they did little more than kick the water can down the muddy road, at what would prove to be a particularly inopportune time.

BENJAMIN BUTLER'S CLEANUP CAMPAIGN

War is hell on infrastructure—particularly on fragile systems in irresolute hands. When Louisiana seceded from the Union in January 1861 and joined the Confederacy in February, local leaders switched to a war footing, and urban drainage sunk on the agenda. When violence broke out in April—and it was local engineer Gen. P. G. T. Beauregard who fired the first shots at Fort Sumter—drainage lost all urgency. "Military companies paraded through the streets, and a large proportion of men were in uniform," wrote a London war correspondent of Confederate New Orleans. "Walls are covered with placards of volunteer companies [and] tailors are busy night and day making uniforms."[35] As Union forces blockaded the river, defense became the city's top priority. Amid the "whirl of secession and politics," gutters went unmaintained, ditches clogged, and steam engines sputtered.[36] The reason was as plain as the result: "The contractors . . . had utterly neglected to comply with their contracts for cleaning and purifying the streets, and the filth was indescribable."[37] These conditions continued for a year, as battles raged elsewhere, and as Confederate *esprit* soared with a string of surprising victories.

In late April 1862, Rear Admiral David G. Farragut's Union squadron charged past bombardments from two Confederate forts at the mouth of the Mississippi and went on to seize New Orleans. By this time, the city's drainage system was in shambles—along with its wharves, warehouses, dry docks, watercraft, and other infrastructure, much of it torched to deprive the enemy of spoils. But now, for once, the problems of New Orleans were no longer New Orleans's problems; local authorities were happy to dump them on their scorned occupiers, under the command of Maj. Gen. Benjamin Butler. When on May 8 Butler wrote to his commander in chief that "New Orleans . . . is at your command," the city's problems, drainage and all, became President Lincoln's.[38]

Consider the irony: back in 1828 and 1831, a young Abe Lincoln had landed at New Orleans as an anonymous upcountry flatboatman, unnoticed at the lower rungs of the social hierarchy. Now the hierarchy certainly noticed President Abraham Lincoln, and like most white southerners, reserved for him the sort of loathing that war correspondent William Howard Russell saw in "a thin, fiery-eyed little woman [who] expressed a fervid desire for bits of 'Old Abe'—his ear, his hair; [either] for the purpose of eating or as curious relics."[39]

In fact, Louisiana had a friend in the old flatboatman because, as the first major rebellious region to return to federal control, the New Orleans area provided Lincoln with an opportunity to demonstrate the strategic benevolence he would later term "re-construction," guided by "the better angels of our nature." One way to make Louisiana an exemplar of Union magnanimity was to give its filthy metropolis a spring cleaning, an effort that would have the added benefit of keeping occupying troops safe. Under the stern rule of Maj. Gen. Benjamin Butler, "the federals assumed responsibility for the administration of the city," wrote historian Gerald M. Capers Jr., "and thus involved themselves in endless troubles of a non-military nature." But, he added, "fundamentally the responsibility was Lincoln's."[40]

Exercising their newfound control of city hall as well as the press, Butler and military commandant Gen. George Shepley staged an exchange of public communiques in June 1862 that got the cleaning project rolling. Knowing full well his power over any impenitent rebel, Shepley issued a deferential entreaty to the "gentlemen" of the city council, that they might consider to put to work "the vast numbers of laborers in this city [who] are unemployed, and suffering from the want of the necessities of life," toward "strengthening the levees and cleaning the streets," thus protecting "the health of the city from miasma generated in the accumulated filth." Shepley's request appeared in the *Daily Picayune* on June 6, aside an article headlined "What Is to Be Done with Slavery," to which Butler replied, "Because of the sins of their betrayers, a worse than the primal curse seems to have befallen [New Orleanians]; the condition of the streets of the city calls for the promptest action for a greater cleanness and more perfect sanitary preparations."[41]

Butler then set forth a series of proposals—orders, really—to create a workforce of two thousand men, properly equipped, to be paid fifty cents a day plus rations for a period of thirty ten-hour days, pending an oath of allegiance to the United States. Their charge: to clean "the streets, squares and unoccupied lands of the city," rid it of "miasma" and "epidemic," and put "those places in such condition as, with the blessing of Providence, shall insure the health as well of the city as my troops." The city council adopted Butler's "resolutions," upped the pay to a dollar a day, and put the work gangs under the direction of the city surveyor, the same position in charge of drainage.[42] Quarantine practices were put in place for river vessels, and the boards of commissioners of the four drainage districts got back to work.

Some old problems persisted: limited funding, insufficient pumping power, confusion over who owned what property and who owed what taxes, even a complaint from fishermen over drainage-caused pollution in Bayou St. John. But the cleansing yielded public health improvements and brought about new progress, following the lakeward-ejection design of the 1857 Pilié plan. New canals were dug by steam dredges on Orleans Avenue to Taylor and Harrison avenues in what is now City Park and Lakeview, where "the water . . . will be pumped and conveyed in a tall race to the Lake" (a "tall race" meaning an outfall canal with guide levees). "Every citizen of New Orleans is interested in the final completion of this work," wrote the board's president, George Ingham, who promised to make "vegetable gardens" of fifteen hundred wet acres. "Converting a swamp into dry land . . . will increase the tax rolls of the city many thousands of dollars, and will contribute much to the health of New Orleans and create room for increased population."[43]

A map dated one year after the Civil War shows the progress that had been made particularly in the Second District (now Uptown), where the Claiborne Canal captured runoff and sent it down four ditches connecting with the New Basin Canal to discharge into Lake Pontchartrain. It was probably this wartime system that brought an end to the tributary bayous in what are now the Fontainebleau and Broadmoor areas, including the legendary Mare à Boré, that "large pond . . . known far and wide[,] sheltering a multitude of wild game." But it would take many more years before subsurface drainage would be installed to ready these neighborhoods for urbanization, and a pond of open water remained on a field in Broadmoor as late as the 1920s.[44]

GEORGE BAYLEY'S PRESCIENT VISION

In October 1864, George Willard Reed Bayley, a highly qualified engineer who succeeded Pilié as city surveyor and aided the wartime drainage efforts, alerted Acting Mayor Stephen Hoyt of his concerns about using the lake as an outfall. "The area to be reclaimed," Bayley wrote, "has to be drained several feet below the level of the receiving reservoir, Lake Pontchartrain, consequently there can be no 'direct drainage to the lake.' The water required to be removed must be elevated, and discharged, either directly or indirectly, into the lake."[45]

Bayley's assessment was among the first recorded public statements since the 1830s to warn that the drainage of swamplands would cause their elevations to drop below lake (sea) level. To address this pending problem, Bayley offered dual proposals. One was to turn Bayou St. John into a sort of elongated harbor, with a navigation lock at the lake, whereby the channel could be widened, deepened, and lowered in its water level. In this manner, adjacent runoff could flow downhill into this new inland harbor-basin, while at the other end, lift pumps would draw the water from the basin and eject it into Lake Pontchartrain.

How to prevent the lake water from slopping back upon the swamp? Bayley's second proposal was as revolutionary as it was prophetic: "A sea-wall might be constructed, say half a mile from the shore line . . . and the space within be filled up from the bed of the lake itself, by means of dredging machines. . . . Such a front on the lake would be very desirable for private residences, or for public gardens. Suitable basins for yachts or other sailing vessels, could be made at intervals, [and] City Park ought to be extended to the lake, and have this sea-wall front upon it."[46] In that 1864 correspondence, City Surveyor George Bayley foresaw the modern geography of New Orleans and the engineering motivations behind it. Drainage would cause land elevations to fall below lake/sea levels; water would therefore have to be lifted to be ejected, which would require that the lakeshore be walled, reclaimed, and raised, creating scenic and valuable new land that would double as a seawall—today's Lakefront. Although his idea about reengineering Bayou St. John did not come to fruition, Bayley did foresee that the twisting rivulet would eventually be gated from the lake and stabilized in its channel. He also correctly predicted that City Park would be extended lakeward, with nearby yachting marinas and lakefront amenities. Bayley died in 1876, exactly a half-century before his ideas would start to come to fruition, and he never received credit for his visions.

All the while, as Civil War battles raged and causalities mounted, the federally administered street-cleaning improved health conditions in New Orleans like never before. By the time the Confederates surrendered in 1865, the city had completed five of its heathiest years on record, four of them coinciding with the bloodiest war in American history. Only fifteen people had died of yellow fever in 1860, and in 1861, largely on account of the quarantining effect of the blockade, the city "enjoyed its first year of entire exemption from yellow fever." After the capture, according to an 1892 northern account, "the thorough cleansing of the city, and improved system of drainage by the federal army of occupation prevented the spread of the disease[;] during these years 1862–3–4–5, only nine deaths being officially reported."[47]

Southern partisans writhed to give Yankee invaders such credit, least of all to Maj. Gen. Benjamin Butler, who became a favorite *bête noire* of unreconstructed rebels. Generations of white New Orleanians would love to loathe Butler for his execution of a civilian accused of desecrating the American flag, and for his infamous General Order No. 28, which enraged southern honor by declaring that any lady caught affronting federal soldiers would be regarded as "a woman of the town, plying her vocation." But the cleansing campaign Butler had organized, with President Lincoln guiding him and Gen. Nathanial Banks replacing him, saved many local lives from yellow fever and other diseases. "Upon the resumption of the civil government," read the 1892 account, "the disease again made its appearance, and in 1867 an epidemic occurred, which caused the death of 3,107 persons."[48] Like the natural disaster of the Carrollton Crevasse in 1816, the human disaster of the Civil War may have actually saved lives in New Orleans, because of how it affected urban water management.

5

Capers and Consequences, 1860s–1870s

COLONEL BROTT'S CAPER

The political turmoil of the Reconstruction Era had a hydrological corollary in the form of a bizarre, high-stakes drainage drama that was as dangerous as it was consequential. It began in September 1868, when the newly formed New Orleans and Ship Island Canal Company (aka Mississippi and Mexican Gulf Ship Canal Company) unveiled to the Republican-controlled state legislature an astonishing plan to excavate a 150-foot-wide, 12-foot-deep navigation channel through the entire metropolitan area. One version of its route had it starting in present-day Old Metairie and ploughing eastward across the city to Bayou Bienvenue through Lake Borgne toward Ship Island, whose depths could accommodate deep-draft vessels (thus the company name). An 1869 update had it starting at a lock with the Mississippi River at today's Shrewsbury, after which it headed north to West End, paralleled the lakeshore eastward, and then cut clear across present-day New Orleans East to the Rigolets and thence to Ship Island.[1] Still other descriptions infer an inner harbor to be dug in what is now the Fontainebleau neighborhood, which would connect with the aforementioned routes. The company's goals were to outcompete high pilotage fees on the lower Mississippi, reduce travel time to Mobile and other coastal cities, "and to afford the dwellers upon the river above the city direct and uninterrupted water communication with the Gulf."[2]

Connecting the river and lake (Pontchartrain, that is) with a narrow longitudinal canal was one thing; folks had speculated about that since colonial times. But cleaving the entire East Bank *latitudinally*—with a 35-mile-long seaway and plans "to widen to 300 feet and to deepen [to] 25 feet" to connect with

Lake *Borgne*—was quite another. Here was a company that came out of nowhere to propose replacing the Mississippi River while turning New Orleans into a coastal city with an inner harbor. It also sought powers of expropriation to attain the needed rights-of-way, as well as leasing privileges on the state-run New Basin Canal. No one seemed to realize that a storm surge pushed westward from Lake Borgne would be funneled directly into the city by this seaway, leading to catastrophic flooding.[3]

The man seeking to redesign the South's largest city was a curly haired Minnesotan of medium build who went by the name of "Col." George F. Brott. Later described as "a born speculator and an ideal promoter [with] restless eyes, a sanguine disposition, winning ways, and a volubility which has seldom been equaled," Brott arrived at New Orleans during the federal occupation, when the town proliferated in crafty opportunists. He managed to pull together six other investors to formulate the plan, after which he paid visits to state legislators, who had been convening in New Orleans rather than Baton Rouge during the Civil War. One by one, Brott and his collaborators won them over, and the project gained traction.[4]

The press had its suspicions, calling Brott's venture a "scheme," questioning the legislative processes behind it, and pondering why "our people should be invited to pay [for] the diversion of trade from our port?"[5] Yet all that the city and government officials seemed inclined to do was kowtow to the company's wishes, starting with tasking their best engineers to accommodate the madcap mission.

LOUIS SURGI'S PLAN

To wit, engineer Louis Surgi, who became city surveyor after George Bayley, dutifully developed a drainage plan to accommodate Brott's state-backed navigation plan, and released it in 1868 to the city council. Surgi's report is something of an archival mystery; we know of it only from secondary sources, and as early as 1869, a journalist noted that "though some pamphlet copies have been distributed, so few are they that we with difficulty obtained the loan of one for our edification." Modified that same year by a committee of seven other engineers, including Bayley, Surgi's plan recommended building an inland levee from "where the Metairie Ridge approaches nearest to the river; thence in a direct line to said ridge; thence [to] the Jefferson and Pontchartrain Railroad; thence . . . to the lake shore . . . easterly to Lafayette Avenue [now Peoples Avenue in Gentilly]; thence in a direct line to the intersection of the Fisherman Canal with Florida Walk [in present-day Arabi, St. Bernard Parish] to the river bank, and along the river bank [back up] to the initial point."[6]

What Surgi and the committee had in mind was a sort of super-polder, in the shape of a thick addition sign (+), built around the urbanized portions of Jefferson, Orleans, and parts of St. Bernard parishes. Its perimeter would take the form of two natural ridges shored up to become artificial levees, connected with a Bayley-style (1864) lakefront seawall, and sealed off to the west and east with lateral protection levees through the swamps, each paralleled with open outfall canals. Everything inside the "bowl" would be drained; everything outside would be left wet. The Ship Island Canal would enter from the east, serving not only for navigation but to carry swamp water eastward into Lake Borgne—"the first time that this route was suggested" as an outfall.[7] To eject street runoff in the city proper, the super-polder would have two additional open outfall canals running straight north, on which steam pumps would propel runoff on their three-mile route into Lake Pontchartrain, rather than the much longer eastward route to Lake Borgne.[8]

Here we see the genesis of today's three main outfall canals in Lakeview and Gentilly. An 1869 map stored in the archives of Tulane University plots their inception, with a "N.O. & S.I. Canal" (that is, the New Orleans and Ship Island navigation channel) drawn where today's Seventeenth Street Outfall Canal is—and "dotted lines represent[ing] draining canals" marking where today's Orleans Avenue and London Avenue outfall canals run.[9] These three seemingly innocuous ditches, all birthed of Brott's folly, would bedevil drainage engineers for generations to come, and two would breach catastrophically during Hurricane Katrina in 2005—a classic case of path dependency, in which carelessly made initial decisions steer future decisions in bad directions.

The 1869 committee went further by adding locks and gates on junctures with the New and Old Basin canals, and recommending that "the ultimate plan of drainage, should be by means of underground sewers, collecting the water and delivering into the [Mississippi] river." That uphill battle—literally—was later judged to be "compiled without any accurate knowledge either of the topography, or of the volumes of water to be handled," and ultimately deemed "impracticable."[10] With the pump-to-the-river concept off the table, the ejection emphasis shifted to pump–to–Lake Pontchartrain, via the outfall canals, and drain-by-gravity to Lake Borgne, via the Ship Island Canal.

In effect, the city's best engineers had become Brott's toadies and, willingly or otherwise, set about designing a public drainage system to accommodate Brott's private shipping interests. The state legislature's groveling was even more egregious. On October 4, 1868, it granted Brott's Ship Island Canal Company a charter to incorporate, and made a point of saluting its effort to "prevent the overflow of the city of New Orleans from the rear." Five months later, the state legislature acted "to repeal all laws and parts of law creating drainage districts in certain portions of the parishes of Orleans and Jefferson, and providing the mode and means of draining the same." With these words, the state killed its own progressive 1858 drainage law, dismissing it as "ineffectual," its funds "misapplied or squandered," its engineering "erroneous in principle and unsuccessful in experience . . . and the health of the citizens endangered." In its stead, the new act went on "to provide for the disposition of the property," along with all taxing and funding mechanisms, "to aid in the construction of the New Orleans and Ship Island Canal." As for Brott's company, it capitalized at $6 million, through stock sales, and later amassed $10 million, the equivalent of nearly $200 million today.[11]

Critics cried foul. Advocates counterattacked. Accusations of corruption flew in each direction. Drainage had gotten caught up in Reconstruction politics, and it came at the expense of the public interest. Drainage districts were abolished, and their commissioners were ordered to turn over all paperwork and assets (ditches, canals, steam engines, dredges, pumps) to whichever political jurisdiction they fell in—the City of New Orleans, the City of Jefferson, the City of Carrollton, or the police juries for unincorporated areas—so that each may transfer them over to Brott's company. The law replaced all those on-the-ground district chiefs with a single governor-appointed commissioner of drainage, to whom would be entrusted a $75,000 bond, and directed that official to aid the Ship Island Canal Company as it reworked regional geography. Signed by Lieut. Gov. Oscar Dunn (Gov. Henry Clay Warmoth having vetoed the bill) and passed by the Republican-controlled House and Senate, the bill became law on March 2, 1869.[12]

On October 4 of that year, a ceremony was held "about a mile above Carrollton" in which the Ship Island Canal Company president, Col. George F. Brott, "dug the first dirt and delivered a brief address." One of the speakers at the event was Governor Warmoth himself, despite his veto.[13]

When corruption charges were later leveled against Governor Warmoth, Brott was called to testify whether, as president of the company, he had ever offered the governor "a sum of money [or stock] to induce him to use his influence with the Legislature to pass the bill over his veto." Brott replied "I did not."[14] Perhaps, but something seemed very fishy.

WATER AND POLITICS

Water does not recognize political boundaries, but political boundaries can certainly affect water. At the same time the Ship Island Canal project made its way down the channels of power, Democrats battled Republicans over the political geography of greater New Orleans. The region's two riverbanks were crisscrossed with the borders of three cities and four parishes, all with their own drainage districts, syndics, and commissioners. New Orleans proper spanned from Toledano Street down to Fishermen's Canal, below which was St. Bernard Parish, and above which were the Jefferson Parish municipalities of Jefferson City and Carrollton City, plus unincorporated enclaves out to Kenner. Across the river was unincorporated Algiers, which was in Orleans Parish but outside the City of New Orleans, plus a necklace of villages amongst plantations and truck farms, from McDonoghville and Gretna to points upriver, and downriver to Belle Chasse in Plaquemines Parish. A single raindrop could flow through multiple jurisdictions as it went from gutter to outfall.

Racially rooted postbellum politics would redraw this map. On one side were Democrats, mostly white southern partisans and former Confederate supporters; on the other was the biracial Republican-controlled state government, headed by Governor Warmoth, backed by federal authorities, and supported by Unionists and recently emancipated slaves. In an attempt to consolidate his power, Governor Warmoth pushed to install his political appointees in unfriendly municipalities, namely Jefferson City, and if that didn't work, subsume these jurisdictions into friendly neighbors, namely New Orleans.

For its part, the City of New Orleans was more than happy to absorb Jefferson City, which would dispense of a potential rival while padding its own tax rolls with valuable real estate. With the Ship Island Canal and its wharf-lined basins slated to bring great economic value to Jefferson City and Carrollton City, all the more did New Orleans covet its neighbors' land.

Toward this end, and following a violent 1869 melee in Jefferson City, the state legislature on March 16, 1870, passed a law annexing Jefferson City into New Orleans, becoming today's Uptown, from Toledano to Lowerline Street. This was essentially a hostile takeover, at a time when communities had little say in their own annexation. *Now the Ship Island Canal harbor would mostly be within New Orleans's city limits.* But it didn't stop there. To counterbalance Jefferson City's Democratic (white) voters now coming into Republican-controlled New Orleans, the annexation of Algiers was included in the 1870 law, under the presumption that its substantial African American population in Freetown would vote Republican.[15]

The 1870 legislation was a terrible loss for Jefferson Parish, and it cost Algiers its independence. The City of New Orleans, meanwhile, more than doubled its urbanized footprint, expanded its share of the now-valued backswamp, and, for the first time, straddled the Mississippi River. The law also updated the charter for the city to create "what was generally known as the Administration system," according to a political scientist writing in 1889. The new model put administrators in charge of specialized bureaus or departments—of Finance, of Commerce, of Assessments (taxation) and Public Accounts, of Police, and of Water-works and Public Buildings, as well as a Department of Improvements, "charged with the construction, cleansing and repair of streets, sidewalks, wharves, bridges and drains."[16] Under this new charter, a board of administrators would oversee drainage projects.

With the legal and political geography now optimized, the state legislature resumed its exertions to lay prostrate for the Ship Island Canal Company. It passed various acts during 1869–71 to provide aid to Brott's company, to reprimand former commissioners for not turning over documents and assets to company officials, and to legally clear the way for work to proceed. On February 24, 1871, the legislature passed "An Act Providing for the Drainage of New Orleans," fawning in its preamble that the company is "prepared to immediately undertake the work [of] the proper and efficient drainage of the city of New Orleans . . . with the only kind of machinery adapted thereto and now ready for use . . . with energy and economy to completion." The act then declared that "the Mississippi and Mexican Gulf Ship [Island] Canal Company . . . is hereby authorized and empowered to excavate drainage canals and protection levees within the present corporate limits of New Orleans and Carrollton . . . and fully drain the area bounded by the protection levees." So as not to burden the company with quotidian operations, the act charged the New Orleans Board of Administrators "to build and run all the pumps and drainage machines necessary to lift the drainage water from said canal or canals over into Lake Pontchartrain."[17] Property owners would be assessed "two mills per superficial foot upon the lands" affected, and "all moneys so collected should be placed to the credit of the Mississippi and Mexican Gulf Ship Canal Company." The city would also pay the company fifty cents for every cubic yard excavated.[18]

Sometime in late 1871, the city started to get cold, wet feet about Brott's caper. Perhaps engineers had privately bent some ears in city hall, or perhaps leaders began to realize just how much power they had ceded to the curly haired Minnesotan who called himself "colonel." With skepticism growing, the city council deemed the company's state-granted powers unconstitutional, on the grounds it had deprived the city of control over its own drainage system (as indeed it had). All along, there had been intense legal challenges to the processes that brought the company into being, with its suspiciously multitudinous names; one court petitioner went on record saying, "there is no such an organization as the New Orleans and Ship Island Canal Company, the pretended corporation being a mere myth gotten up for fraudulent and illegal purpose."[19] The prevailing opinion was that this was a speculative venture, of the "carpetbagger" variety, in which "Col." George F. Brott and his cronies would launch a company, issue stock, win over a sympathetic legislature eager to revitalize its postwar economy, and cash out as the stock price rose.[20] We are left to wonder if Brott ever truly intended to dig that seaway, and apparently, so did his contemporaries. The project lost momentum; bonds were devalued; and revenue problems mounted. In May 1872, Brott's shady enterprise, nearing bankruptcy, transferred its rights to a New York capitalist named Warner Van Norden, and went on to fight years' worth of court battles.

Ostensibly, the Ship Island seaway was still a "go," even though its prospects looked dim. As for the drainage system, which had also been authorized over to Brott and now transferred to Warner Van Norden, it would definitely move forward because, unlike the seaway, the city really needed drainage.

WILLIAM BELL'S PLAN

The task of designing a drainage system around Brott's mess fell to the city's board of administrators, and specifically to William H. Bell, who held the position of city surveyor previously occupied by Surgi and Bayley. In May 1872, the same month that Brott transferred his company's rights to Van Norden, Bell built upon the ideas of his predecessors and released his *Chart of Draining Sections of New Orleans, Showing Present Canals, With Protection Levees and Reservoir*

Canals. He estimated that the construction to be carried out by Van Norden's outfit would cost $3 million.[21]

Bell's plan brought topographic exactitude to the important questions of where to place the pumps and where to eject the outfall. Regarding pump locations, Bell strongly dissuaded the belief "that the drainage of this city can be drawn from the river bank and thrown into the lake by a single line of draining machines placed upon the lake shore." Rather, he recommended "rehandling" the water—that is, installing a sequence of pumps, interior and perimeter, to relay, lift, and push the water as needed, past the Metairie-Gentilly Ridge, over the sedimentation that often clogs canal bottoms, and up and over the proposed lakefront levee.[22] This rehandling approach is embedded in our modern system.

To the question of outfall receptacle, Bell made a distinction between *urban* drainage, meaning of densely populated areas, and *sub*urban drainage, meaning the lakeside swamps. For the former, he called for underground "sewers or large iron pipes, below the city into Bayou Bienvenue, the pipes passing under the navigation canals," and outflowing eastward into Lake Borgne. For the latter, he recommended open outfall canals into Lake Pontchartrain.[23] Bell later advised that the pumps on the Pontchartrain-directed outfall canals be placed along the lakefront perimeter, as opposed to the interior of the basin.[24] In time, the outfall canals that Bell designed to connect with Lake Pontchartrain would make him a *de facto* advocate of Pontchartrain disposal, despite his earlier distinctions in using Borgne or Pontchartrain depending on urban-versus-suburban. The Borgne-versus-Pontchartrain debate, and the perimeter-versus-interior debate, would persist for decades to come, with major consequences for our modern system.

In a subsequent *Plan of Proposed Improvements for the Lake Shore Front,* released in 1873, Bell reprised Bayley's 1864 lakeshore breakwater idea, envisioning a broad reclaimed landmass protruding into Lake Pontchartrain, complete with scenic roadways, promenades, inner harbors, locks for navigation, drainage infrastructure, and siphons to eject discharge farther offshore to prevent shoreline pollution.[25] In time, this plan would be Bell's legacy, as it would inspire the Lakefront Improvement Project fifty years later, just as Bayley's idea from 1864 probably informed Bell in 1873. Bell also deserves credit for articulating the notion of rehandling the water through the framework of what would become our three open outfall canals in modern-day Lakeview and Gentilly. And it was Bell's 1872 plan that Warner Van Norden, filling in for the court-mired Brott, would dig into the landscape.

WARNER VAN NORDEN'S BIG DIG

However chimerical its predecessor, Van Norden's outfit made impressive progress in executing "the Bell Plan."[26] With a legislated commitment to excavate
up to 50,000 cubic years of dirt per month, its steam dredges and work teams
managed to move over 5 million cubic yards of muck from twenty miles of major
drainage canals and another sixteen miles of ditches, completing two-thirds of
its entire obligation. The team also installed seven steam-operated waterwheels,
each with capacities of up to 3 million gallons per hour and a lift of seven feet.
That amounted to a grand total of 780 cubic feet per second (cfs)—seven times
more than the 111 cfs total system capacity from 1840, but a pittance in the face
of the 171 million cubic feet that a heavy all-day rain typically dumped on New
Orleans. The new system could remove 0.125 inches of rainfall per hour, more
than ever before, but it would still take over two and a half days to handle a heavy
all-day rain of roughly 6 inches. And it would do next to nothing to remove
swamp water, much less the groundwater.[27]

Van Norden's outfit could go no farther. Costs by 1875 had totaled $1,713,635,
more than the city could pay, especially after the state forbade it from increasing its municipal debt. Imagine how much worse matters would have been had
the Ship Island seaway also been dug! Instead, the drainage portion of Brott's
scheme, having gone first, ended up killing the seaway portion, which had been
the linchpin all along. The dog had died, so to speak, but the tail kept wagging it.[28]

The great Ship Island Canal escapade came to an ignominious end on February 24, 1876, when the state legislature passed an act authorizing the city to
retake control of all drainage work, and if it so wished, buy back all the digging
tools and pumping apparatus from Van Norden.

One might have hoped that this mercy killing might mark a new day, after
which the progress made by Van Norden on Bell's good plan would resume under new auspices. Alas, it did not. After the 1876 return of drainage to city control, wrote a later court, "little, if any, work was done thereafter by the city, and
the abandonment of the work resulted in largely destroying the value of that
which had been done, the rusting and decay of the machinery and tools, and the
inundation and overflow of the portion of the lands attempted to be drained."[29]

Brott's caper had been an utter fiasco—lost time, wasted money, an em-
barrassment to Reconstructionists, and a relearning of the lesson that drainage should not be privatized. Lawsuits between the city and transferee Warner

Van Norden went all the way to the US Supreme Court, twice, and persisted into 1897, nearly thirty years after Col. George F. Brott first conceived the Ship Island Canal.

While Van Norden got dragged through the courts, Brott continued to peddle his cockamamie brand of infrastructure entrepreneurism elsewhere in the nation. Was he a crook? Evidence seems to suggest the cheerful busybody—part P. T. Barnum, part Rube Goldberg, part drainage king—may have actually believed his fantasies, at least enough to hawk them convincingly to powerful men, which was his true talent. His dubious reign started with that questionable "Col." title, and ended in 1896, when he unveiled his idea for an "ocean to ocean . . . bicycle railway over which you can ride two hundred miles per hour." Reported a gullible journalist for the *Philadelphia Inquirer,* "Colonel Brott says not only will his bicycle railway be built, but over a comparatively short time it will extend all over the country. . . . A bill is pending in before Congress granting permission to construct the road from Washington to New York."[30]

Brott's Ship Island Canal was never built. But as a proposal rolling forward during politically volatile times, it created a set of assumptions that affected Bell's engineering of 1872–73, and Van Norden's execution of 1872–75. Had the proposal never arisen, the drainage decisions of the 1870s would have played out otherwise, and we'd have a different drainage system today.

What resulted instead is well-documented in Thomas Sydenham Hardee's 1877 *Topographical and Drainage Map of New Orleans and Surroundings,* which shows all four of Surgi's and Bell's planned outfall canals in operation, as they remain today. First, to the west, is the "Upper Line Protection Levee and Canal." As per Louis Surgi's 1868 plan, this outfall canal paralleled the Jefferson and Pontchartrain Railroad, and earned its name from the upper line of the City of Carrollton, soon to be the Orleans-Jefferson parish line. An 1869 map labels it the N.O. & S.I. Canal—that is, the New Orleans and Ship Island Canal Company—making it clear that this drainage outfall canal (the role that it serves today) was initially envisioned to be a shipping canal. Neither name stuck; instead, locals nicknamed it by the street nomenclature of Carrollton, whose original circa-1833 grid started with First Street (now St. Charles Avenue) and proceeded back seventeen blocks to a rear drainage ditch. Today we call this rear channel the Palmetto (Washington Avenue) Canal, but at the time it abutted Carrollton's Seventeenth Street, and thus earned the sobriquet "Seventeenth Street Canal." It retains that name to this day, even though its eponymous roadway is long gone, and it has since become the main drain of all of Lakeview and much of Uptown and Old Metairie. The Seventeenth Street Outfall Canal made world news in 2005 when its federal floodwalls ruptured, unleashing a catastrophic deluge in neighborhoods built on drained, sunken swamplands. Why was this particular canal originally dug here? Because of Brott's caper, this being the only section of his proposed shipping channel actually excavated.[31]

Next was the "Orleans Canal," today's Orleans Avenue Outfall Canal, an extension of the drainage apparatus that had been created at the Faubourg Tremé end of Orleans Street during the tenure of Joseph Pilié. It had been positioned to parallel the Canal Street, City Park, & Lake Railroad, later the streetcar line to Spanish Fort. This excavation decision would later define the western edge of City Park, now one of the largest urban parks in the nation.

Then there was the London Avenue Canal, which together with the trajectory of Dubreuil's 1740s mill race—today's Elysian Fields Avenue—would undergird the geometry in nearly every modern subdivision in Gentilly. London Avenue is now gone, having been renamed for A. P. Tureaud, but its name lives on in the London Avenue Outfall Canal, which, too, breached during Hurricane Katrina in 2005, deeply inundating Gentilly.

Finally there was the Lower Line Protection Levee and Canal, the eastern edge of the polder as defined by Surgi in 1868. He had pegged it for Lafayette Street, today's Peoples Avenue, and it too would become an antecedent axis framing the future development of Gentilly. While the lower levee has since been moved farther eastward a number of times, there is still an internal drainage canal on Peoples Avenue.

As millions of cubic yards of muck were being dug throughout the lakeside swamps during the early 1870s, the City of New Orleans continued its expansionist drive. After the 1870 double-annexation of Jefferson and Algiers, city leaders next eyed Carrollton City, seat of Jefferson Parish and home to a scenic community with fine river and rail access. Politicians had long savored this area; an earlier version of the 1870 annexation bill had included Carrollton in the booty, though it was later removed, probably because its Democratic voters might have upset the expected Democrat-Republican balance that Jefferson and Algiers would bring to New Orleans ballot boxes.[32] But in February 1874, a Republican senator representing New Orleans reintroduced the idea, this time targeting not just "all the city of Carrollton" but also "Jefferson Parish up as far as St. Charles Parish."[33] Had that bill passed, the City of New Orleans today would encompass the entire urbanized East Bank, from Kenner to the Rigolets. Instead, the bill was trimmed back to the original Carrollton limits, and on March 23, 1874, that Jefferson Parish city became the New Orleans neighborhood of Carrollton. With it came a whole slice of the backswamp, all the way to the Lake Pontchartrain shore.[34]

The 1870–74 changes of political geography, along with the capers of 1868–76, set the scene for all subsequent decisions of drainage engineering affecting the heart of the city. On the positive side, they simplified water management, in that more polders now pertained wholly and solely to one political jurisdiction, the City of New Orleans. But they complicated what to do with the water in one key polder—Hoey's Basin, between the Metairie Ridge and the river's natural levee—which was now sliced between Jefferson and Orleans parishes by the Upper Line Protection (Seventeenth Street) Canal, causing interparish headaches to this day. The changes also forced Algiers into awkward collaborations with Jefferson and Plaquemines parishes, as well as the communities of Gretna and Belle Chasse, in managing water in a major shared polder on the West Bank. Water and politics, it seemed, just could not agree on their geographies.

EDWARD FONTAINE'S EKMUZESIS PLAN

Despite the Brott fiasco of 1868–76, the idea of replacing the Mississippi River with a manmade seaway continued to bedevil the civic brain trust, as if some demonic little imp kept whispering it to men who had more power than wisdom. In 1878, for example, Congress authorized the Barataria Ship Canal Company "to construct and operate a ship canal from New Orleans to the Gulf of Mexico" across the West Bank, "and to grant to said company the right of way for that purpose."[35] A southward-oriented version of Brott's idea, this deep-draft seaway to Grand Isle proposed to "be the only gate through which the largest ships will reach the port of New Orleans, [i]f the Eads jetties should prove a failure," a reference to the serious shoaling problem at the mouth of the river. Instead, jetties designed by nautical engineer Capt. James Eads, built at South Pass to constrain the flow, increase velocity, mobilize sediment, and scour out the channel, worked wondrously, and the West Bank was spared the schism that had nearly befallen the East Bank.[36]

Still the imp whispered, and the ideas got wackier. In 1879, a professor of theology and natural science named Edward Fontaine published a thesis on "ekmuzesis," a term he coined from the Greek "*ek,* out of, and *muzeo,* to suck." Fontaine proposed to turn New Orleans into a *de facto* valley, by cutting a deep thirty-six-mile-long canal across the entire metropolis and connecting two river meanders with a steeper, shorter, straighter channel to the Gulf. The swift flow of the canal's diverted river water, Fontaine argued, would create a sucking force, and once the city connected drainage and sewerage pipes to the canal bed at acute angles, *ekmuzesis* would "drain the noxious fluids of the city and the contiguous marsh-lands" and discharge them far downriver. Fontaine got so far as to convince the federal government to publish his sucking treatise at taxpayer expense.[37]

That same government, in the century to come, could not put down the idea of replacing the lower Mississippi with a straighter tidewater channel. In time, tens of millions of dollars would be spent on this folly, and it would cause the deaths of hundreds of people, with damages in the billions of dollars.

6

The Progressive Era,
1880s–1890s

It arose in part from postbellum industrialization, which swelled the ranks of the middle class and gave more Americans disposable time and income to engage in civic affairs. It benefited from railroads, which birthed a leisure-travel industry and opened eyes and minds to new places and peoples. It paralleled the rise of higher education and advancements in science, communications, and engineering. It would later become known as the Progressive Era, and while New Orleans remained perfectly regressive in matters such as race relations, its push for other social and structural improvements equaled or exceeded those of many other American cities in the late nineteenth century. Throughout the nation, citizens initiated the progress, and in time the body politic caught up.

What gave folks in New Orleans an incentive was yet another yellow fever epidemic, the worst in twenty years. As in 1853, it likely started with an Irish-born laborer, Frank Walsh, who had been working on Eads's jetties at the mouth of the Mississippi. Walsh returned to New Orleans on July 10, 1878, felt sick, and checked into Charity Hospital, where he developed symptoms of "intense pain[,] nausea, vomiting; tongue coated with white and yellowish fur[,] temperature 105°[,] capillary congestion of face, [and] delirium." He died on July 19, his corpse an "intensely yellow jaundiced hue."

Cases soon multiplied, and when deaths increased twenty-fold by August, it was clear 1878 would be a bad epidemic year. The board of health still held the view that "intensely warm . . . atmosphere loaded with moisture" was the ultimate cause, but thanks to Dr. Edward Barton's research from 1854, it was now understood that poor sanitation and stagnant urban water were key proximate determinants. Targeting those problems, the board distributed thousands of gallons of pure carbolic acid, a toxic phenol extracted from coal tar, to disinfect "whole areas of the city."[1]

In all, 4,056 New Orleanians perished in 1878, mostly during August through October, the vast majority white, with immigrants and transients overrepresented. True numbers were understood to be significantly higher, due to misdiagnosis or oversight; total cases and recoveries were even less precisely known.[2] The Yellow Fever Epidemic of 1878 ended up killing as many as 20,000 people throughout the Mississippi Valley, including 5,000 in Memphis and 3,200 in Mississippi, both record highs. While New Orleans probably introduced the scourge domestically, it likely originated in Havana, the Spanish colonial "Pearl of the Caribbean" to which *Nueva Orleans* once answered, and with which it maintained rigorous commercial and cultural ties. Maybe some help could be found there.

STANFORD CHAILLÉ, RUDOLPH MATAS, CARLOS FINLAY, AND WALTER REED

The saffron scourge being endemic to both cities, the US Havana Yellow Fever Commission decided in 1879 to unite researchers from New Orleans with their Cuban counterparts to compare notes. Chairing the commission was Dr. Stanford Emerson Chaillé, a physiologist at the University of Louisiana School of Medicine who specialized in microscopic methods. Chaillé brought along a medical student named Rudolph Matas, a Louisianian of Catalonian descent, to serve as the delegation's scribe.

In Havana, the American delegates met a commanding figure by the name of Dr. Carlos Juan Finlay, a forty-six-year-old Cuban scientist of Franco-Scottish ancestry who Matas later described as "the model of exemplary wisdom, of the laborious worker wealthy in strength of knowledge, in rectitude of principles, in conscientiousness and intellectual integrity."[3] Finlay had been studying yellow fever since the 1860s, and his more recent work had subverted the conventional wisdom. He knew, of course, that urban sanitation was important for various health reasons. But did filthy stagnant water actually *cause* yellow fever? "Sanitary measures generally adopted to prevent the spread of yellow fever," wrote Dr. Juan A. Del Regato, himself a famed Cuban physician, actually "were inconsistent with a number of observed facts." Yellow fever, it seemed, was not contagious in a person-to-person manner, nor in an environment-to-person manner, as miasmic theory had posited. Rather, it was ported. "An independent agent was needed to transmit the disease," wrote Del Regato; thus "to prevent the transmission[,] this agent would have to be destroyed."[4]

From Chaillé's published research, Finlay gleaned the idea to examine victims' tissues and lesions though microscopy. What he found made him doubt miasmas, and focus instead on vectors—namely, something associated with poor sanitation which had the ability to tap human blood vessels—"all of which conditions," Finlay later wrote, "the mosquito satisfied most admirably through its bite."[5]

Finlay began testing in December 1880, starting with the *Culex* species, which was known to lay eggs in puddles and flourish in urban environments. Determining that the females sucked blood, Finlay carefully harvested their eggs, hatched and nurtured the offspring to maturity, and exposed them to a human infected with yellow fever.

Now came the experiment. Having secured "necessary authorization" from the Spanish military to recruit twenty volunteers for the study, Finlay designated fifteen of the soldiers to serve as experimental controls and subjected the other five to the female *Culex* which had bitten the sick patient. "The first experimental subject," wrote medical historian Enrique Chaves-Carballo, "was inoculated on June 30, 1881. Nine days later, he developed fever, jaundice, and albuminuria."[6] The results convinced Finlay that *Culex* was a vector, and he wrote up his findings. On August 14, 1881, Finlay presented "The Mosquito Hypothetically Considered as the Agent of Transmission of Yellow Fever" to the prestigious Spanish Royal Academy of Medical, Physical, and Natural Sciences at its meeting in Havana.

The response was devastating—"incredulity and ridicule from his colleagues," according to Chaves-Carballo.[7] That was the prevalent view throughout the medical community, if it had noticed at all. But there were exceptions, such as Louisiana-born Dr. Rudolf Matas, that young scribe who had met Finlay in 1879 and later advocated for Finlay's "mosquito hypothesis." Among those Matas had convinced of Finlay's finding was Alabama-born Army physician Dr. William Gorgas. Americans of the Gulf Coast kept the Cuban's research alive.

By the late 1890s, the United States had more direct interests in the Caribbean basin. Capitalists invested in fruit-growing regions; imperialists eyed possible island colonies; and engineers envisaged a canal dug across the Central American isthmus. Then, in 1898, war broke out with Spain, in which more

American soldiers would die of tropical disease than in battle. Two years later, the US Army Fourth Yellow Fever Commission formed under the leadership of Maj. Walter Reed. Once again, American doctors set off for Cuba, which was now on its path toward independence following the defeat of Spain in the Spanish-American War.

Having heard impassioned praise from admirers like Matas and Gorgas, the commissioners began their Havana sojourn with a visit to 110 Calle Aguacate—home of Carlos Finlay, now sixty-seven and still busy with his experiments. "Finlay was elated," wrote Chaves-Carballo of the meeting, "and regaled the visitors with a detailed exposition of his ideas, reprints of his publications, experimental notes, and documents related to his work." Finlay also showed them the eggs of the *Culex* mosquito, which an entomologist on the commission confirmed to be *Culex fasciatus.* The right people were finally all in the same room.

Standing on Finlay's shoulders, the American commissioners got to work at Quemados, Cuba, where an epidemic had been breaking out. This was truly heroic medicine: one team member got infected by a trial *Culex* and fell severely ill, while another got bitten by a wild *Culex* and later died. But under the perseverance of Major. Reed, the commissioners managed to conduct rigorous tests and obtain critical information from tissue, blood, and organ samples, which Reed analyzed in his lab in Washington. He soon proved scientifically that *Culex fasciatus* was indeed the vector of the virus that caused yellow fever—just as Finlay had suspected in 1881.

On October 23, 1900, Reed presented his findings to the American Public Health Association in Indianapolis. He shared authorship of "The Etiology of Yellow Fever—a Preliminary Note" with three colleagues, one of whom, he wrote in a footnote, had died of the very disease he helped demystify. Reed also acknowledged that the mosquito hypothesis had been "first advanced and ingeniously discussed by Dr. Carlos J. Finlay," to whom the commission wished to "express our sincere thanks" for his critical research "during the past nineteen years."[8]

The New Orleans medical community, forged by yellow fever, had played a vital role in solving the riddle, and key figures went on to illustrious careers. Stanford Chaillé, chair of the 1879 commission, became known as "the Father of Hygiene and Health Education" and served as dean of the University of Louisiana School of Medicine. That institution later became Tulane University of Louisiana, and its School of Medicine today is home to the Rudolph Matas Library of Health Sciences. William Gorgas went on to establish mosquito-control campaigns in Florida, Cuba, and Panama, saving thousands of lives. Finlay remains a national hero in Cuba today, and Reed's memory lives on through the Walter Reed Army Institute of Research.

As for *Culex fasciatus,* that invasive pest was later reclassified as *Stegomyia fasciata* and is now known as *Aedes aegypti.* It's still rife in New Orleans today, a product of the Middle Passage, the cause of unimaginable suffering, a vector of dengue and Zika today, and the inadvertent motivator of two great thrusts for swamp drainage in New Orleans. The first followed the city's two worst epidemics, in 1853 and in 1858, and the second followed the third-worst epidemic, in 1878, at the dawn of the Progressive Era.

THE NEW ORLEANS AUXILIARY SANITARY ASSOCIATION

What the 1850s plagues did to spur government investment in public health, the 1878 scourge did to engage civil society in the effort. At the same time the Yellow Fever Commission set off for Cuba in 1879, a group of "thoughtful and intelligent merchants" and other citizens formed the New Orleans Auxiliary Sanitary Association, "for the purpose of promoting public health," wrote a political scientist in 1888.[9] "It was felt that, in the condition of the finances of the City, it was necessary to invoke private subscription. The appeals of the Association met with a liberal response." In the years to come, determined volunteers pushed for urban improvements, criticized derelictions of public duty, and even got involved in day-to-day operations. For example, they arranged for garbage and excrement to be loaded onto river barges and dumped midstream well below city limits, rather than at the nuisance wharves right by the downtown water source.

The Auxiliary Sanitary Association understood that water could be a cleansing friend of public health; only when it stagnated did it become a foe. To that end, the association raised funds to install two giant steam pumps on the levee, "with a daily capacity of about 8,000,000 gallons each, and arranged a system of pipes by which the principal streets at right angles to the river are supplied with flushing water." Through its members, many of them prominent business leaders, the association was able to build the system for only $75,000, whereas the same effort by the city would have run closer to $200,000. Little wooden hydrants with spigots were installed on selected arteries, and while they lacked sufficient pressure to hose away muck or extinguish a fire, they served well in flushing gutters.[10] Members of the association's Flushing Committee went further in 1881, issuing technical recommendations to improve both sewerage and drainage via "the ability to 'store' . . . surplus water in the vast 'reservoir canals' . . . to 'flush' the draining canals, thus avoiding the introduction of additional water from the navigation canals, and to that extent relieve the draining wheels."[11]

Relieving pumps, storing water, flushing stagnation: such enlightened thinking would thrill a modern-day urban water manager, and it came from a bunch of amateurs with day jobs. Indeed, the Auxiliary Sanitary Association was pure Tocquevillian America, in which smart autodidacts formed private associations to address local problems where government was too ineffectual to act. In the words of one admirer, the citizens behind the association "were actuated by the most patriotic motives—the rehabilitation of the city."[12] But if asked, members would probably make clear their goal was to fill in where government fell short, and that the true solution was better government, not volunteers manning pumps and flushing gutters. They would probably also acknowledge that public health was good for business.

Officials took note. The years 1879–80 saw a new state constitution, a slate of new government officials, and a reorganized Louisiana Board of Health, with Dr. Joseph Jones as president. That was good news for New Orleans because Dr. Jones, already esteemed for his mastery of medical record-keeping, had been closely involved in the city's 1878 epidemic. He treated sixty patients in his own office, most of whom recovered, and later documented the experience in the *New Orleans Medical and Surgical Journal.*[13]

Attention from Jones's board of health and civic leaders soon returned to the issue of drainage, which by now took the form of what people called "the Bell Plan." As it currently stood, that system could remove at most 0.125 inches of rainfall per hour, about 90 percent short of what was needed, and it dried only a slender perimeter of the swamp. The city's population, meanwhile, had grown stridently. "Some 70,000 French and Creoles, 30,000 Germans, 60,000 Negroes and mulattoes . . . 10,000 Mexicans, Spanish, and Italians [and] 80,000 or 90,000 . . . Anglo-Americans" lived cheek-by-jowl upon the limited space of the natural levee.[14] Pressure came from real estate interests as well as public health experts to turn the backswamp into living space and enable the Queen of the South to reign from bank to shore. Commercial interests and progressive forces such as the Auxiliary Sanitary Association spoke with one voice to finally solve the drainage problem, and the city, with so much potential tax revenue at stake, heard the call as well.

BENJAMIN HARROD'S ELEVEN-POINT REPORT

In 1888, the city council moved forward on two parallel drainage efforts. For one, it requested the state legislature to create a Commission of Public Works for the city, to be led by the mayor and a commissioner, staffed with the chairmen from five city council committees, and including "six citizens to be selected from different districts of the city."[15] In a related second effort, on November 13, the city council called for a report on the status of the current Bell Plan system, to be written promptly by the newly appointed city surveyor, civil engineer Maj. Benjamin Morgan Harrod.

Harrod, born in New Orleans in 1837, had graduated from Harvard in 1856 and returned home to study engineering and architecture before enlisting in the Confederate Army. Serving as an engineer during the Vicksburg Campaign and at the final defenses at Richmond, Harrod attained the rank of major by the time of the surrender at Appomattox. Afterwards, Major Harrod's career, like that of Barthélemy Lafon a half-century earlier, made him seem omnipresent—as state engineer of Louisiana, designer of levees, member of the Mississippi River Commission, surveyor of tributaries, architect and engineer in private practice, later a member of the Panama Canal Commission, and, starting in 1888, chief engineer of the City of New Orleans.[16]

On November 22, 1888, within ten days of the city's request, Harrod released his *Report on Drainage, to the City Council of New Orleans.* This important document examined the polders of the main populated portions of the East Bank of Orleans Parish, which at the time comprised four drainage districts. The First District (downtown, between the New Basin and Old Basin canals), spanned 2,218 acres and would amass 29.5 million cubic feet of runoff in a heavy rain of six inches over 24 hours. The Second (Uptown, from the New Basin Canal to Carrollton), which was the largest, at 8,650 acres, would collect 72.3 million cubic feet in such an event. The Third (the lower faubourgs) had 4612 acres and 42 million cubic feet; and the sparsely populated Fourth (today's Ninth Ward), had 3,724 acres and 27 million cubic feet of runoff.[17] Harrod's report handled Algiers separately, and excluded "the lands between the ridges and Lake Pontchartrain," meaning most of today's Lakeview, Gentilly, and New Orleans East.[18]

Harrod recounted the hydrological regrets he had inherited from New Orleans's past—projects dating to the late 1700s that made a tough job tougher. If he had his druthers, Harrod wrote, he would have capitalized on the natural flow of Bayou Bienvenue for the eastern wards, discharging into Lake Borgne, and particularly on Bayou St. John, the natural outlet for most of the city, flowing into Lake Pontchartrain. Yet, despite a court ruling that Bayou St. John "was not a naturally navigable stream[,] the strange result has occurred that, while a private corporation has acquired the right to improve its navigation, the City has lost its original and normal servitude on the only natural use the stream possessed, that of drainage." This explains why, to this day, Bayou St. John is not used for municipal drainage.

Then there were the circa-1830s New Basin Canal and the circa-1790s Carondelet (Old Basin) Canal, which, "however important as arteries of commerce, were so located as to interfere seriously with the natural drainage of the City," their guide levees imposing "apparently permanent obstructions."[19] Here we see evidence that, on a deltaic plain, a manmade intervention like a canal (or track bed, or roadbed) is the equivalent of a natural ridge, with hydrological consequences to match. But none of this was going to change, Harrod understood, so the dutiful civil servant resigned himself by declaring "any plan of improvement should accept and deal with the[se] conditions."[20]

Now for solutions, Harrod laid out an eleven-point strategy in his report. First, he needed better maps and data—$10,000 for surveyors to collect accurate topographic measurements, rainfall estimates, canal capacities, and more.

Next, levees needed to be higher and stronger; main canals needed to be sufficient in number, length, and capacity; a wider array of intercepting canals was required to draw the gutter flow into the main canals; the gutters themselves could be better designed; new "drainage machines, of proper number, power and location" had to be installed, and old-fashioned water wheels had to be replaced with better-lifting centrifugal pumps; outfall canals had to be extended and better designed; a means to flush the canals of debris and stagnation was needed; and "subsoil or tile drainage of streets and premises," meaning underground pipes for subsurface drainage, had to be installed.

Revealingly, Harrod also called for the "the extension of *permanent and impervious pavements* as rapidly as possible." Anathema to planners today, this directive reflected an ethos that called for every drop of rain to be sent to the pumps as quickly as possible. If water was the problem and drainage the solution, why dilly-dally? What resulted was a call for a hardened, "closed" drainage system—what today would be called "gray" infrastructure, and a rejection of "blue" (water retention) and "green" (vegetated permeable surfaces).[21] Harrod went further, rejecting any notion of storing water in the canals, and recommending quite the opposite: "It is of extreme importance in the drainage and sanitation of the City that the water surface in these canals *should be kept down by pumping to the lowest possible level*—six feet below the surface of the soil, and more, if practicable."[22] Why? Because otherwise, "the adjacent lands are . . . in a state of such thorough saturation as to be unable to absorb and retain any important proportion of the rainfalls." Groundwater was viewed as an obstacle to the riddance of surface water, and the way to dispense of both problems was by drawing down the canal stage "to the lowest possible level." Thus, pump, pump, pump. Draining New Orleans was no longer a job for gravity; it was a job for machines. Use them!

As for green infrastructure, Harrod set aside "the lands between the ridges and Lake Pontchartrain"—today's Lakeview, Gentilly, and New Orleans East—leaving their full drainage for another day. Rather, he wrote that this area "should be leveed, cleared and [partially] drained, and its surface exposed to the beneficent influences of the sun and air," implying that sunlight would cure the dampness of what in times past would have been called miasmas. "The sanitary benefits of this would extend over the City, while the local economic and agricultural results could be made to pay for this part of the improvement."[23]

Where should the outfall be? Harrod opted to discharge some lakeside areas northward through the Bell-era outfall canals, for which steam-driven water wheels might suffice. But the bulk of the urban runoff, Harrod thought, would go eastward into Bayou Bienvenue and Lake Borgne, entrained by powerful centrifugal pumps. *Perhaps if Harrod included today's Lakeview, Gentilly, and New Orleans East in his calculations, he would have seen more need to pump north to Pontchartrain, and less need for the long trip eastward to Borgne.* This counsel, which would be reiterated in the 1890s, would lead to our present-day pumping stations being placed along the Broad Street–Florida Avenue corridor, with enormous consequences. "An incidental benefit of this part of the plan," Harrod went on, is the "preservation of Lake Pontchartrain in fit condition for pleasure and business," a recognition that the lakeside resorts of West End, Spanish Fort, and Milneburg were important economic interests at the time.[24]

Total cost estimate: a surprisingly modest $961,600, less than Raymond Thomassy's colmates plan of 1858, unadjusted. To pay it, Harrod recommended the issuance of bonds backed by a real estate tax millage, perhaps with the jumpstart of a special tax.[25]

Harrod's 1888 report was not intended to be a technical plan, but rather an assessment of current circumstances and roadmap to a better system. It paralleled various drainage bills being discussed in the state legislature, amid a growing public discourse on water management. Indeed, drainage was the talk of the town in the late 1880s; city papers published 154 drainage stories in 1886; 206 in 1887; 329 in 1888; and 326 in 1889. In those two last years, around the time the Harrod report came out, drainage made headline news 137 times.[26]

THE ORLEANS PARISH LEVEE BOARD

Adding to the momentum was the State of Louisiana's creation of "the Orleans Levee District and the Board of Commissioners thereof" in July 1890. Seven appointed board members were "charged with the construction and repairs . . . of all levees in said Orleans district," to be funded by a tax of up to one mill on assessed real estate value—itself an incentive to levee and drain.[27] As its name implied, the Orleans Levee District Board was created to shore up embankments and not to drain bottomlands, but everyone understood that drainage depended on building rear levees for polderization, and this would be the agency to build them. "Beginnings of a Drainage System," read a *Daily Picayune* headline on January 4, 1891, under which was reported the Orleans Levee Board's new charge to build "a complete and unbroken embankment or dyke around the city from the river to the lake, above and below [to] convert the inclosed area into what in Holland would be called a 'polder.'" The journalist understood this was only an initial step. "It would then become necessary to use enlarged means for raising the water out of the basin so formed, and something would have to be done to that end."[28]

A letter to the editor published that same day, titled "Agitating for a Drainage System," responded to a prior conversation titled "The Only Way to Get a Boom Here," and pointed to the Harrod report as the way to get it done. Private engineers also weighed in, proffering their own drainage plans. One "Mr. Joseph Jouet" submitted his vision to the mayor in 1881, and a "Mr. J. L. Gubernator" did the same to the city council in 1889; the former recommended discharging to Lake Borgne, the latter to Pontchartrain. A third plan, from Mr. S. D. Peters, agreed with Jouet.[29] The suggestions were in response to the Orleans Levee Board's 1890 offer to citizens of "a premium of $2500.00 for the best plan of drainage for the City of New Orleans." But the board admitted it could only provide citizens with a rudimentary city map, with "no exact knowledge . . . as to topography, areas, hydrography, or other data essential for the formulation of an efficient plan." As a result, "although there were several plans submitted in response to the advertisement, none could be accepted."[30]

The lack of data did not bode well for drainage, but no one could fault citizens for lack of enthusiasm. Indeed, the zeitgeist of the 1880s–90s might remind modern New Orleanians of the aftermath of Hurricane Katrina in 2005, when myriad entities—mayoral commissions, the city council, out-of-state nonprofits, neighborhood associations, even private citizens—took it upon themselves to release their own recovery plans, rebuilding plans, resiliency plans, sustainability plans, and so forth. One wag called it "plandemonium," and that could well describe the civic spirit for drainage improvement in circa-1890 New Orleans.[31]

Yet despite the groundswell, City Surveyor Benjamin Harrod failed to convince city hall to fund his eleven-point strategy, not even that initial topographic survey. Reason: Mayor Joseph Shakespeare had signed numerous contracts for pressing concerns like the fire department and criminal justice, and he was apparently disinclined to dig the fiscal hole deeper for a long-term investment like drainage. An attempt by "public spirited citizens" to raise private moneys for drainage also fell short.[32] What was needed was political leadership to match the progressive public zeal.

Into that role stepped a new mayor, John Fitzpatrick. Born in 1844 and orphaned in childhood, Fitzpatrick came of age in St. Mary's Asylum in what is now Bywater, and worked as a newsboy and carpenter in the Third Ward before going into politics. Allying himself with working-class folks and the public interest, he became commissioner of public works in 1884, a position that exposed him to water-management issues. After defeating Mayor Shakespeare in the run for city hall and assuming office on April 25, 1892, "Mayor Fitzpatrick showed an enlightened interest in the problem of the drainage of the city," wrote historian John Smith Kendall in 1922, "and under his auspices the movement looking to the solution thereof was encouraged and developed."[33] Drainage was finally poised to become an executive priority.

Late that year, the city council debated allocating funds for that long-overdue topographical survey. So convoluted had drainage politics become that some councilmen were certain such dollars had already been allocated, and that proper maps already existed. "A rigid search was made for records of this nature," read a later report, "but without result." The council had to be convinced that, despite twenty years of piecemeal jury-rigging and patchwork repairs—twenty years of talk—"no work of any magnitude ha[d] been done since 1872 and 1873 under what is known as the 'Bell Plan.'"[34]

So convinced, on Tuesday night, January 31, 1893, the city council passed Ordinance No. 7170, allocating $17,500 "for the making of a thorough and complete topographical survey of the city of New Orleans, and to obtain all necessary data [for] drainage"; for directing the city engineer to "prepare and submit . . . a complete and comprehensive plan; and for appointing "an advisory board of engineers for the purpose of approving or disproving" the aforesaid map and plan. A subsequent Ordinance No. 8327 appointed members to the advisory board on drainage, to be led by the city engineer, Linus Weed Brown.[35]

LINUS BROWN AND THE ADVISORY BOARD'S
TOPOGRAPHIC SURVEY

The advisory board on drainage brought forth the next generation of drainage kings, and the most qualified to date. Among them were ten engineers, a draftsman and "computer" (mathematician); three esquires; and dozens of field crew and support staff. Many, even the newcomers, had apprenticed under local mentors and learned the idiosyncrasies of local geography, presaging a noble trait of the city's modern drainage community: deep-rooted in-house expertise, institutional devotion, and love of city. Mechanical engineer Linus W. Brown, for example, had originally come down from New York to work for the Southern Pacific Railroad, but soon found himself drawn to the challenge of New Orleans drainage. He became the assistant and eventually the successor to City Engineer Benjamin Harrod, the native-born New Orleanian now considered the dean of Louisiana engineering. Harrod and Brown, mentor and protégé, would later mutually credit each other for the drainage breakthroughs to come.[36]

Besides Harrod, Brown's advisory board included Maj. Henry B. Richardson, current chief of the Louisiana State Board of Engineers; Albert Baldwin, whose kin would later make engineering history; and everyone's mentor, Rudolph Hering, an American-born water engineer who trained in Germany and honed his expertise in Philadelphia, Chicago, New York City, and beyond. As for "buy-in" from city hall, the ex-officio chairman of the advisory board was none other than Mayor John Fitzpatrick.[37]

Ordinance No. 7170 seemed rather quotidian at the time, lost amid the seven-hundred-plus drainage articles published by the local press during the previous year. The only coverage it got came two days later in the *Daily Picayune,*

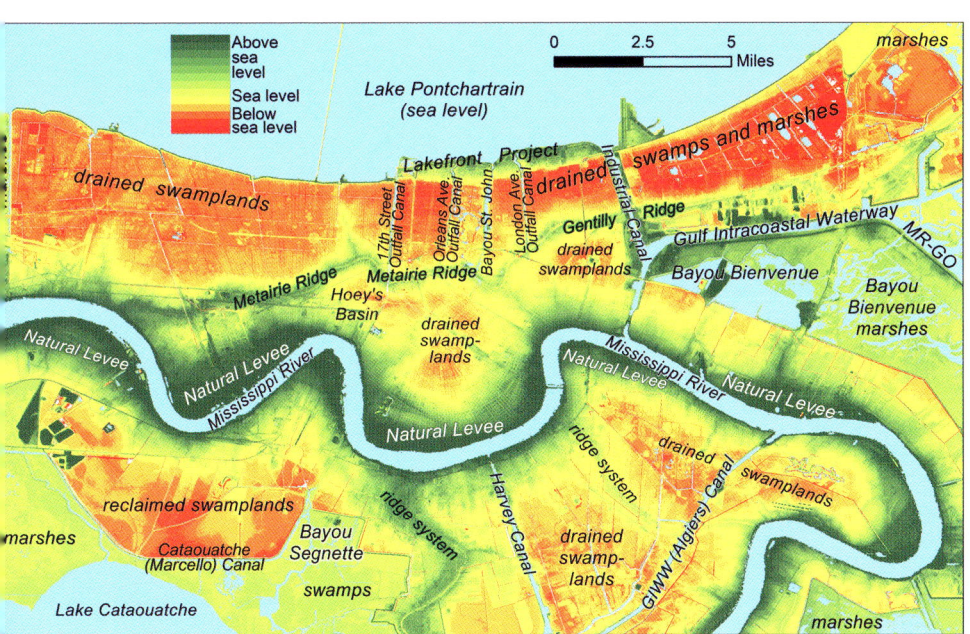

Urban/political geography, *top,* and topography/hydrology, *bottom,* of greater New Orleans. Maps by Richard Campanella, using Landsat satellite imagery and FEMA LIDAR data.

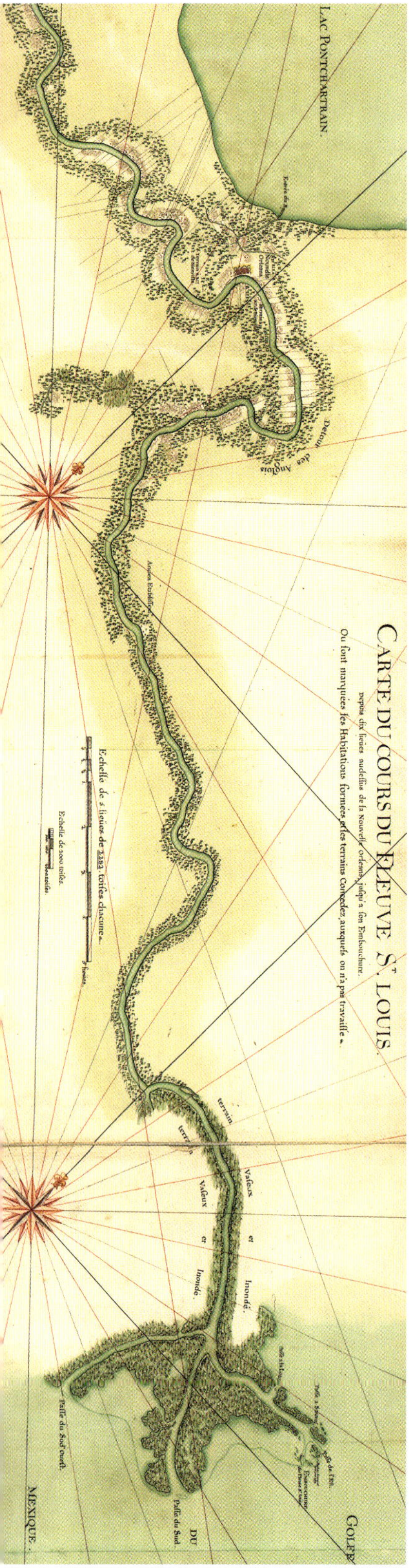

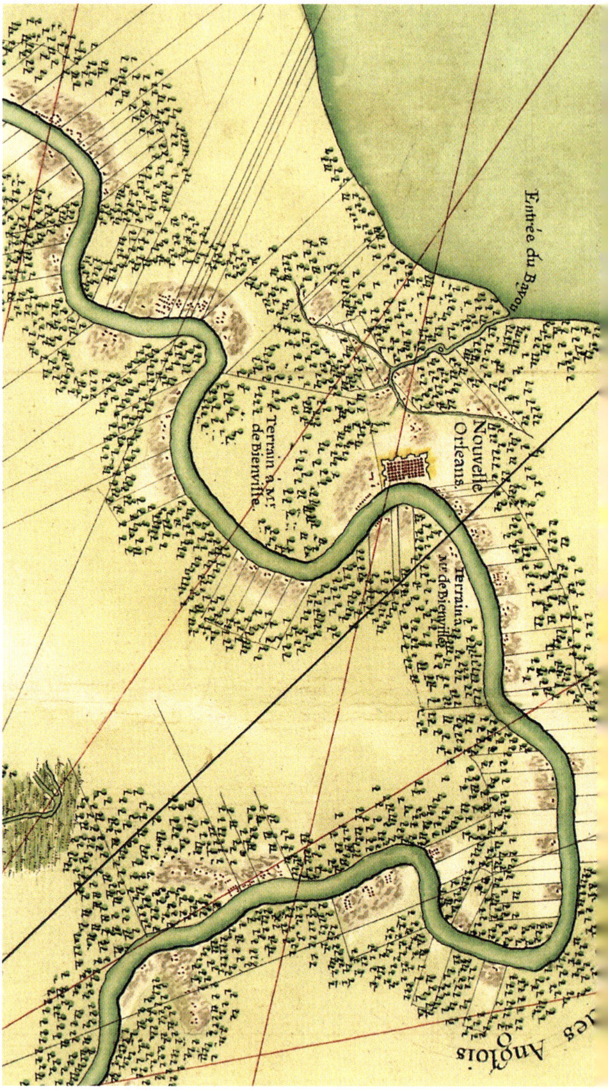

Left: The lower Mississippi depicted in 1732, capturing the prograde
nature of this fluvial delta as it extends into the *Golfe du Mexique*
astride an apron of swamps and coastal marshes. *Above:* Detail of New
Orleans area, illustrating the higher elevations of the natural levee
(areas closer to Mississippi River, demarcated with French long-lot
plantations). Details of *Carte du Cours du Fleuve St. Louis,* courtesy of
Library of Congress.

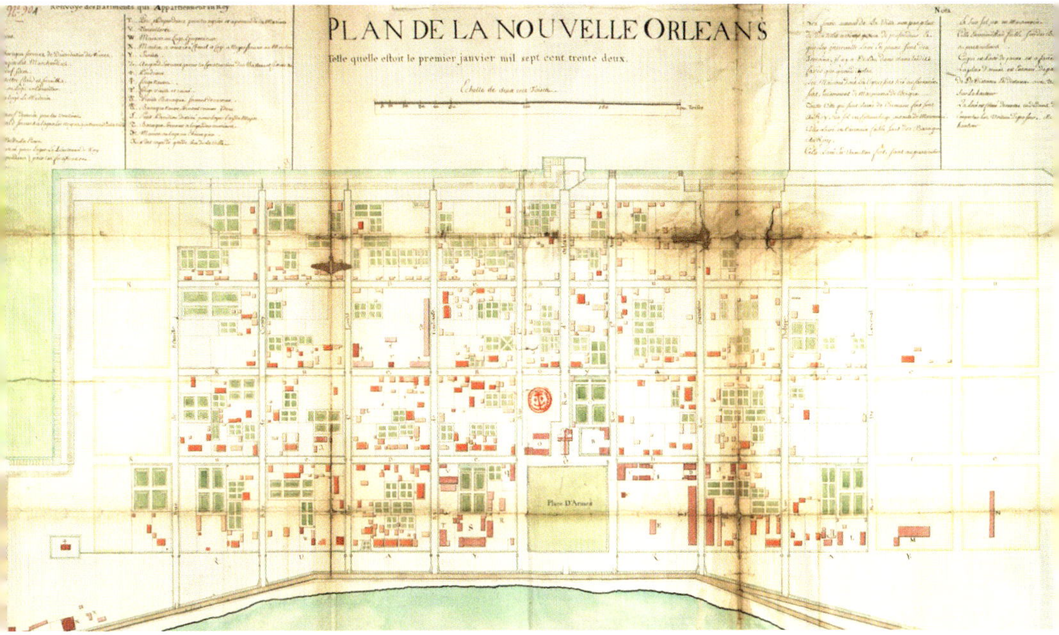

Ignace-François Broutin's 1721 map of New Orleans, *top,* shows houses, gardens, and drainage gutters reaching no further back than today's Dauphine Street, where runoff discharged into a moat (detail, *below*). Courtesy of Dépôt des Fortifications des Colonies, Archives Nationales de France.

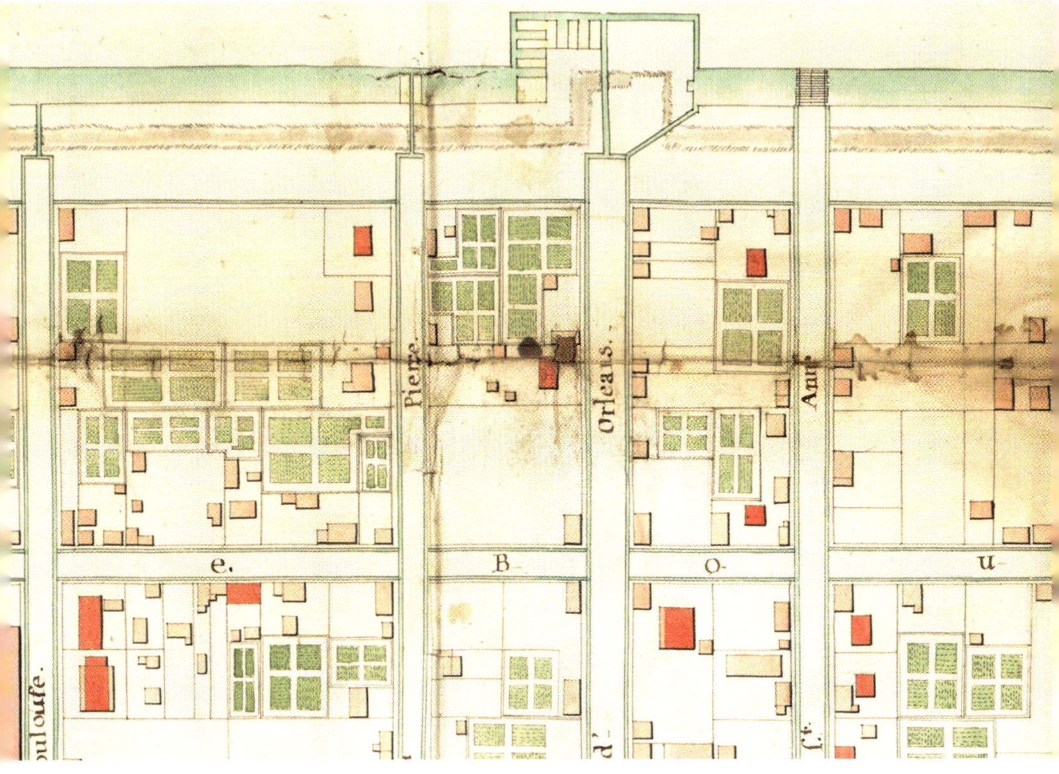

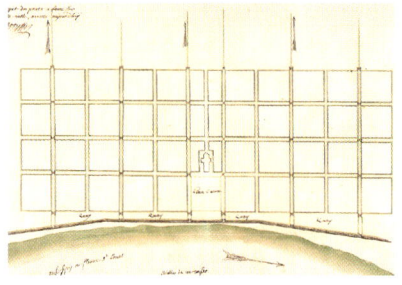

Ignace-François Broutin's 1732 sketches show how run-off would flow through *fossez* (ditches) across intersections, *below*, beneath brick pedestrian bridges, *bottom*, and out toward the backswamp (arrows in map *at left*). Never fully executed, such plans nonetheless illustrate the acumen and intentions of early French engineers in managing water in New Orleans. Courtesy of Collection Moreau de Saint Méry, Archives Nationales de France.

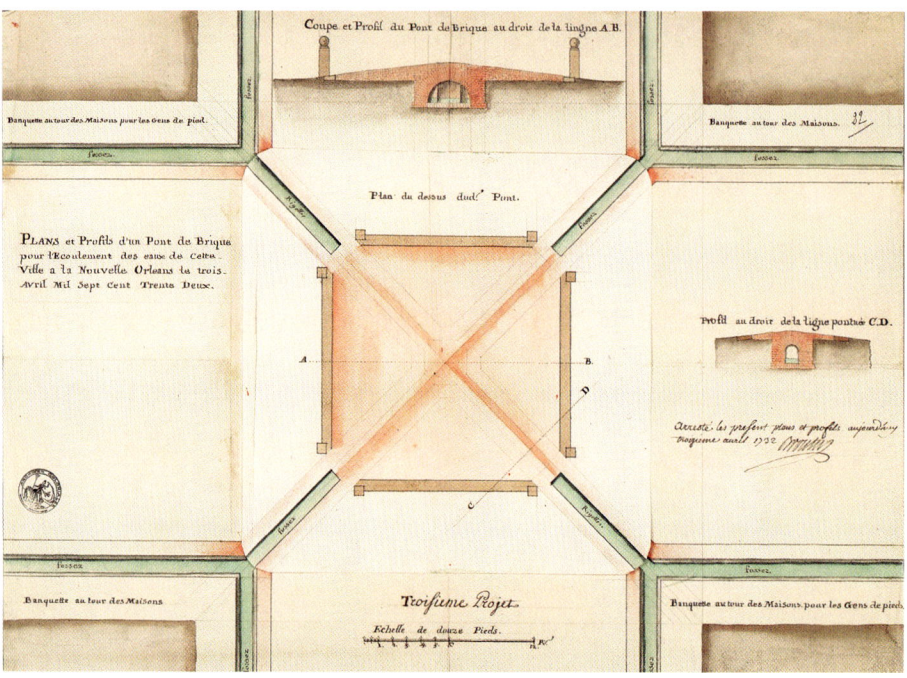

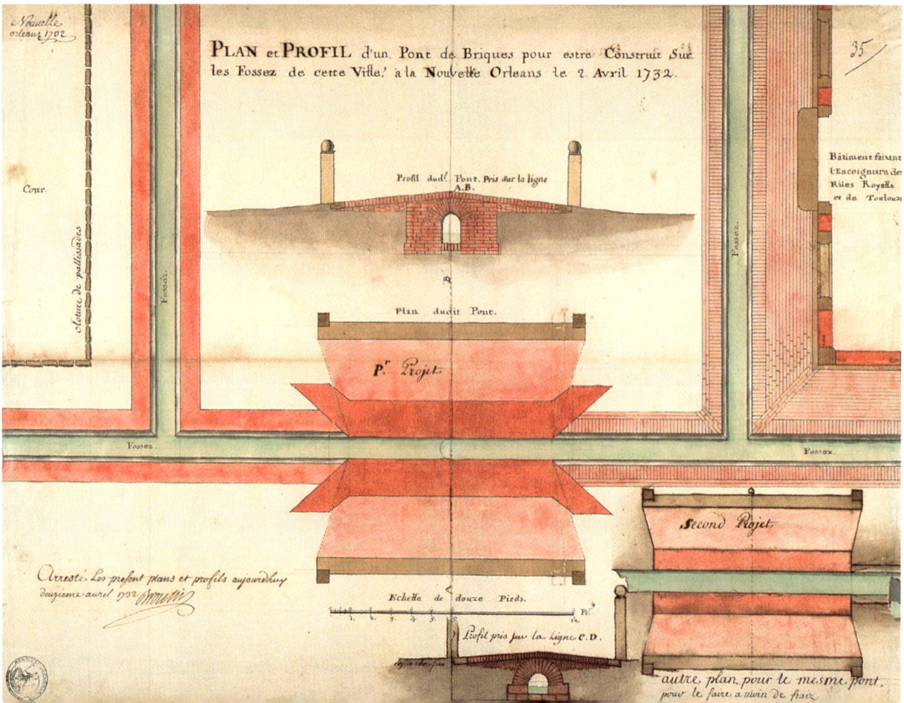

This unusual 1775 Spanish map, which positions the West Bank *at top* and New Orleans (today's French Quarter) *at lower left,* shows how long-lot plantations on the natural levee of the Mississippi River drained gravitationally through backswamp tributaries (bayous) to adjacent bays (*lagos,* or lakes). These cadastral patterns are readily recognizable in the modern-day urban morphology. Courtesy of Archivo General de Indias, Sevilla, Spain.

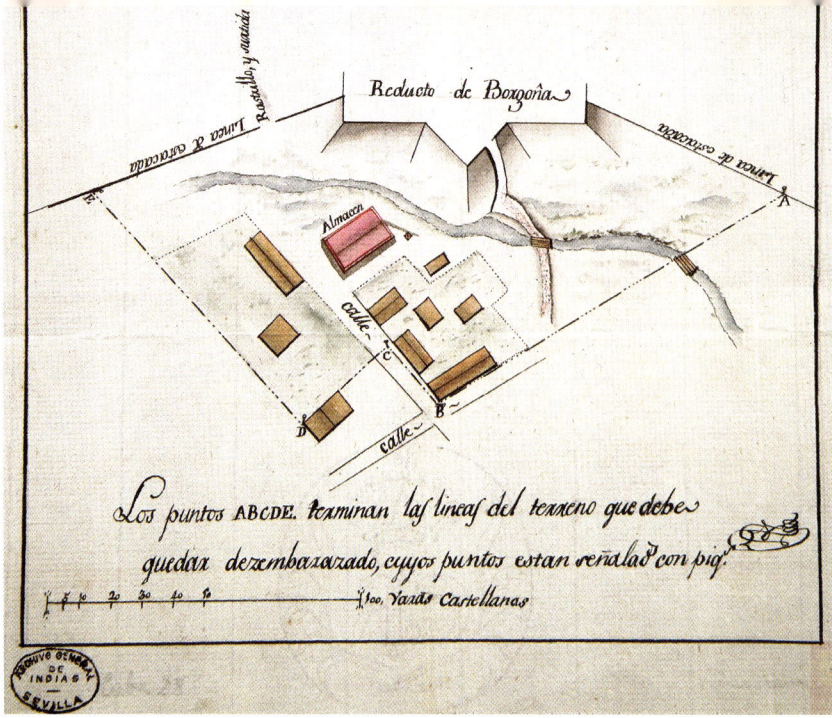

Above: 1790s Spanish drawing shows how runoff from the city proper, *top,* discharged through the bastion *Reducto de Borgona* and into a moat and eventually to Bayou St. John. Today, this area is bounded by North Rampart, Iberville, Canal, and Burgundy streets. *Plano del Sector del Reducto de Borgoña,* courtesy of Archivo General de las Indias, Sevilla, España. *Below:* Detail of 1809 map shows how a moat along the rear fortification (Rue Rampart, today's North Rampart) drained out the Carondelet Canal, today's Lafitte Greenway. *Plan Dressé en exécution de l'arrière du Conseil de Ville de la Nve. Orléans,* courtesy of the American Geographical Society Library Map Collection, University of Wisconsin at Milwaukee.

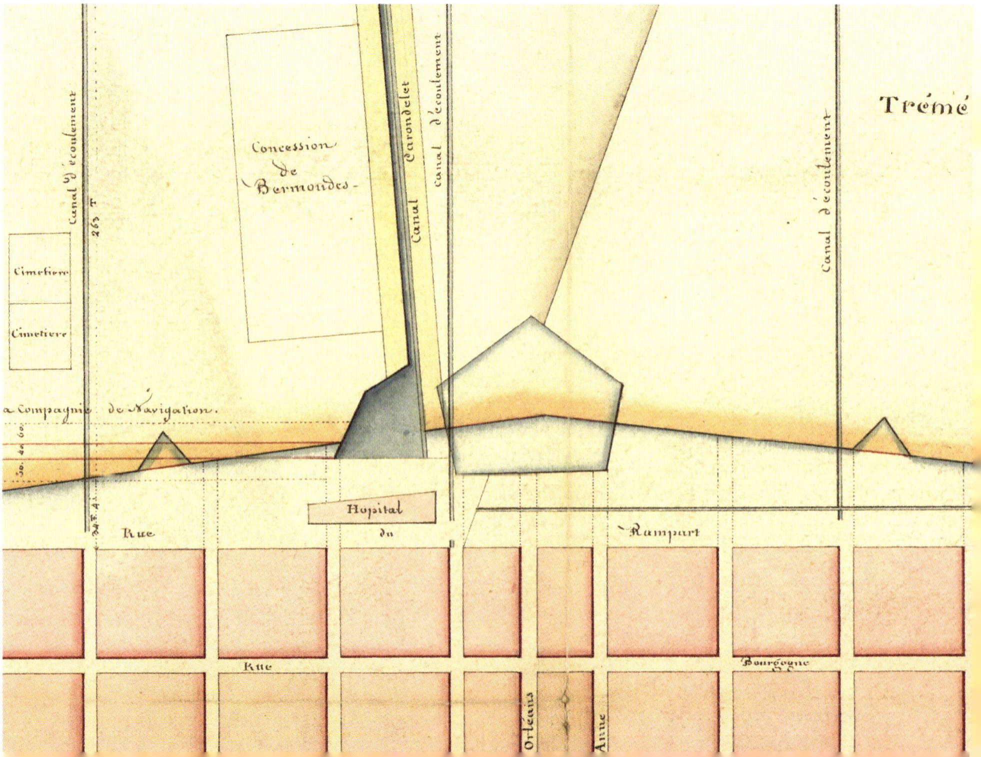

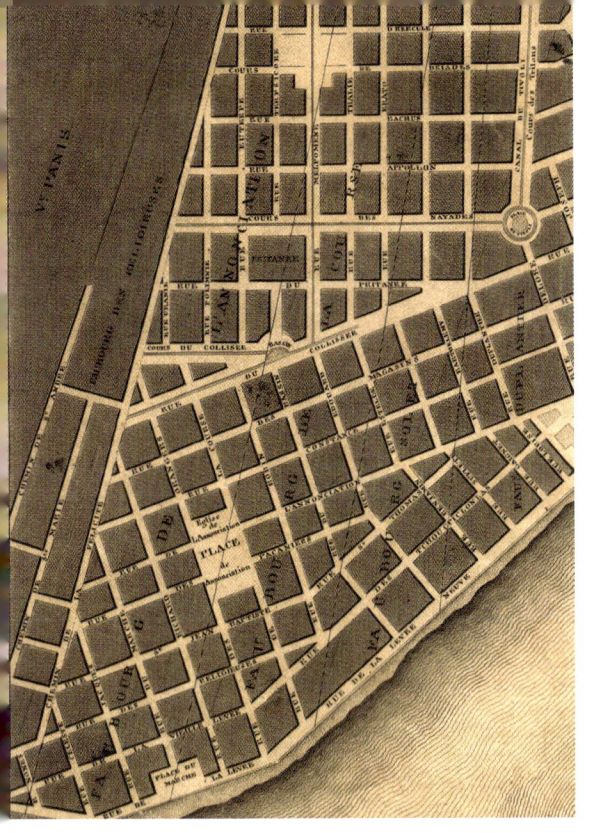

Left: Barthélemy Lafon's plan for today's Lower Garden District, laid out 1806–10 across four former plantations, whose angled boundaries are visible in this detail of Jacques Tanesse's *Plan of the City and Suburbs of New Orleans* (1815). Note Lafon's landscaped drainage features along broader arteries *at top.* Courtesy of Library of Congress. *Below: Plan of the City and Environs of New Orleans, taken from actual survey by B. Lafon,* 1816, shows how runoff flowed gravitationally into the backswamp and out Bayou St. John, *top.* Courtesy of Wikipedia Commons.

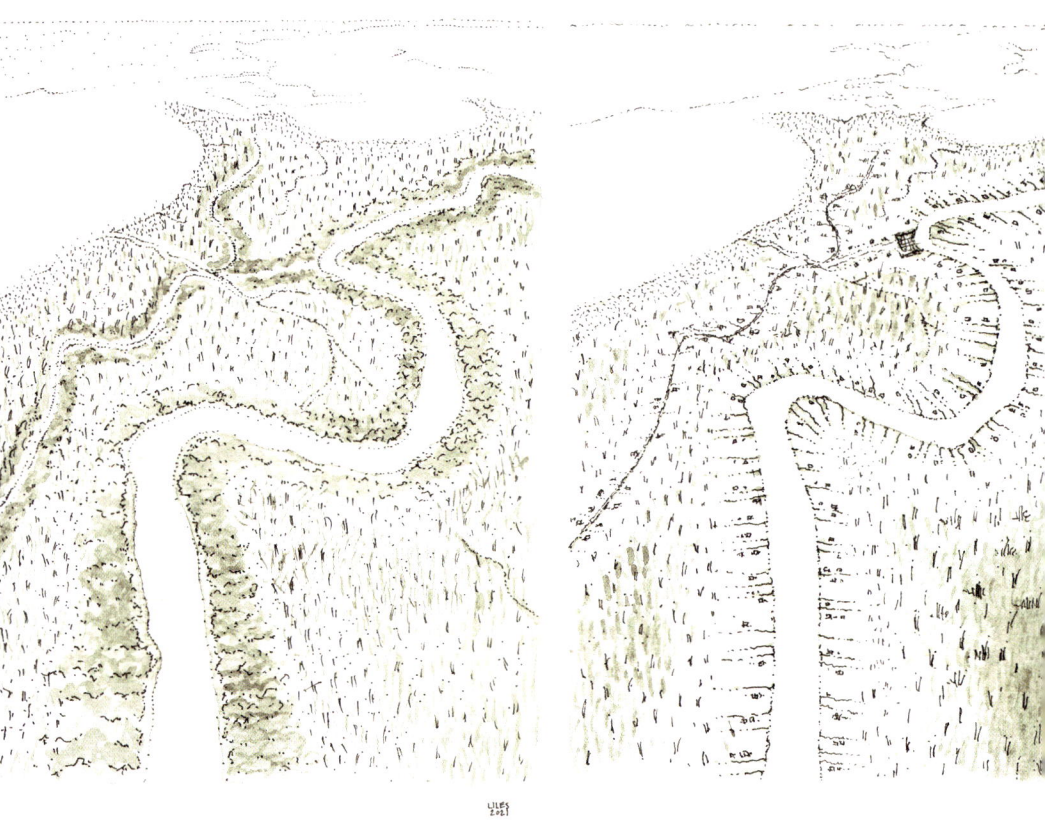

These drawings depict what the New Orleans region looked like circa 1700, *left,* circa 1750 (*right;* note city *at top right* and plantations fronting river); circa 1830 (next page), as urbanization spread upriver; and circa 1870 (next page, *far right*), by which time the backswamp had been largely cut over, and early mechanized drainage systems had been installed. Drawn by New Orleans architect Andrew Liles, based on information provided by author.

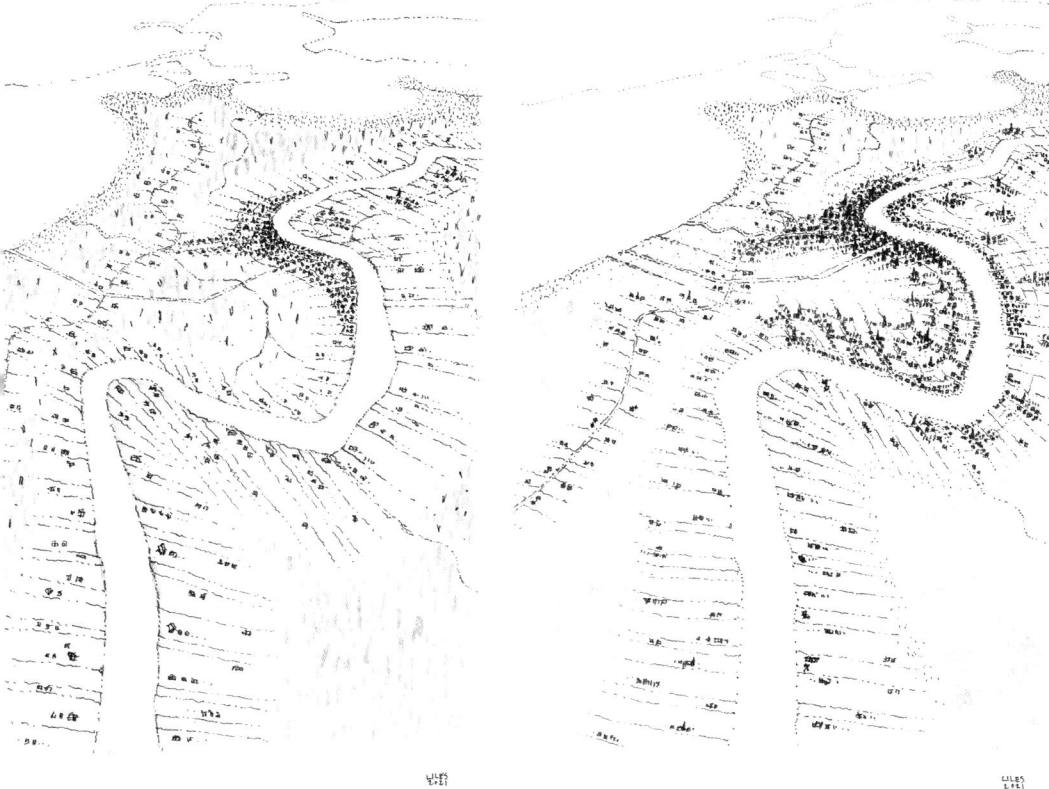

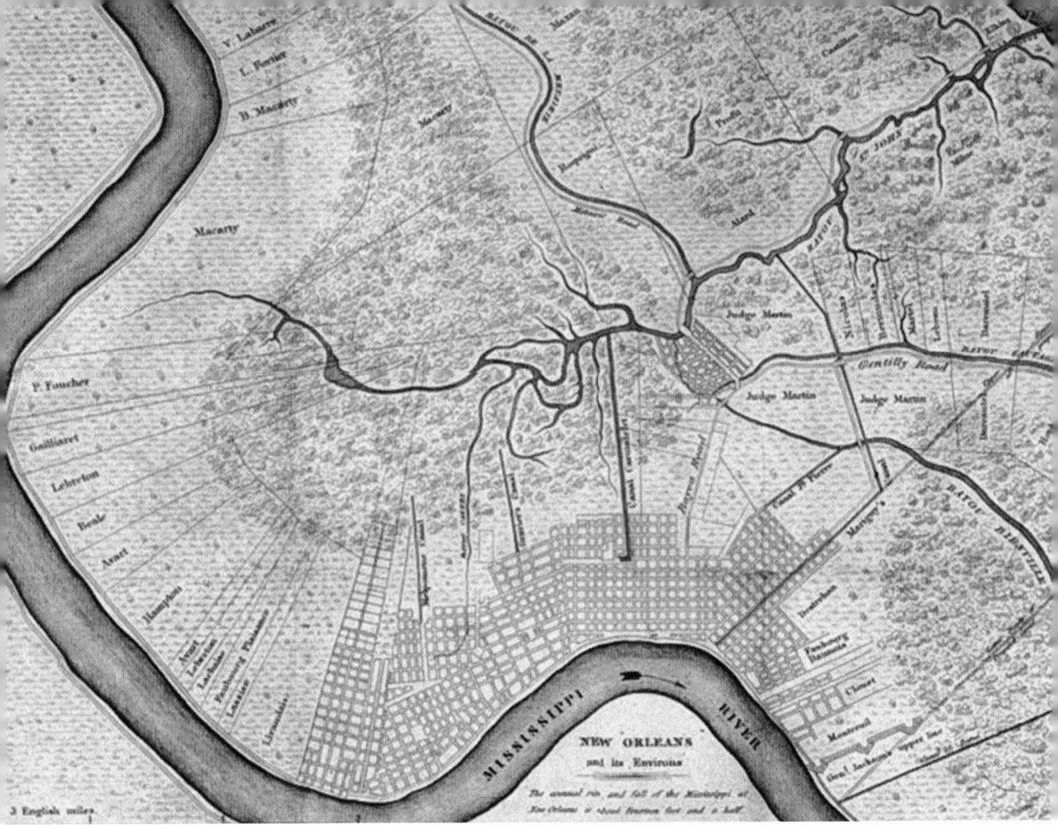

NEGROES HIDING IN THE SWAMPS OF LOUISIANA.

Top: 1829 map by Francis P. Ogden shows urban expansion into adjacent long-lot plantations from original city (French Quarter, *at lower center*). Note how uptown runoff drained through tributary networks flowing into Bayou St. John and out Lake Pontchartrain, *top right,* whereas in lower part of city, runoff drained out Bayou Bienvenue, *right,* to Lake Borgne. Detail, *City of New Orleans,* courtesy of Library of Congress. *Bottom:* Escaped slaves, known as maroons, often established furtive communities in the swamps and marshes that were later drained to become today's Metairie, Lakeview, Gentilly, New Orleans East, and much of the West Bank. This sketch dates to 1837. Courtesy of Library of Congress.

Dr. Edward H. Barton's influential yellow fever report, *left,* and sanitary commission map, *below,* researched during 1854 to 1857, advanced scientific understandings of public health and furthered efforts to drain the swamplands. Author's collection.

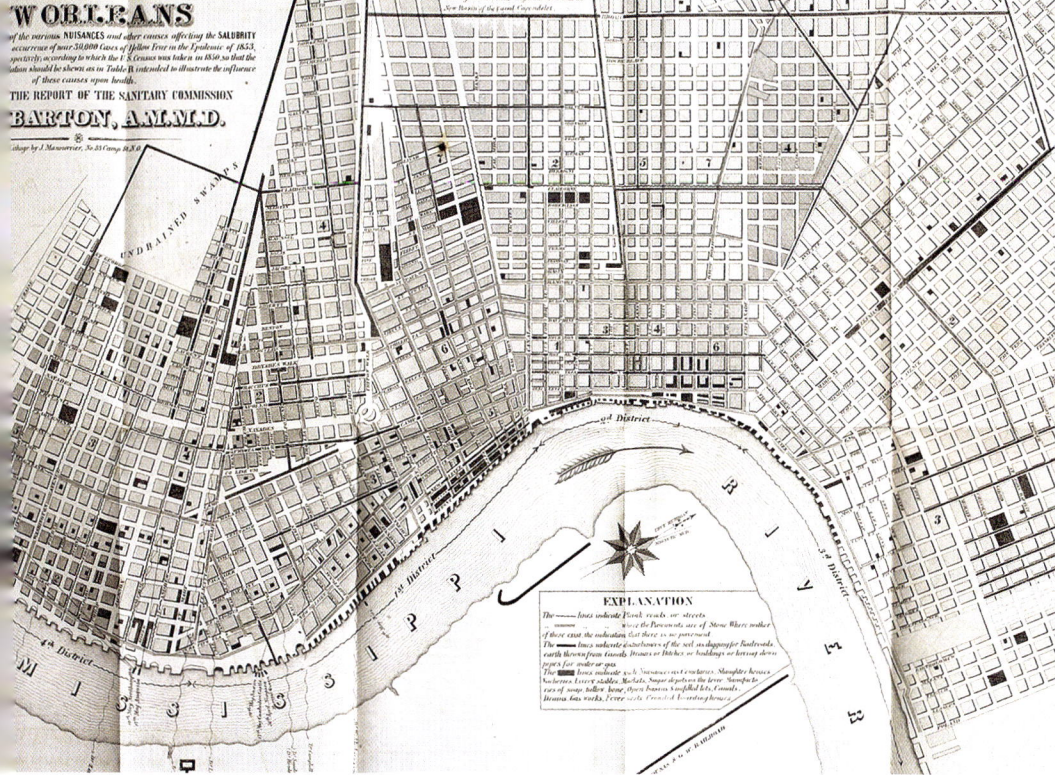

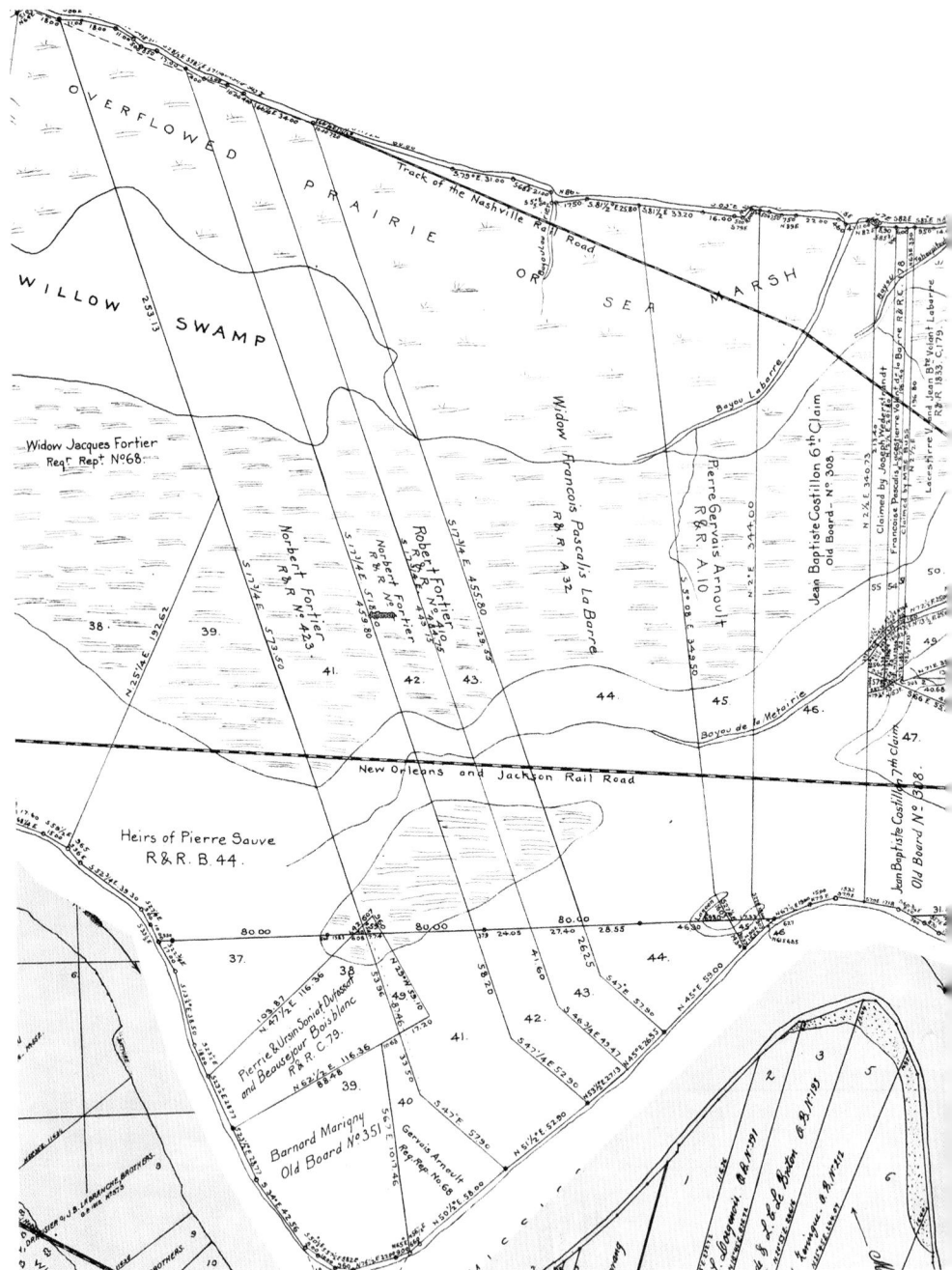

This state lands map shows 1850s conditions in today's East Jefferson Parish. Note Bayou de la Metairie, *at center* (today's Metairie Road, running along the Metairie Ridge), Hoey's Basin to its south, and various wetlands descriptors throughout, including "willow swamp" in interior and "overflowed prairie or sea marsh" by Lake Pontchartrain. By 1915, all this would be reclaimed, and by late 1950s, fully drained for suburbanization. Courtesy of Louisiana State Lands Office.

Top: First District Drainage Bond, City of New Orleans, 1858, depicts steam-powered paddle, *detail below,* like that operating at Bayou St. John since 1835, city's first mechanized method of water removal. Seal *at left* shows paddle at another angle. Courtesy of Tulane University Louisiana Research Center (Jones Hall), New Orleans Municipal Papers, 1782–1925, Public Works–Streets, box 7, folder 8.

Detail of *New Orleans, La. and Its Vicinity*, published in 1863, captures how urbanization petered out as elevations diminished toward the backswamp, *center*, and marshes (*top, distance*). This basic geographical pattern persisted throughout the 1700s and 1800s, but radically changed with modern drainage in the early 1900s. Courtesy of Library of Congress.

Genesis of a bad idea: this 1869 map shows how Colonel Brott's caper, to build the New Orleans & Ship Island navigation canal (*top*, green arrow), spawned an interlocking drainage plan entailing two additional outfall canals to be dug (yellow and blue arrows). Detail of Map and Profile of the New Orleans and Ship Island Channel courtesy of Tulane University. By 1877, the navigation canal idea faltered, but the complementary drainage plan got executed, giving us today's Seventeenth Street, Orleans Avenue, and London Avenue Outfall Canals (*middle*, marked in green, yellow, and blue arrows, respectively). All three would bedevil future drainage planning and engineering, to this day. Detail of Topographical and Drainage Map of New Orleans courtesy of Library of Congress. Two of the three outfall canals, none of which were ever gated, saw their federal floodwalls breach in three spots (*bottom*, red circles) during Hurricane Katrina in 2005, causing catastrophic flooding throughout neighborhoods which had subsided four to eight feet below sea level (dark blue areas). Landsat satellite imagery courtesy of USGS and FEMA.

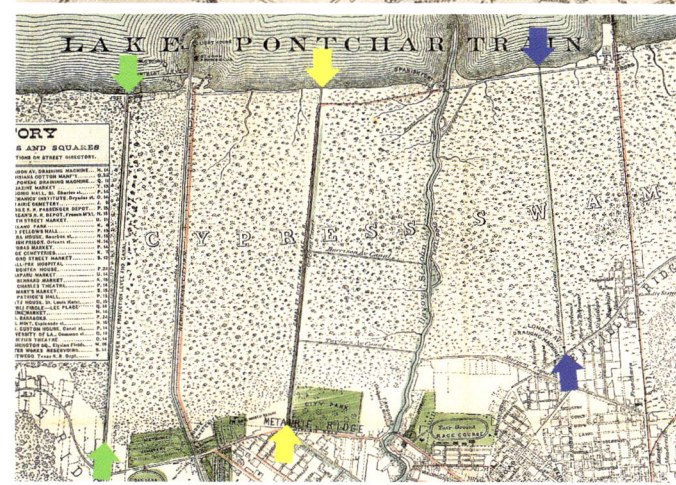

Scenes from New Orleans–area swamps captured by George François Mugnier in the 1890s. *Top* four photographs are from the Des Allemands area on the West Bank; *lower* two scenes are from New Orleans proper. George François Mugnier Photo Collection, courtesy of New Orleans Public Library.

Top row, left to right: Linus W. Brown, Benjamin M. Harrod, and Kate M. Gordon; *middle row, left to right:* George A. Hero, Marcel Garsaud, and Albert Baldwin Wood; *bottom row, left to right:* Marcia St. Martin, Ghassan Korban, and David Waggonner. Courtesy of *Harvard Graduate Magazine,* New Orleans Sewerage & Water Board, National Council of Examiners for Engineering and Surveying, Hero Family Records, New Orleans Public Library, Wikipedia Commons, and Waggonner & Ball Architects.

24   DRAINAGE OF THE CITY OF NEW ORLEANS.

The advantages of and objections to Lake Borgne as a rece the drainage are as follows :

The borders are mostly uninhabited and a slight polluti water has no disadvantages. It is open to the Gulf, and the and fall more rapidly. The greater fluctuations cause a r plete dispersion and a more rapid removal of the drainage wa mean level is several inches lower than Lake Pontchartra

Final disposal into Lake Borgne. distance from the city to the lake is somewhat greater, b Bienvenu, which by dredging can readily be made of sufficie convey the drainage water, runs from the lake nearly to the the city and can be utilized, without detriment, to receive the flow. It is, also, the natural outfall for the drainage of a lar the city.

Conclusion as to point of final disposal. In our opinion, the proper place therefore to discharge drainage from the City of New Orleans, and the more or less water from light storms, is Lake Borgne. Although farther city, it possesses, as indicated, a number of material advant Lake Pontchartrain as a drainage receptacle, which make it preferable.

Left: The 1895 drainage report that would forever transform New Orleans, and its key decision (*right,* from page 24) regarding the selection of an outfall. But how to fund it? The answer came with a June 1899 tax referendum, in which women in particular, *below,* enthusiastically supported drainage improvements. Author's collection.

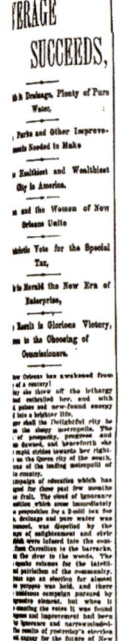

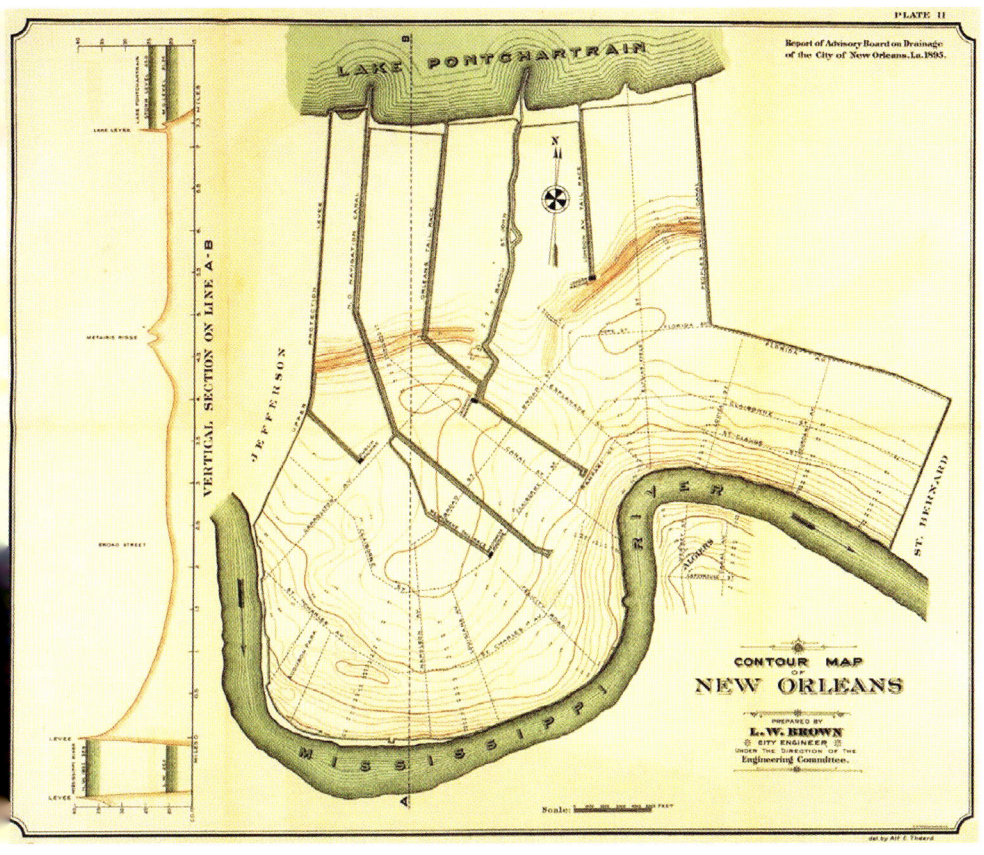

Linus W. Brown's Contour Map of New Orleans, created for and originally published in the 1895 drainage plan, would influence countless engineering projects for decades to come. Note the elevation profile, *at left,* and the very first areas to have subsided below sea level (oval-shaped contours *at center*). Courtesy of New Orleans Sewerage & Water Board.

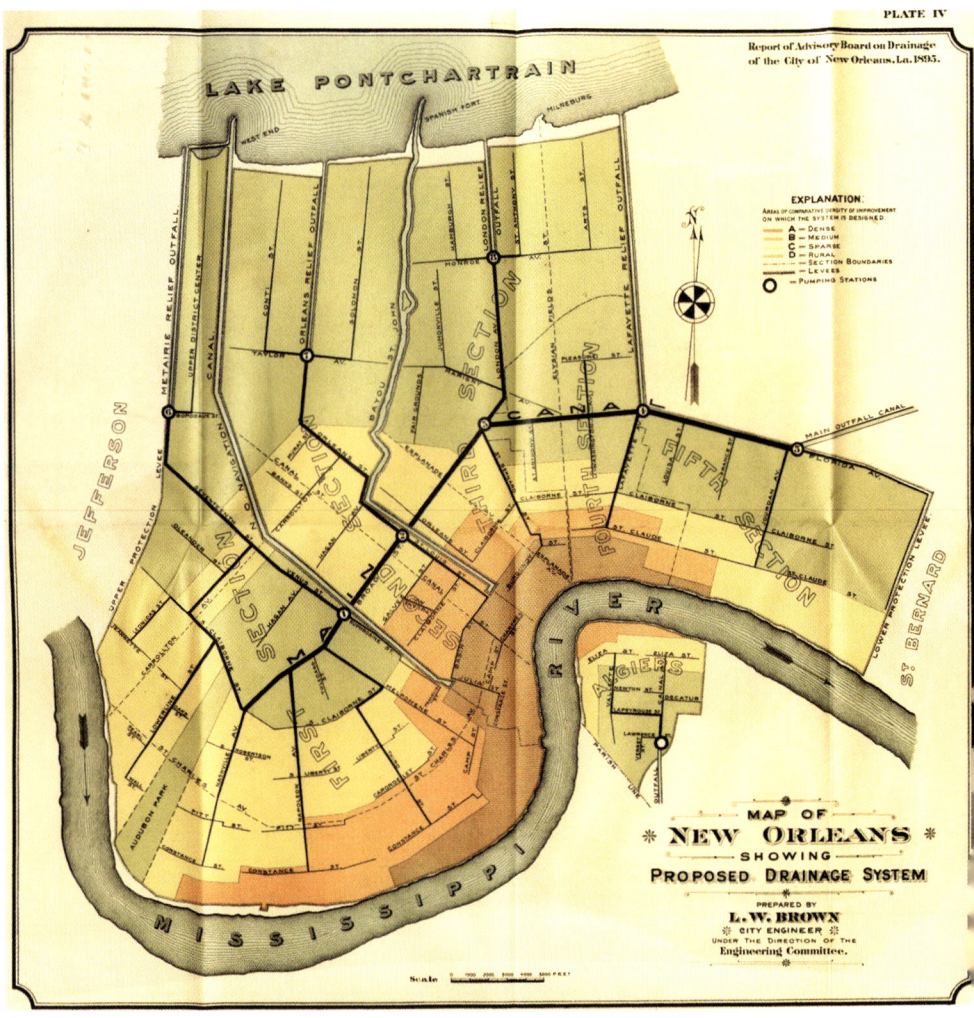

Proposed drainage system, based on the 1895 plan led by Linus W. Brown. Note the Main Canal, *at center,* heading eastward to the Main Outfall, meaning Bayou Bienvenue and ultimately Lake Borgne, a decision that necessitated placing the main pumps (labeled 1 through 5) in the interior of the city, along Broad Street and Florida Avenue. Courtesy of New Orleans Sewerage & Water Board.

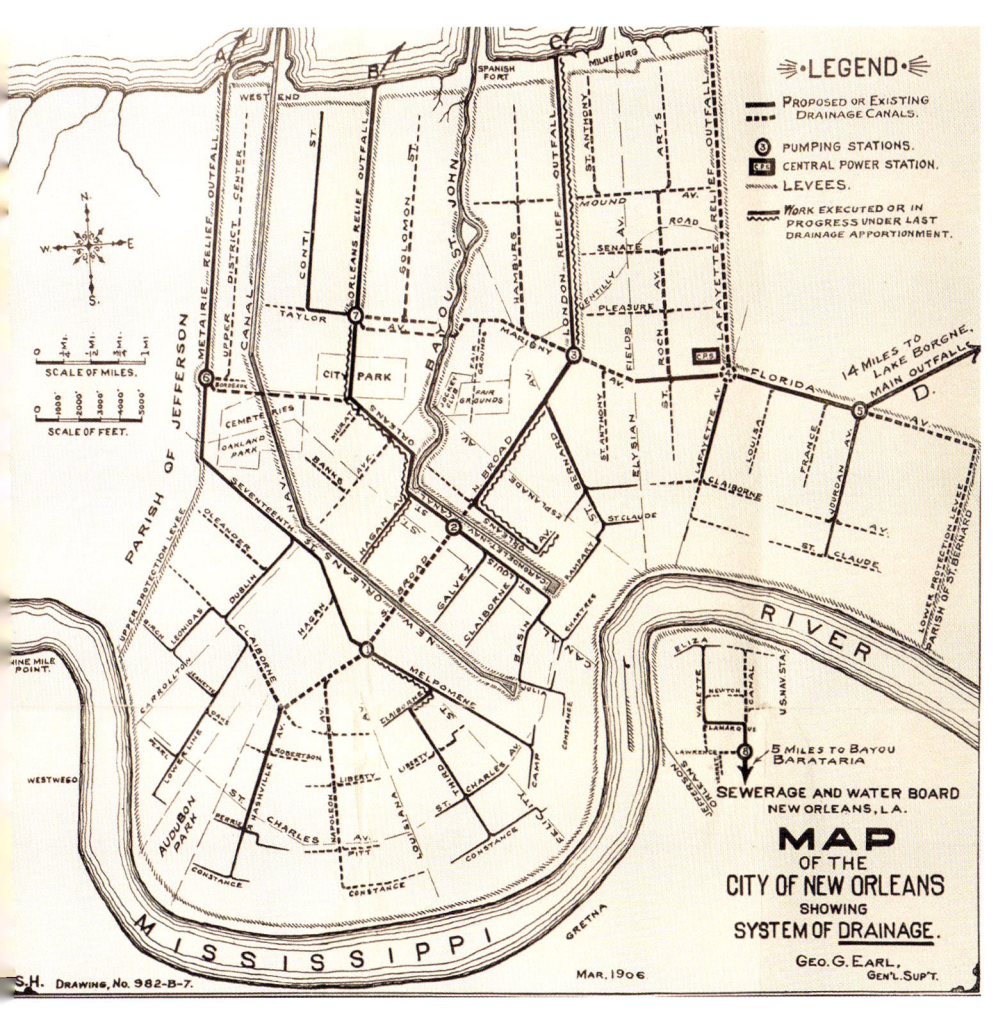

Build-out of drainage system as of 1906, generally following the 1895 plan. Note "14 miles to Lake Borgne, Main Outfall," *at right*, reflecting an outfall decision that was later largely changed to Lake Pontchartrain. From the *Semi-Annual Report of the New Orleans Sewerage & Water Board*, 1907.

Bayou Bienvenue today, looking into the Central Business District. Drone photograph by Marco Rasi, 2021.

Scenes from the installation of the canals, pumping stations, and electrical generators of the New Orleans drainage system, 1900–1912. From New Orleans Sewerage & Water Board, *Semi-Annual Reports*, 1901–13.

George Hero, *left,* "the Drainage King," at work with his engineers installing the West Bank system, *lower right,* in time for the 1915 ceremony in Plaquemines Parish, *top,* in which arrangements were made for President Woodrow Wilson to activate the pumps from the White House. Courtesy of Hero family records.

New Orleans Item headlines from the Great Storm of 1915, *top*, from the first major local news article on soil subsidence (1919, *center*), and from the novel engineering solution to the drainage problem caused by the excavation of the Industrial Canal (*bottom*, 1920).

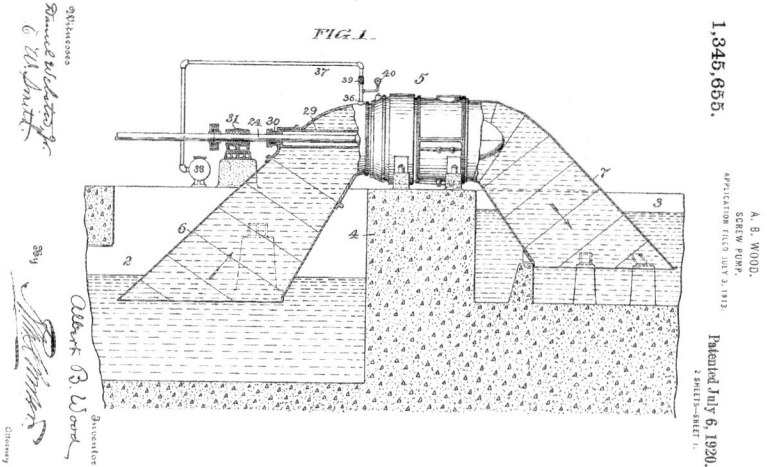

Top: Sketch of Wood Screw Pump in patent application filed in 1913 and granted in 1920 by the US Patent Office. *Middle and bottom:* Engineers and staff pose outside and inside Baldwin Wood's fourteen-foot-diameter screw pump, larger than the world's largest pump he had previously invented, and still in service today. Courtesy of American Society of Mechanical Engineers and New Orleans Sewerage & Water Board.

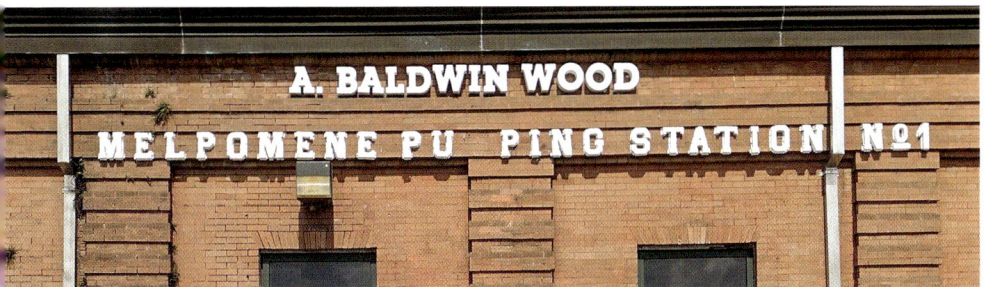

Outside and inside the Melpomene Pumping Station, built in 1900, enumerated as No. 1, named in honor of Albert Baldwin Wood, and still the linchpin of the world's greatest urban drainage system. Photos by Richard Campanella, 2018–21.

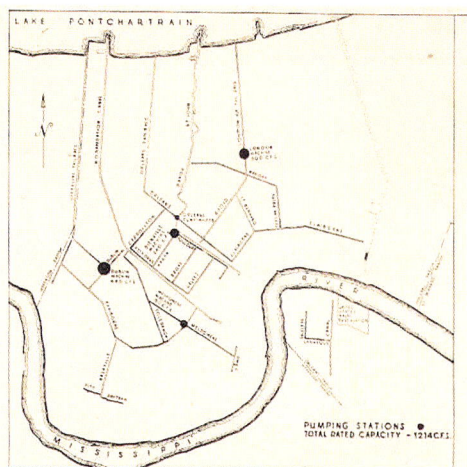

1894

"DURING THE PAST FORTY YEARS, EFFORTS HAVE BEEN MADE TOWARDS THE PROPER DRAINAGE OF THE CITY, BUT, WITH THE EXCEPTION OF THE WORK DONE BETWEEN THE YEARS 1871 AND 1873, NOTHING OF ANY PERMANENT VALUE, OR OF A NATURE EMBRACING THE COMPLETION OF ANY COMPREHENSIVE OR GENERAL SYSTEM, HAS BEEN EXECUTED"

PAGE 47, REPORT ON THE DRAINAGE OF THE CITY OF NEW ORLEANS BY THE ADVISORY BOARD, 1895

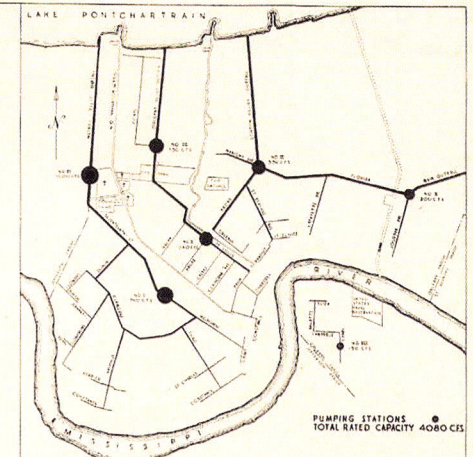

1904

"THE DRAINAGE SYSTEM A NOW CONSTRUCTED, SHOWS 20 MILES OF LINED AND COVERED CANALS, 3 MILES OF WOOD LINED CANALS AND 17 MILES OF OPEN AND UNLINED CANALS —— THERE IS COMPLETED ABOUT 44 PER CENT OF THE SYSTEM AS ORIGINALLY DESIGNED IN 1895, AND LEAVES TO BE CONSTRUCTED, 26 MILES OF LINED AND COVERED CANALS, — 8½ MILES OF WOOD LINED AND 25½ MILES OF OPEN CANALS".

PAGE 9, TWELFTH SEMI-ANNUAL REPORT OF THE S. & W. BOARD, 1905

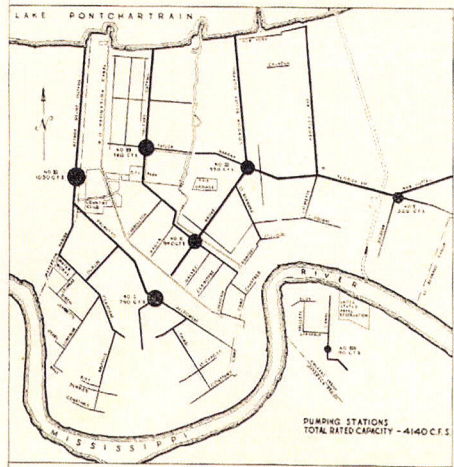

1913

THE DRAINAGE COMMISSION ESTIMATED IN 1900 THAT THE TERRITORY BOUNDED BY METAIRIE RIDGE AND FLORIDA WALK ON THE NORTH, JEFFERSON PARISH ON THE WEST AND PEOPLES AVENUE ON THE EAST, INCLUDING ALGIERS, WOULD ULTIMATELY HOUSE APPROXIMATELY 855,000 PERSONS. THE AREA CONSIDERED WAS WELL SERVED BY DRAINAGE CANALS BY 1913.—

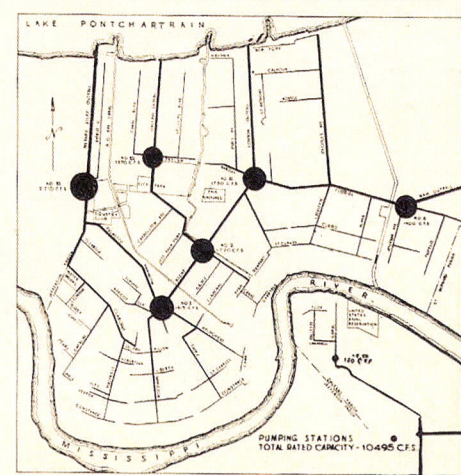

1926

"SANITARY NECESSITY HAS COMPELLED THE EXTENSION OF THE MAIN DRAINAGE SYSTEM TO ELIMINATE IMMEDIATELY SURROUNDING SWAMP AREAS WITH CHEAPER OPEN CANAL DRAINAGE, THIS IN TURN HAS —— INDUCED THE SCATTERED BUILDING DEVELOPMENT WHICH, FOLLOWING THIS DRAINAGE, HAS CREATED SOME TAXABLE PROPERTY VALUES, WHICH, IN TURN, HAVE BEEN RECOGNIZED BY SUCH SCATTERED WATER MAIN EXTENSIONS AS HAVE BEEN MADE. THE DRAINAGE OF THESE LANDS AND THESE SCATTERED WATER EXTENSIONS GENERALLY REPRESENT A GREATER PROPORTIONAL EXPENDITURE THAN THE TAXABLE VALUE OF THE LAND AND IMPROVEMENTS AT PRESENT WARRANTS".

PAGE 15, FORTY-SIXTH SEMI-ANNUAL REPORT OF THE S. & W. BOARD 1922

GROWTH IN AREAS SERVED BY
STORM WATER DRAINAGE SYSTEM
NEW ORLEANS 1894-1926 LOUISIANA
COMPILED FROM THE RECORDS OF THE SEWERAGE AND WATER BOARD
BY THE CITY PLANNING AND ZONING COMMISSION-HARLAND BARTHOLOMEW & ASSOCIATES-CONSULTANTS

First reclamation, followed by subsurface drainage; then infill, grading, and urbanization; and finally lakefront reinforcement: these aerial photos, from 1936, *top*, and circa 1950, show the extent to which former swamps and marshes became cityscape—and subsided below sea level. From author's collection of photographs from the US Army Air Corps and New Orleans Public Service, Inc., now Entergy.

Drainage begets drainage, because its prior success in spurring urbanization yields more runoff to be removed—and higher expectations for dry streets. Here we see expansions of the subterranean drainage canal on Orleans Avenue, and on South Claiborne at Lowerline, in the late 1930s. Courtesy of New Orleans Sewerage & Water Board.

Metairie in 1957, *above*, four decades after its reclamation and just prior to subsurface drainage; note the fragments of old bayous. Within ten years, nearly every acre would be urbanized, as it appears *below* in 2005. Photo mosaic by Del Hall / US Air Force; 2005 scene courtesy of Jefferson Parish.

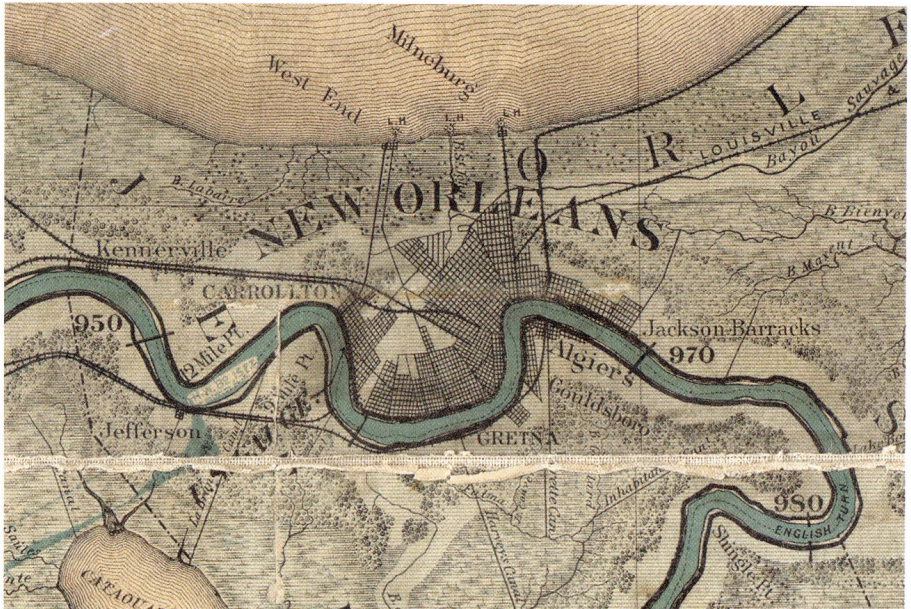

Greater New Orleans around 1890, a decade prior to the installation of the modern drainage system, at which time nearly the entire region sat above sea level, as nature had created this fluvial delta. Courtesy of Tulane University Digital Library. Greater New Orleans in modern times. Areas still above sea level are shaded green, and those that have subsided below sea level—thanks almost entirely to artificial drainage—are shaded red. Satellite image / LIDAR elevation composite by Richard Campanella, using Landsat and FEMA/State of Louisiana source data.

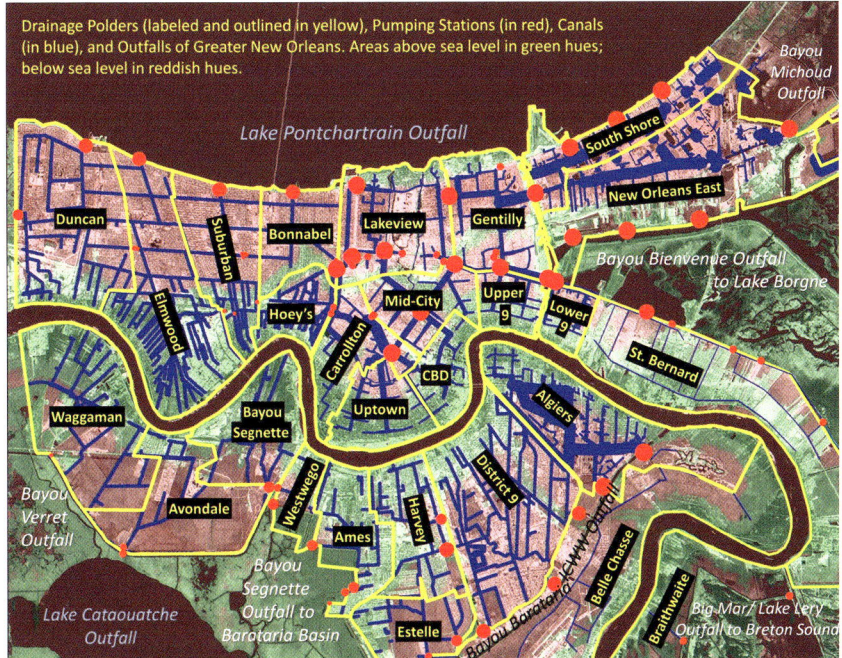

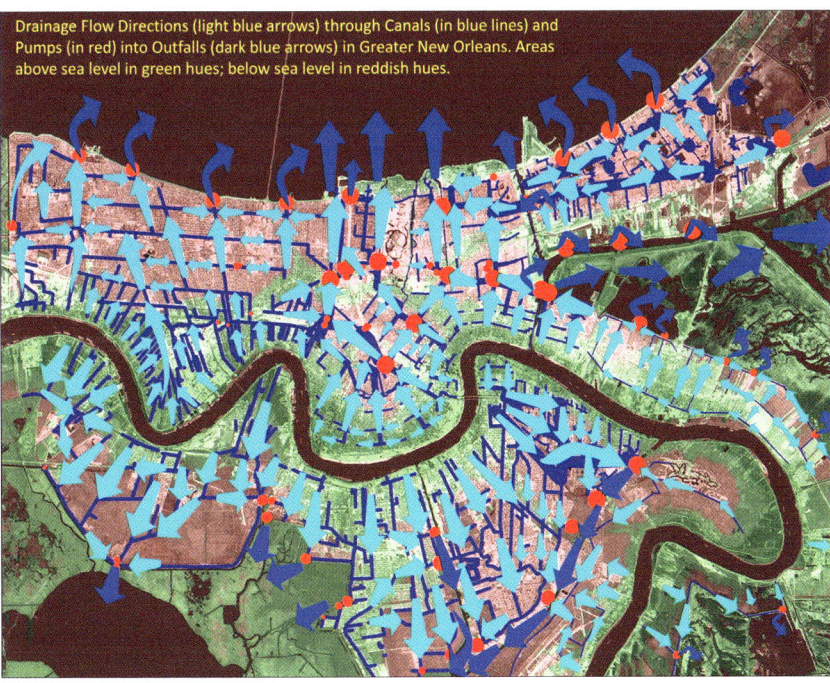

Polders (yellow outlines) and drainage systems (blue lines indicating canals, with red dots showing pumping stations) of greater New Orleans. Areas shaded red have subsided below sea level. Analysis and map by Richard Campanella. How runoff flows through metro-area drainage systems, by gravity or pump (cyan arrows, with red dots showing pumping stations), and where it gets ejected (dark blue arrows). Background areas in reddish shade have subsided below sea level. Analysis and map by Richard Campanella.

Uptown runoff flows gravitationally into Pumping Station No. 1, *bottom,* which pushes it through the Washington/Palmetto Canal (seen here during a dry spell). That discharge then arrives to Pumping Station No. 6 in Lakeview, *center,* whose pumps raise the water, from below to above sea level, so that it may flow through the Seventeenth Street Outfall Canal, *top,* to Lake Pontchartrain, which is roughly at sea level. As it does, the discharge flows above homes in adjacent Metairie, Bucktown, and Lakeview, all of which have sunk to four to eight feet below sea level. Drone photos by Marco Rasi, with author, 2021

A closer look at Pumping Station No. 6 in Lakeview, as it raises runoff from below-sea-level stages, *at right,* to above sea level, *at left,* in order to eject it into Lake Pontchartrain, two miles off *to the left.* Note the raking devices *at right,* designed by Albert Baldwin Wood to remove debris from the catch basin before the runoff enters the lift pumps. Drone photo by Marco Rasi, with author, 2021.

Pumping Station No. 3, *bottom,* in Gentilly lifts runoff drained from polders in the Seventh and Eighth wards and expels it through the London Avenue Drainage Outfall Canal, *top,* into Lake Pontchartrain. Like the Orleans and Seventeenth Street canals, this channel's water stage flows well above the below-sea-level elevations of adjacent neighborhoods, all of which flooded deeply due to Hurricane Katrina–induced floodwall breaches in 2005. Drone photos by Marco Rasi, with author, October 2021.

Whereas engineers in New Orleans proper decided in 1895 to place their main drainage pumps in the interior of the city (which later proved problematic), those in suburban Metairie learned from this mistake and positioned their pumps on the lakefront perimeter. Here at Bonnabel, the pumping station acts as a storm-surge gate, while the pumps inside raise stormwater runoff and eject it into Lake Pontchartrain. Drone photos by Marco Rasi, 2021.

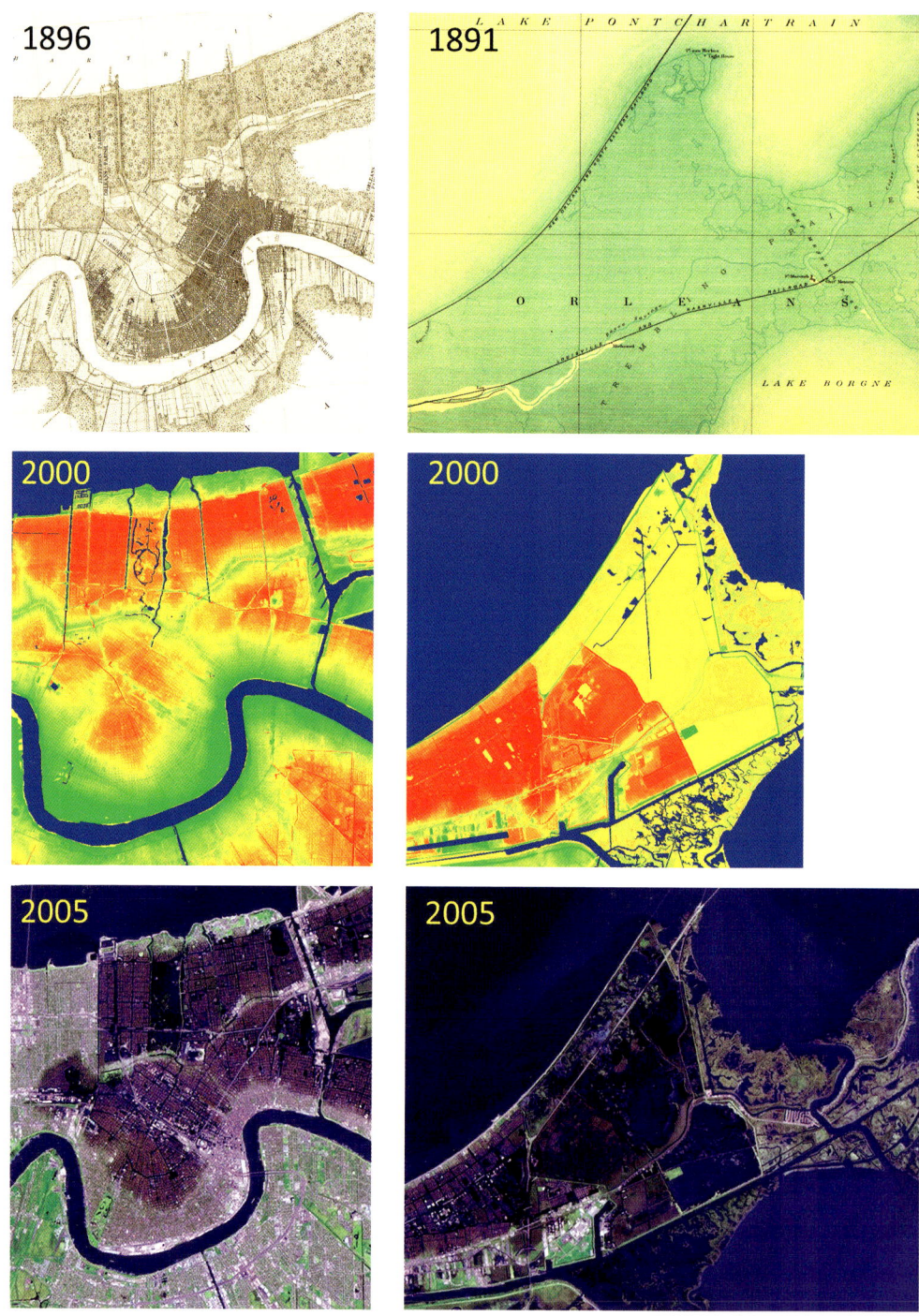

Areas that were historically swamp and marsh, *top images,* were artificially drained in the early 1900s, causing them to subside into topographic bowls (*middle images,* with areas below sea level shown in red), which made them vulnerable to catastrophic flooding when levees breached during Hurricane Katrina in 2005 (*bottom images*). Maps and analysis by Richard Campanella using USGS, FEMA, State of Louisiana, and Landsat data.

Inside and outside the billion-dollar West Closure Complex, the world's largest pumping station (eleven five-thousand-horsepower pumps capable of ejecting nineteen thousand cubic feet of water per second), astride North America's largest sector gates. The complex is located within two miles of the site of the famous Hero Day drainage ceremony of February 1915. Photos by Richard Campanella, 2019.

Dutch Dialogues in action, 2010. Visible in *upper left* photo are New Orleans Mayor Mitch Landrieu, *left;* David Waggonner, back to camera; and Dale Morris, in blue shirt; in *upper right* photo is Tulane architect John Klingman presenting a design to Sen. Mary Landrieu. Three years later, Waggonner & Ball released its Greater New Orleans Urban Water Plan (cover *at lower left*), aimed to add "green" and "blue" approaches to New Orleans's traditional "gray" stormwater-management system—with the hope of reducing dreaded flash floods like that of May 8, 1995 (*lower right*). Photos courtesy of Waggonner & Ball.

bottom of page four, and it mostly fussed about the allegedly "lost" maps and the price tag of the new survey. "If the city is to spend $18,000 or $20,000 in making surveys every time a new ditch is dug," the editorialist pontificated, "it will finally be ditched into bankruptcy."[38]

In hindsight, Ordinance No. 7170 would prove to be an almost literal watershed moment for New Orleans. Its success rested on three strengths: it invested in science (exceeding the $10,000 Harrod had requested in 1888); it made drainage someone's job (the city engineer); and it backed up that responsibility with the rigor of peer reviewers (the advisory board on drainage), making this a team project and not a solitary vision (like most of those prior plans, of Dunbar, Pilié, Surgi, Bell, and Harrod). The *Picayune* editorialist missed these key distinctions, but waxed lofty nonetheless: "The age of improvement is just commencing," he intoned. "The city must sooner or later have a complete system of sewerage and of drainage. . . . It must have the entire area of the swamp . . . converted into dry ground, covered with residences, gardens and truck farms."[39]

It turned out to be sooner. Members of the advisory board got right to work understanding the terrain, figuratively and literally. Being pragmatic engineers, they realized they could not revert to the *tabula rasa* of the 1700s, with its unobstructed watersheds and natural drains, and accepted the sunk costs of the 1800s. The engineers would work within the polderized, canal-crossed framework left by their predecessors, and not call for the removal of guide levees, the filling of canal beds, or the reopening of natural drains. But neither would they be handcuffed to existing assumptions and conditions; they were quite willing to buck convention and think anew.

The engineers began by conducting an exhaustive review of all prior drainage documents and, finding extant data to be "meagre, crude, and unreliable," concurred that a state-of-the-science topographic survey was "absolutely required to properly solve the problem." They next spelled out all assumptions, definitions, standards, and metrics, as good scientists are wont to do. As good government ought to do, the advisory board also held a series of public meetings, from 1893 through 1895, for citizens to contribute ideas and concerns—the forerunners of today's ongoing water-management hearings and community charettes. Then, in the hot summer of 1893, the advisory board dispatched its Topographical Survey Department crews to the field.[40]

Under Brown's guidance, the surveying crews set out to measure an interlace of baselines, offsets, profiles, and spot elevations, focusing on the main river-to-ridge basin of the East Bank of Orleans Parish plus the urbanized portion of Algiers. The baselines, 150 linear miles' worth, were trajectories radiating outwardly from known benchmarks, from which exact distances (measured by chains), directions (azimuths), and elevational changes (measured by quadrant or theodolites) were recorded. Baselines usually followed streets and avenues—"Elysian Fields street, from the River to the Lake . . . Lowerline street, from the River to Claiborne . . . General Taylor street, from the River to St. Charles avenue"—and aimed to hit ridges, lulls, and other hypsographic extremes.

Next, lines known as offsets, which radiated perpendicularly from the baselines like branches from a tree trunk, were measured for the same information. To those were added elevational profiles from an additional 270 streets, block after block, front- to back-of-town, crisscrossing the urban footprint. As a separate set of spot measurements, the teams "obtain[ed] the elevation[s] of the curb, gutter, center of street and property line at intersections, and in center of block, . . . of all culverts, bridges and . . . railroad tracks at intersections" as well as "existing drainage canals and tail-races," meaning trench bottoms on the ejection side of the pumps. The raw field data were then passed to "computer" B. Shall and draftsman A. F. Theard to triangulate into isopleths.

The resulting one-foot contours were referenced not to sea level, but to the Cairo Datum, a now-obsolete vertical baseline pegged to the Army Corps of Engineers benchmark at the Cairo, Illinois, confluence of the Ohio and Mississippi rivers. This theoretic plane had been used since 1879 to reference most Mississippi Valley elevation-mapping projects, at a time when the best geodetic control emanated from the continental interior rather than coastal perimeters. The advisory board adopted the standard definition of the Cairo Datum as being "a plane 21.26 feet below Mean Gulf Level," which had the curious (and often misunderstood) effect of jacking up contour postings to magnitudes New Orleanians could only dream of, such as 37 feet at the French Quarter riverfront and 25 feet along the Metairie-Gentilly Ridge. But of course, these were not heights above Lake Pontchartrain or the Gulf of Mexico (sea level); they were above an imaginary plane some 21 feet *below* those watery surfaces. That put the French Quarter riverfront at a more reasonable 16 feet above sea level, and the ridges around 4 feet.[41]

What resulted was what the advisory board modestly described as "a map on a scale of 600 feet to the inch [with] one foot contour lines."[42] In fact, the yearlong effort yielded ten beautiful linen charts, large enough to cover a wall, titled *Topographical Map of New Orleans,* credited to L. W. Brown, and superior to all previous local elevation maps.[43]

The city got its money's worth from that $17,500. Brown and the engineering staff had done such a fine job that their topographic survey would get adduced, adapted, modified, and ingested repeatedly for years, for everything from drainage projects and roadwork to levees and real estate projects. Developers in 1909, for example, cheerfully described their new Gentilly Terrace subdivision as being "twenty-seven feet above the Cairo Datum Line," a figure drawn from the 1895 survey but erroneously construed to mean "that it is the most elevated residential section in the City of New Orleans."[44] Incredibly, the S&WB still uses the circa-1879 Cairo Datum today, even though mapmakers elsewhere moved on to the North American Vertical Datum of 1927, and later to the satellite-based North American Vertical Datum of 1988 and World Geodetic System of 1984, to standardize hypsography.

The survey teams did far more than measure elevation. They computed slopes and polder volumes, set out gauges to record rainfall, computed its relation to runoff, and analyzed US Weather Bureau records to understand broader seasonal patterns. Staff members compiled the findings in neat tables for the engineers to compute the velocities for different water volumes to be mobilized across varied distances and heights, according to pump capacities in variously sized canals. None of this had ever been done before locally, and the keystone dataset was Brown's topographic map.

In and among the graceful contours of that map are three spaces warranting closer attention. First are the swampy lowlands between the Metairie-Gentilly Ridge and the lakeshore, where no contours appear. This was understandable; their flatness would have fallen within the one-foot contour interval, and their wetness probably precluded entry by the field crew, whose instruments required stable footing. But these polders' topography would soon radically change, from a wet plane to a dry concavity, and in doing so, would throw off all the engineers' assumptions of just how much runoff would accumulate here, how much would have to be pumped out, and where it should be ejected.

Which brings us to two other curious spots on the Brown map, one in the swamp basin riverside of the Metairie Ridge, and the other in the riverside of the Gentilly Ridge. Historically these were the subbasins of the now-extinct Bayou des Cannes and the now-filled upper reaches of Bayou Bienvenue; today they would be in the Broadmoor/Mid-City area, and in the central Seventh, Eighth and Ninth wards. Brown's map shows both of these low spots to be 20 feet in elevation—above the Cairo Datum, that is. Subtract 21.26 from that number, and for the first time, *we have documented evidence of parts of the New Orleans land surface falling below the level of the sea.*

What likely happened in these cutover backswamps was that just enough artificial drainage, going back to the 1830s, had drawn down just enough swamp water to allow the crews to set up their instruments and collect surface measurements. What they apparently revealed was that the soil body, robbed of its water volume, had begun to consolidate and subside. Being low to begin with, the compaction put them below the level of the sea and lake. Drainage-caused subsidence would only get deeper, faster, and more widespread, and in time, it would subvert the advisory board's best-laid plans.

PONTCHARTRAIN OR BORGNE?

The kingpin decision of the advisory board, affecting all subsequent choices, was the question of outfall. "There are," the engineers summated, "three localities where [runoff] can be disposed of, namely: the Mississippi River, Lake Pontchartrain, and Lake Borgne." They swiftly rejected the river option, citing the costs of the uphill lift, and focused on the two tidal lagoons (bays): Lake Pontchartrain to the north, or Lake Borgne to the east.

Their predecessors had grappled with this same vexing choice. Earlier systems had used Lake Pontchartrain, via Bayou St. John and its tributaries, except for the lower faubourgs, which used Lake Borgne, via Bayou Bienvenue. Louis Surgi in 1868 was the first to suggest Lake Borne for the entire city, in that era when drainage got engulfed in the Ship Island seaway project—itself envisioned to penetrate through Borgne. Surgi's successor, W. H. Bell, proposed a "both" approach, but eventually designed his outfall canals for Lake Pontchartrain.

Now the advisory board was revisiting all that, and the stakes were high because the final decision could reorient the entire system. Pontchartrain certainly had its attributes: "*nearer to the center of the city[;] at present used for an outfall[;] less expensive.*" But it came with a cost. To a surprising degree, the engineers worried about polluting the lake and ruining the "numerous pleasure resorts" at West End, Spanish Fort, and Milneburg. They sensed that Lake Pontchartrain, nearly "land-locked" with only two narrow straits to the sea, did not have adequate tidal action to flush out runoff pollutants, leading to overloading.

Lake Borgne, on the other hand, opened widely to the Mississippi Sound and Gulf of Mexico, and its marshy shores were "mostly uninhabited." It was more clearly a bay, with brackish tides that "rise and fall more rapidly," and water levels "several inches lower than Lake Pontchartrain," creating a flushing action with "a more complete dispersion and a more rapid removal of the drainage water." To be sure, Borgne was farther away than Pontchartrain, but it had the helping hand of Bayou Bienvenue, "which by dredging can readily be made of sufficient size to convey the drainage water."

After weighing the pros and cons, the advisory board decided resolutely. "In our opinion," it wrote in its *Report on the Drainage of the City of New Orleans* (1895), "the proper place therefore to discharge the daily drainage water from the City of New Orleans, and the more or less polluted water from light storms, is Lake Borgne."[45] Much like Bell had done, the advisory board hedged its bet by saving Lake Pontchartrain as an auxiliary relief discharge, for use "only occasionally."

So Lake Borgne it was, and for a rather enlightened environmental reason for this era—to prevent water pollution. As a result of this decision, the nation's most herculean urban drainage project got a 90-degree pivot.

brown's advisory board plan of 1895

First to reflect the pivot was the advisory board's new delineation of five "drainage sections," numbered First through Fifth in sequential upriver-to-downriver order, plus an Algiers Section on the West Bank. This polder-by-polder approach broke a big project into smaller interlocking parts, and made each one "capable of sectional construction," buildable incrementally.[46] All surface runoff within each East Bank section would end up in a "main canal[,] a fundamental feature of the plan," to be installed beneath "Broad street . . . down Florida avenue . . . to Bayou Bienvenue" and eastward into Lake Borgne. How the runoff would get there was the next major change, as *this would be a largely closed, underground system.* "Branch canals start from the main canal, and then diverge with constantly decreasing capacity, separating into main drains, branch drains, and surface gutters, which extend to the limits of each section."

The design mimicked dendritic streams flowing through a natural watershed, only this was for urbanized polders with barely perceptible slopes.[47] The "surface gutters" would be the equivalent of fifth-order streams, the "branch drains" forth-order streams, the "main drains" third-order, the "branch canals" second-order, and the "main canal" being the river. Precise measures of runoff, in cubic feet per second, were determined for each tier, along with slopes and velocities, all of which drove the metrics for the main canal.

Even on a dry day, the engineers expected 38 cubic feet of soil water, mostly from Mississippi River infiltration, to flow into the system every second. During a heavy rain, that figure swelled to an estimated 12,234 cubic feet per second, across all five East Bank sections, with another 671 cfs in Algiers. That larger figure would drive the sizing of the main canal.[48]

To call it a mere "main canal" was engineering understatement. What was being proposed for the South Broad–to–Florida Avenue corridor was a cavernous seven-mile-long underground river, seventy feet wide, its bottom fifteen feet below grade level, its "entire bed and sides [lined] with smooth permanent material in order to insure cleanliness, and a good velocity and discharge."[49]

Unlike a river, however, gravity could not mobilize this water, as the entire tunnel was "practically level."[50] Instead, the engineers proposed colocating enormous pumps along Broad to relay the water eastward along Florida Avenue and into Bayou Bienvenue, which would have to be dredged, widened, and "cleared by the removal of sunken logs" to get the discharge to Lake Borgne. From ancient trunks mired in the *cyprière,* to smooth, clean, fast, mechanized drainage: such were the ambitions of the advisory board.

The pumps were the system's beating hearts, sucking up runoff, sucking down groundwater, and mobilizing both for expulsion. "For the purposes of obtaining the fullest advantage of the proposed system," the advisory board iterated, the water "*must be kept pumped down,* throughout all its branches and mains, *to the lowest possible level.* Constant vigilance and care must also be exercised to keep [all] canals *free of deposits of any kind,*" because any volume lost to debris could have been more water removed.[51] "The resulting effect of . . . removing the ordinary flow of water," the engineers concluded, "will be to materially lower the water level over all parts of the territory under consideration, including the swamp areas, and to keep the entire territory permanently dry and available for improvement."[52] One need not read between the lines to understand that every drop of water was deemed problematic, solvable by bringing to bear every modicum of system capacity. In today's parlance, this was an all-out prioritization of gray infrastructure over blue or green.

To exert the sucking power, the advisory board designated "main pump stations" and "auxiliary pumping stations," based on their locations, destinations, and capacities. There would be five main pumps. Pumping Station No. 1, slated for what is now South Broad at Melpomene, would have a capacity to relay and lift up to 2,738 cubic feet per second. Pumping Station No. 2, at South Broad at St. Louis Street, would handle up to 3150 cfs; Station No. 3, originally for Hope Street and later moved to Florida Avenue at London (now A. P. Tureaud) Avenue, would have 3,250 cfs; Station No. 4, Florida Avenue at Lafayette (now Peoples) Avenue, 3,000 cfs; and Station No. 5, at Jourdan and Florida, 2,500 cfs. Runoff quantities, augmenting as they moved eastward across rain-soaked neighborhoods, would get "rehandled" from pump to pump, as Bell put it in 1874, like a gigantic bucket brigade. Once the runoff reached Bayou Bienvenue, courtesy of a diagonal levee-lined open canal dug to connect with the wending natural channel, gravity would take over and discharge into Lake Borgne.[53] This diagonal Main Outfall Canal remains salient today in the triangular shape of the wetlands directly behind the Lower Ninth Ward.

For Station Nos. 1, 2, and 3, the pumps would steer most water eastward down the main canal. Roughly a quarter of their respective capacity would be reserved to relay the remaining water northward, through "branch relief canals" (the advisory board's term for W. H. Bell's four outfall canals, dug in the early 1870s by Warner Van Norden) and helped along by those "auxiliary pumping stations." Only "when the capacity of . . . the main canal becomes insufficient to carry off the water [eastward to Lake Borgne], then it is necessary that the pumps deliver the excess into the branch relief canals, to flow into the auxiliary pumping stations . . . and thence into Lake Pontchartrain" to the north. In this way, the advisory board gracefully integrated the problematic but still useful old system into their superior new system, as a supplemental alternative during the heaviest rains.[54]

There would be three of these auxiliary, northward-pointing pumps, to be numbered 6, 7, and 8. Auxiliary Station No. 6 would relay and lift runoff along the Seventeenth Street Outfall Canal (which the advisory board called the Upper Line Canal because it formed the Jefferson-Orleans parish line). Auxiliary Station No. 7 would do the same on the Orleans Avenue Outfall Canal in what is now Lakeview by City Park, and Auxiliary Station No. 8 for the London Avenue Canal in Gentilly. The advisory board called these extant channels "relief outfall canals," clarifying their Pontchartrain-discharging role as subsidiary to (that is, relieving of) the main outfall canal's eastward thrust to Bayou Bienvenue and Lake Borgne. In this manner, the Lake Borgne decision begot all subsequent project-design decisions.

The new system would have ninety-five miles of canals, of which thirty would be lined with concrete beds and covered, and its eight pumping stations would have a total capacity of 18,991 cubic feet per second—16 times more than the existing system's 1,214 cfs capacity, and over 171 times more than the first steam-powered system.[55] Total estimated construction costs: $7,933,691, or nearly $250 million today.

The advisory board's *Report on the Drainage of the City of New Orleans,* released in 1895 after a year and a half of research, represented American engineering at its best, done by local engineers using local resources, "tackling the world's toughest drainage problem."[56] The report had some notable omissions, mostly regarding scope. It did not cover most of what is known today as New Orleans East, largely unoccupied at the time, and lower Algiers, still mostly agrarian. It spoke little of levee protection from river or lake, except as extant topographic conditions—this was an internal drainage plan, after all, and not an external flood-protection plan. Relatedly, there was zero talk of lakefront seawalls or land reclamation, à la George Bayley or William Bell, because the Lake Borgne decision steered attention away from Lake Pontchartrain. It also saved pump design and power generation for another time, as the board members were mostly civil and not mechanical or electrical engineers.

There was also one major oversight, which came into relief only with hindsight. The engineers failed to foresee that their drainage interventions would trigger soil subsidence—massive, fast, deep sinkage below sea level, which would soon render obsolete their topographic data and subvert their design plans. Yet they did notice that one spot which had dropped below sea level. On page 17 of their report, in a passage mostly devoted to general topographical description, there appears the sentence, "*A large portion of the basin between the foot of the [natural levee of the Mississippi] and Metairie and Gentilly Ridges is below mean gulf level.*"[57] But the authors did not flag that important observation as an omen, except to call such spots "swampy and practically waste land, affording very unsanitary conditions." The engineers had read a twentieth-century problem through a nineteenth-century lens, and in doing so, overlooked a colossal geo-engineering accident that would befall future New Orleans.

THE NEW ORLEANS DRAINAGE COMMISSION

"Drainage Plan Before the Council," read the papers on Wednesday, May 1, 1895. It wasn't exactly front-page news—top of page 6—but this time the *Daily Pica-*

yune gave drainage its due, summarizing the advisory board's "voluminous document" with a headline commensurate to the lengthy article: "The Advisory Board Making a Complete Report, In Which the Requirements Are Fully Set Forth, And a Plan of Canals and Pumping Stations, To Cost Eight Million Dollars, Capable of Sectional Construction, is Recommended."[58]

For the rest of the month, city councilmembers digested the complex plan, batted around comments, and quizzed the authors. Some questioned the Lake Borgne discharge decision. Some questioned the cost. Others pondered the fate of the Old and New Basin canals. Property owners on Broad Street formed an association to fight the huge main canal. But the general sentiment in the council chambers, in the neighborhoods, among commercial leaders, and certainly in city hall—Mayor John Fitzmorris had chaired the advisory board—ran supportive, even enthusiastic, about the plan.[59]

Nature lent a hand to their effort. At dawn on May 24, a lightning storm with near-record rainfall put streets "beneath sheets of water." The next day, with headlines reading "Rain Makes a Plea for Drainage," members of the advisory board went before the city council, armed with two years' worth of maps, graphs, and data. The ensuing debate went on for weeks, mostly on account of councilmembers whose constituents might be inconvenienced, namely along Broad Street.[60]

Finally, on Wednesday evening, July 10, 1895, the city council put the advisory board's plan up for a vote. Hands went up eighteen in support versus four votes against. Ordinance No. 10991 declared "that the plan of drainage as submitted by the Advisory Board be . . . approved and accepted and made the plan of drainage to be put in execution in the city of New Orleans."[61] The century-long effort of so many drainage kings had finally culminated with the city officially committing democratically to a professionally designed comprehensive drainage system.

Now came the tough part: managing, funding, and executing the advisory board's good work. Because of the legal authority needed and sheer scope of the effort, the state legislature intervened in city affairs by creating the New Orleans Drainage Commission through Act No. 114 of its 1896 session. That act set forth the new commission's composition, its contractual and financial machinations, its "cooperation" with the Orleans Parish Levee Board (OPLB), and its mission, to "have charge of the construction and administration of the system of drainage of the city, [as] one of the most important public bodies to which the future welfare of New Orleans has been committed."

But the drainage commission was not quite the dedicated agency that drainage had long needed; it was an uncompensated panel of officials with other jobs, including the president of the Orleans Parish Levee Board. Only that agency could levy taxes (one mill on the dollar) to get its job done, and it was "construction, repair, control and maintenance of all levees in the district, where on the river, lake canal or elsewhere." It was not drainage.[62]

KATE GORDON AND THE WOMAN'S SEWERAGE
AND DRAINAGE LEAGUE

What saved the day for drainage was sewerage service and potable water, the two parallel improvements for which progressive citizens had also been clamoring. A sewerage system would enable indoor plumbing, sending human waste to a modern plant at the lower end of the city for treatment and disposal. Meanwhile, at the upper end of the city, fresh water could be drawn from the top of the Mississippi River, purified at a modern treatment plant, and pumped to thousands of domestic kitchens and baths. With sewerage and water systems in place, New Orleans could leapfrog from the eighteenth to the twentieth century, and with a drainage system, it could finally expand into a modern metropolis.

Progressives mobilized. Some formed the Citizens' League; others, the Citizens' Party. The groups soon merged and helped elect Walter C. Flower mayor in April 1896, after which a new reform-friendly city charter was put in place. The Citizens' League also enabled Abraham Brittin to get elected to the city council, where he articulated three important considerations for the forthcoming improvements. In the words of historian John Smith Kendall writing in 1922, Brittin argued that "with drainage should go water-supply and sewage"; that "the city itself should undertake the work, and not leave it with a private corporation"; and that "these various activities should be *concentrated under the control of a single board,* and not left to the management of the city council."[63] Upping the ante were three yellow fever outbreaks in 1897, 1898, and 1899—none calamitous, but alarming enough to demonstrate "that a general house-cleaning was necessary for the salvation of New Orleans, and that the methods which had been depended on for drainage . . . for 200 years, would have to be immediately abandoned."[64]

Dovetailing with Brittin, progressive citizens formed a coalition known loosely as the "sewerage, water and drainage campaign committee." They joined others petitioning city officials to form a "sewerage and water board," and hold a public referendum to fund the board with a special tax of up to 2.5 mills of assessed real estate value. "The choice," Brittin made clear, "lies between a special tax and city ownership on the one hand, and private control on the other."[65]

With these goals in mind, the Woman's Sewerage and Drainage League, led by suffragist Kate M. Gordon, aimed to drum up support for the special tax in a referendum in which some women would have the rare opportunity of voting. Ladies over age twenty-one had to certify that they were on the tax assessor rolls, meaning they were landowners, and had resided in their respective ward for at least six months, in New Orleans for one year, and in Louisiana for two years.[66] Both the opportunity as well as its restrictions were reminders that the Progressive movement was largely a crusade of upper-middle-class whites, and in New Orleans, that mostly meant well-educated denizens of the Uptown "silk stocking wards." Not many Black women were landowners, and they as well as Black men faced racist and classist barriers at the ballot box. As for blue-collar whites, the Woman's Sewerage and Drainage League condescendingly presumed "the same old barrier of indifference" would prevail among the working-class wards of the lower parts of the city. Gordon, president of the league, adjusted her message to appeal to those communities, but did not spare them her exasperation. "Can any one tell me what is the trouble with the women of New Orleans?" she huffed in a March 1899 appeal. "Haven't they any civic pride? . . . Why should not the women be as much interested in the cleanliness of the streets as they are in their back yards? . . . The women of the League for Sewerage and Drainage beg the women of this city to join them in this struggle . . . for by getting this tax voted, $14,000,000 will be put in circulation among our working men."[67] Roughly three-quarters of property owners in the city were women, due largely to widows' inheritances, and advocates knew their vote would be key.[68]

The big "drainage and sewerage election" was scheduled for Tuesday, June 6, 1899. Rules regarding who could vote, per gender, were so complicated that the *Daily Picayune,* which strongly supported the measure, decided to spell out everything clearly in a special column. No mention was made of race, but the heavy favoritism of property owners, plus the intensifying statewide disenfranchisement of the Black population, made this mostly a referendum of the white populace. The main question on the ballot: FOR or AGAINST "the special tax of two mills for forty-three years, to be devoted to water, sewerage, and drainage."[69]

The Uptown elites had grossly underestimated their working-class counterparts. Voters in every ward overwhelmingly supported the special tax, resulting in a 6,089-to-389 landslide. This time, drainage made banner headlines.

HATS OFF TO OUR PATRIOTIC WOMEN!
Sewerage Succeeds, And With It Drainage, Plenty of Pure Water
The Men and the Women of New Orleans Unite
How the Glorious Victory Was Nobly Won

Centered on the front page was an illustration of an allegorical woman named *New Orleans,* astride a black stallion named *Progress,* leaping over a loathsome chasm labeled *Sanitary Neglect, Commercial Stagnation,* galloping toward distance hills "to a greater city," with a winged angel glowing "Victory" in the sky above.[70] The latest, greatest generation of drainage "kings" now had thousands of "queens" in their realm, and it was they who finally secured the hard dollars to dewater New Orleans.

THE NEW ORLEANS SEWERAGE & WATER BOARD

On June 22, 1899, the New Orleans City Council readily passed Ordinance 15,391, officially levying the tax that voters had approved on June 6. On August 8, a special session of the state legislature in Baton Rouge passed Act No. 6 to create the Sewerage & Water Board of New Orleans. On August 18, the legislature amended the Constitution of the State of Louisiana to carry into effect the tax and the issuance of bonds "to establish therein public systems of sewerage and water."[71] In March 1902, the state passed the "Merger Bill," amending the 1899 act by consolidating the circa-1896 Drainage Commission with the Sewerage & Water Board.

New Orleans would thenceforward have a single, tax-funded, professionally staffed, state-backed public utility for sewerage, water, *and* drainage. Oddly, no one ever made a point to update the name, and to this day, the Sewerage & Water Board of New Orleans, "tackling the world's toughest drainage problem," still does not have "drainage" in its name.[72]

With speed and efficiency not normally associated with New Orleans, work got underway on installing the three separate systems—drainage, water, and sewerage—beneath the same streets, and managed by the same utility. Initial contracts were let two years before the tax referendum, starting in August 1897, followed by a slew after the June 1899 vote. Maj. Benjamin M. Harrod, unquestionably New Orleans's drainage king at this key moment, now served as chief engineer of the Drainage Commission, under Chairman R. M. Walmsley, who would later oversee the commission's merger into the Sewerage & Water Board.

Speaking of mergers, engineers did consider whether drainage (that is, removal of stormwater runoff plus swamp and soil water) and sewage (toilet waste, bathwater, and sink water from residences, businesses, and industry) ought to be built separately or combined into a single system, as many cities have. They opted for separate systems, for three reasons. Discharging sewage into Lake Borgne and especially Lake Pontchartrain would be deleterious to public and environmental health, even on circa-1900 standards. Additionally, positioning sewerage lines, with their weak gravitational flows, inside of the big stormwater conduits could lead to waste fermentation beneath city streets, creating stenches and health risks. Finally, the two systems had two very different construction timetables, and no one wanted the complex drainage system to slow the installation of the top-priority sewerage system. So the two systems were built separately. The third system, for the treatment and distribution of potable water, had a whole different geography, originating from a point source (top of the river in Carrollton, with a treatment plant farther back in Carrollton) and pumped through a distribution network to each household, and it too would be built separately on its own priority schedule. The linear guts of all three systems, however, ran more or less adjacently to each other—mainlines and pipelines, within the same cavities, dug beneath the same streets, maintained by the same utility.[73]

7

Dewatering New Orleans,
1900s–1910s

As the new century dawned, construction progressed on the premier nodes of the drainage system, the shedlike stations for the Broad Street pumps. Blueprints came from the desk of the versatile Benjamin Harrod, who was an architect as well as engineer, surveyor, and planner. What he designed were masterpieces of municipal architecture—functional yet beautiful, monumental without being ostentatious, simple yet surprisingly detailed, and at once Modern and Classical. In their original form they each had eight bays, one per pump, their foundation stones inscribed "1899" along with the names of Harrod and others. Attuned to the heroic ethos of civil engineering in this era, Harrod succeeded in rendering the sheds into stately neighborhood landmarks. Their unassuming facades of reddish-brown brick were subtly complicated with banding, pilasters, entablatures, and beveled corners, and their broad hipped roofs were topped with ventilating monitors, giving them a distinct profile. Harrod's pumping stations, the beating hearts of a vascular system of manmade hydrology, exuded gravitas.[1]

The vessels of the system were a different story—dug expediently and built to work, not beautify. Streets were torn up all over town, and as today, neighbors complained of disruptions, expropriations, and impacts on property values and quality of life, which authorities heard out in public hearings and addressed as they could.[2] By 1901, over $3 million had been spent building out roughly 20 percent of the entire system. By early 1902, canals had been dug, interconnected, and covered on stretches of Canal Street, Basin Street, Claiborne Avenue, St. Charles Avenue, Julia Street, Chartres Street, and Constance Street. "We entered all of these canals," wrote a team of inspectors, and "walked through quite a length[,] carefully making personal examinations of . . . material and workmanship, as well as the alignments, connections, their general condition and operation." Laborers progressed on other stretches of these arteries, as well as Perrier Street and Nashville Avenue uptown, where engineers monitored soil conditions and made on-the-spot construction modifications.[3]

Everything done for East Bank drainage had to be replicated in smaller form for that one part of the City of New Orleans across the river. "The drainage of Algiers on the right [west] bank of the Mississippi," declared the advisory board in 1895, will entail a "system of gutters, branch and main drains, and branch canals leading into a main or intercepting canal located on Canal [now Whitney] Avenue. From a main pumping station [at] Lawrence Street and [Whitney] Avenue, the total run-off . . . will be lifted . . . and delivered into the outfall . . . canal, leading into Bayou Barataria."[4] Gravitational ditches to be dug on Vallette, Eliza, and Lapeyrouse streets would get runoff to the pump station. As on the East Bank, unexpected site conditions sometimes arose, and contractors usually responded with ad hoc adjustments.

THE BOARD OF INQUIRY

Such tweaks were par for the course in a complex project like this—to a point. But as contractors translated the 1895 plan into field execution, they made ever more adjustments, on matters such as concrete types, drain locations, rainfall assumptions, and canal designs and capacities. Even some pumping sites were modified. Because these changes affected contracts, budgets, and overall project goals, they implicated former city engineer Linus W. Brown, mastermind of the 1895 plan, who was not pleased to see his expertise questioned. On August 2, 1901, he wrote a letter to Drainage Commission Chairman R. M. Walmsley, warning that the changes were "in direct violation of the plan of drainage as adopted by the City Council and approved by the Legislature [and] will most seriously affect the future welfare of this city."[5]

The commission took the charges seriously and formed a Board of Inquiry of the Conduct and Character of the Drainage Works. Brown's indignation would backfire on two accounts. For one, the board's highly qualified engineers found that, in fact, the alterations had "been made in the interest of increased efficiency or economy," were "wise and proper," or otherwise did not have "an intention of altering the original plans."[6] Second, having the podium, the three board engineers took the opportunity to redress a major issue in Brown's design, regarding power source. Times had changed since 1893, when the advisory board had first convened, and a new technology had since revolutionized cities. The "boilers and steam engines" recommended in the 1895 report to power the pumps ought to be changed to "a central station . . . from which the power [will] be distributed by electricity to each pumping station[,] similar to changes that are everywhere taking place in the industrial and mechanical world."

By concentrating power at one central plant, the number of skilled workers could be minimized, realizing over $20,000 in annual savings.[7] This Central Power Station would be located just off the main canal along Florida Avenue at its intersection with Lafayette (now Almonaster) Avenue, ideal for its ample space and centralized position. It also sat at the crossroads of two discharge directions: eastward in the main canal to Bayou Bienvenue and Lake Borgne, as most water would go, and northward to Lake Pontchartrain, in that circa-1870s Lafayette Relief Outfall Canal, now the People's Avenue Canal.[8] Once the water-purification plant opened in Carrollton in 1908, a second power-generation plant was built adjacently; in time, this plant, on Eagle Street off South Claiborne, would become the main generation plant, and the source of most pump energy.

In recommending centralized electrical generation over multiple self-standing steam engines, the Board of Inquiry next found itself planning for contingencies. What if a storm knocked out that central plant? Wouldn't the very lowlands drained dry by the pumps now be residentially developed, and prone to flooding with the pumps off? "We think your engineer," board members told city officials, "should plan and estimate the cost of a reasonable safeguard against such emergency," suggesting an auxiliary power station at either Pumping Station 6 or 7 as the solution. Such was the 1902 version of resiliency planning, aka "future proofing."

Other concerns came up at the board's inquiry. Did the pumps, regardless of power source, have enough capacity? By one account, they were among the largest and best available in the country, eight-foot-diameter vertical-shaft screw pumps with synchronous movement.[9] But this was not the rest of the country; New Orleans's needs were unique in their nature and magnitude. Engineer Charles Louque testified that "the canals are too small [and] the pumps are inadequate," throwing further shade on Brown's designs. Writing to Mayor Capdevielle, Louque stated: "I come to caution you, first as a friend, then as a party interested, thirdly as an expert . . . that the pumps which the contractors of drainage are at this moment putting up at Broad street will prove a total failure. The gigantic proportions of the preliminary work would lead one to suppose that the pumps were to be the largest in the world, while the pumps themselves are simply playthings and toys."[10]

Questions also arose about the main trunk canal beneath Broad Street to discharge to Bayou Bienvenue and Lake Borgne. Contracts to build that all-important water tunnel got delayed, even as work progressed on "pumping stations Nos. 1, 2 & 3 . . . so arranged . . . for operation along the line of the proposed main canal." The reason was because planners had opted for a section-by-section construction approach, starting with the more urbanized areas, which, being closer to the "relief outfall canals" (that is, the Seventeenth Street, Orleans Avenue, and London Avenue outfall canals), would discharge northward into Lake Pontchartrain. This precisely contradicted the famous 1895 advisory board decision to steer most runoff eastward toward Lake Borgne, and it's not surprising that it triggered fears among Brown and others that the 1895 plan was being subverted in a fundamental way.[11] In fact, engineer Benjamin Harrod testified that "the canal on Broad street, the main outfall, [was] left to be built last" by the recommendation of none other than the advisory board itself, the body behind the 1895 plan, "because the city would have benefitted by building the other canals first."[12]

Nevertheless, the scheduling decision, and its effect of (temporarily) flipping the discharge direction, broached questions about pump station locations—indeed, about the whole Borgne-versus-Pontchartrain outfall debate. "Until that [main] canal is built," asked the Board of Inquiry chair to a project engineer, "all of the water . . . *must necessarily go out into Lake Pontchartrain?*" "Yes, sir," came the response. "And this water cannot be prevented from going into this Lake Pontchartrain without the building of the main canal [on Broad]?" the chair pressed. The project engineer replied, "I know of no way it can be done."[13]

Here we see the beginnings of the gradual reorientation of the drainage outfall, away from Lake Borgne and increasingly toward Lake Pontchartrain. It would take decades for this shift to be fully actualized, but already as of 1920, more water went to Pontchartrain than to Borgne, despite all the prior decisions and lingering insistence that Borgne was superior.[14]

Perhaps operational reality was demonstrating that the advisory board had made the wrong decision in 1895. What can be said with certainty is that, had the system been originally designed to discharge into Lake Pontchartrain, the pumping stations *would not have been located in the middle of the city,* at the "bottom of the bowl," along Broad Street. *They would have been located along the lakeshore, at the perimeter of the bowl, and the entire hydrology and engineering of the modern system would be safer and simpler.*[15] We know this because future systems, built in eastern New Orleans, the West Bank, and Metairie, all used perimeter pumps; indeed, the only reason to have put the pumps in the middle of the "bowl" was the Lake Borgne decision of 1895. Had Pumping Stations 1, 2, and 3 been located lakeside, at the heads of the outfall canals into Lake Pontchartrain, there would have been no floodwall breaches during Hurricane Katrina in 2005, because the phalanx of pumps at their mouths would have served as surge-blocking gates, and the outfall canals behind them would have been below-grade and fed by gravity, with no floodwalls to breach.

Too late. By the time the ink had dried on the *Report of the Board of Inquiry of the Conduct and Character of the Drainage Works,* the big pumping stations on Broad were all but complete, and they were hooked up to the already-built outfall canals discharging into Pontchartrain, astride the not-yet-built main canal pointing toward Borgne.

Such inadvertent detours and provisional-cum-permanent fixes would become the norm. "The New Orleans drainage system has never been a static entity," observed the Army Corps of Engineers on the hundredth anniversary of the Sewerage & Water Board. "Changes were made in plans and construction [since] the earliest days of the Drainage Commission, and there was never a point where the system planned in 1895 was in place, as designed."[16]

Like all complex infrastructure, the New Orleans drainage system, then and now, is a network of networks with a thousand moving parts, some of them rather improvisational, most requiring constant maintenance and upgrades, and all with weblike feedback loops prone to unintended consequences. Each would reconfigure the baseline conditions to which future engineers and authorities would have to contend, further inscribing path dependence into the processes and outcomes of decision-making.

"THE LOWERING OF THE WATER"

Steadfastly the work progressed. Contractors did the heavy lifting—thousands of laborers, white and Black, native and immigrant, working among steam shovels and dredges, knee-deep in mud. They removed old cypress trunks, smoothed out canal bottoms, lined tunnel walls with bricks or wood, covered them with steel and concrete, and connected them with pipelines and storm drains. By 1901, about 20 percent of the system was in place, including the main sheds and pumps, powered by synchronous electrical motors running on a twenty-five-cycle three-phase altering current.[17] By 1904, about 44 percent of the system was complete. By 1905, the reclaimed terrain had expanded from 16,000 to 22,000 acres, and the pace of water removal increased from 1,214 to 5,000 cubic feet per second. Over 20 miles of covered concrete-lined canals now crisscrossed the landscape, along with 17 miles of open canals, 3 miles of wood-lined canals, and, by 1909, over 106 miles of subsurface pipelines up to three feet in diameter.[18]

After "subsurface drainage was commenced," wrote US Weather Bureau forecaster Isaac M. Cline of these years, "an immediate result was the lowering of the water in the drainage canals 8 to 10 feet below what it had been previously, thereby effecting a much more rapid transfer of storm and other water from the city through the drainage system into Lake Pontchartrain."[19] Once a two-dimensional city with hardly anything underground except dirt—no tunnels, few basements, not many graves—now New Orleans had grown a third dimension, in the form of three vital infrastructural systems.

The three systems had mirror versions across the river. Algiers streets got torn up, drains and pipes laid, and a pump station installed at Lawrence Street and Whitney Avenue. In 1907, workers began digging the three-mile-long Algiers Outfall Canal to connect to Bayou Barataria, the West Bank equivalent of the main canal on Broad.[20] Concurrently, sewerage and water lines were installed, and when the three systems went online in 1909, Algiers joined the twentieth century. Not to be left behind, the Jefferson Parish Police Jury set about in 1910 creating a drainage district for neighboring McDonoghville, Gretna, and Harvey, matching those under development in Amesville (Marrero) and Westwego, and catching up with its rival across the parish line, Algiers.[21]

As more apparatus went into the soil, more water came out, with ever-greater efficiency. The system yielded astonishing results while still under construction. "Storm water from moderate storms was removed rapidly, and saturated soil and stagnant street gutters were drained by pumping standing water in the canal system to ten to fifteen feet below street level," wrote the Army Corps of the early 1900s. "Mosquitoes decreased noticeably. Land within the city limits that had formerly been too wet for building or agricultural use became available for development, and mortality rates dropped significantly."[22] It was as if a whole new city were emerging from the morass.

Among the first to pounce on the new real estate were "homestead associations"—private land speculators who bought up vacant land as it dewatered. The hottest properties lined the fringes of the topographic bowl, just beyond the natural levees and ridges. For Carrollton and Uptown this meant today's Hollygrove and Fontainebleau; for the Metairie Ridge (City Park Avenue), it meant Parkview, Navarre, and later Lakeview; off the juncture of the Metairie and Gentilly ridges, it meant Mid-City and Gert Town; off the shoulders of the Esplanade Ridge, it meant the interiors of the Fifth, Sixth, and Seventh wards; and off the Gentilly Ridge, it eventually gave rise to the greater Gentilly region.

Developers of Gentilly Terrace, laid out in 1909, barely mentioned how drainage had enabled the lower half their subdivision to be built. Instead, they drew homebuyers' attention to the higher natural ridge of Gentilly Boulevard. Promoters declared its "sandy loam" soils to be "twenty-seven feet above the Cairo Datum Line" (which was technically true) and surmised it thus constituted "the most elevated residential section in the City of New Orleans" (utterly false).[23] Soon Gentilly Terrace's streets would be lined with California bungalows and Spanish Revival cottages, as if transported from Pasadena.

In the same year, the Lakeview Land Company broke ground on major new subdivisions to the west of City Park. "LAKEVIEW HAS ARRIVED," it proclaimed in June 1909. "The progress of 'Lakeview' has been phenomenal" since swamp drainage. "In the space of twelve months, 'Lakeview' has been converted into rich income-bearing residence property. [It's] the logical suburb for a crowded city . . . in one of the most beautiful and healthful spots in Louisiana. . . . Up will go the value of Lakeview lots!"[24]

Similarly, the marshy Uptown bowl, which the 1895 topographic survey had determined to be the first spot to drop below sea level, now looked perfectly enticing. "The ability to drain the moor around Broad Street," noted an Army Corps study, "meant a whole new real estate market for the center of the city." That area gained the name Broadmoor, and over the next generation, "virtually the entire 250-block neighborhood . . . developed for residential settlement," filling in the last open lands in that part of the city.[25] As in Gentilly Terrace, Broadmoor, particularly along "Napoleon avenue and Claiborne . . . running back to Broad street and beyond," drew its architectural inspiration not from the humid Gulf Coast but from the dry West Coast. Real estate agents in 1920 gave it the nickname "Little California" and hailed it as the scene of the city's "greatest building activity."[26]

During the opening decades of the twentieth century, New Orleanians *en masse* moved off higher ground and onto lower ground. "Cellars can now be dug," marveled one reporter in 1903; "the first cellar ever dug in New Orleans will be under . . . a twelve-story addition to one of the large hotels." Pumps had been "rapidly drying out the soil, not only the surface water from drains, but the soil water," and the "supersaturation of the ground has disappeared. Formerly water could be struck two feet below the surface, but now it is necessary to go six feet for it."[27] Beamed writer George Washington Cable in 1909, "There is a salubrity that could not be when the mosquito swarmed everywhere, when the level of supersaturation in the soil was but two and half feet from the surface, where now it is ten feet or more. . . . The curtains of swamp forest are totally gone. Their sites are drained dry and covered with miles of gardened homes."[28]

Drained dry. Consider the hydrological implications of those words in this fluvial delta. Desiccation meant less water volume in the soil body keeping it fat and moist, and more air spaces inviting it to consolidate and shrink.

Covered with miles of gardened homes. Likewise, every home—and rooftop, driveway, sidewalk, and street—represented that much more hardened surfaces, and less porous soil to absorb and filtrate precipitation. Recall Benjamin Harrod's directive from 1888, before he created the 1895 plan, in which he called for the "the extension of permanent and impervious pavements as rapidly as possible" such that every raindrop flowed straight to pumps.[29] That gray-over-green imperative was already exacting its toll. "Increasing numbers of buildings and area of paved land," observed the Army Corps retrospectively, "particularly on the lake side of Broad Street, were reducing the ability of the soil to retain precipitation, thereby overwhelming the drainage system."[30] That meant that, in heavy rains, runoff poured down the drains and into the canals at a pace faster than the pumps could remove it. What could result was inundation of those *miles of gardened homes*—unless, of course, you increase pumping capacity. And so, as we saw how drainage begets flooding, and how flooding begets drainage, now we see how drainage begets drainage.

That's precisely what happened to the 1895 plan. Engineers never truly completed its charge, because the work done to date birthed the need for more work. The baseline had to be reset, and the drainage kings, namely Benjamin Harrod and Rudolph Hering, reconvened in 1910 to reprise their strategic plan. Harrod, aged seventy-three and highly accomplished, and Hering, aged sixty-three and known as "the father of modern municipal sewerage systems," brought world-class expertise to the team. But luckily for them they had a bright young "prince" named Baldwin at their side, and he would change the world of urban drainage.[31]

ALBERT BALDWIN WOOD

Albert Baldwin Wood hailed from Louisiana aristocracy in all its ethnic tap-roots. His anglophone father was a captain from a founding family of Pennsylvania, who, with his brother, had relocated to New Orleans and prospered in finance and commerce. His francophone mother descended from the Bouligny and Fortier families, whose forebears included the Spanish commandant Francisco Bouligny and scholar Alcée Fortier, the very embodiment of the French Creole intellectual elite.[32]

Born in 1879 and raised on Prytania Street, "Baldwin" seemed to inherit his namesake uncle's gift for numbers. The precocious youth devoured all things mathematical and mechanical, and went on to study engineering at Tulane University's new St. Charles Avenue campus. It is said that he and a buddy, having read papers by Guglielmo Marconi, managed to put together a "wireless" and communicate across classrooms—perhaps the first radio in New Orleans.

Baldwin Wood earned a BS in engineering at age twenty and graduated with honors in 1899. Though his pedigree and intellect could have landed him in the most lucrative lines of work, he instead took a civil servant position as an assistant manager with the New Orleans Drainage Commission, just as it became the New Orleans Sewerage & Water Board. "A. Baldwin Wood," as he was known professionally, would devote the remaining fifty-seven years of his life to that city utility, accepting no more than a standard salary and foregoing "a king's ransom" in royalties from his eventual thirty-eight patented inventions.[33]

What the young prodigy saw inside the just-completed pumping stations was a series of nested problems: too much water coming in, with too much debris, needing to be drawn and discharged at a much greater volume and velocity. The existing pumps were vertical shaft screws, which worked on the principle of vertical displacement. That is, their eight-foot-wide threads turned inside a chamber, taking in water at the bottom, raising it up within the spiraling thread spaces, and spilling it out at the top, at which point gravity took it out the canals to the lake. "They were sunk in the basement," noted the American Society of Mechanical Engineers, submerged in a pool of incoming runoff known as a suction basin, "and when anything was out of order the mechanics had to go down in the subterranean chamber to repair them."[34]

Repairs were needed constantly, because, as the 1902 Board of Inquiry noted in a section titled "Carelessness and Wantonness," "drains are choked with a most remarkable collection of garbage and trash," including the "unsightly deposit of paper and sweepings in the gutters" and invasive hyacinth and other aquatic plants in the open canals.[35] Worse yet, every time the pumps had to be taken offline for repairs, canal water back-flowed into the subterranean screws.

There were other problems. "Several pumping stations have been in service now for ten to twelve years," reported S&WB General Superintendent George G. Earl in 1911; "the time is inevitably coming when the Board will [have] a lot of broken-down machinery, entirely unfit for use and incapable of being further maintained in efficient service." Even if the pumps were all running perfectly, they still lacked a capacity commensurate to runoff volume, more of which arrived faster into the suction basins thanks to ever-expanding urbanization and diminishing permeable surfaces. "At times it had to be driven far above its rated capacity," wrote Earl of the laboring machinery, "to overcome . . . the deficiency in power and capacity which exists whenever rainfalls of much more than average intensity occur."[36]

Wood assessed the situation and determined the need for two tiers of water-removal apparatus: "constant duty" pumps, to handle daily soil-water infiltration from the river, and "storm pumps," which needed much greater capacity and speed. Intakes for both types had to be continually filtered of flotsam, while storm-pump motors had to be startable quickly, under a mounting load created as runoff rushed into the suction basin. They needed to run at a constant speed, and when stopped, their discharge pipes could not allow any backflow. Fundamentally, the whole phalanx of pumps, bay after bay in station after station, had to be raised up out of those dank dungeons and brought onto the floors of the sheds, accessible to mechanics.[37]

Baldwin Wood was the right man with the right job. His first inventive response had come in 1906 when, in answer to the shortfalls of the vertical screw pumps, he designed centrifugal pumps to fit into the bays. This device used an impeller spinning at high speeds inside a horizontally oriented chamber, creating hydrodynamic energy to accelerate a torrent of water upwardly from the suction basin in the "basement" of the shed, and outwardly into the external discharge basin, where the now-heightened water would flow by gravity to its outfall. If a vertical screw pump might be analogized to a helicopter, then Wood's new centrifugal pump would be a powerful jet turbine, blasting a torrent of water horizontally. He also invented "'flapgates' to stop water from backing up when the pumps were stopped—a concept that surprised the engineering world."[38]

Wood's next wave of inventions came in the early 1910s, as the Sewerage & Water Board wrangled with updating the 1895 plan. Some upgrades involved the "nouns" of the system—drains, pipelines, canals; their capacities, materials, and so forth. But most focus was on the "verbs"—that is, the pumps, in all their complex engineering physics.

Wood experimented with models. He took his 1906 designs, modified them on paper, and built a one-to-ten-scale model with a twelve-inch diameter, which he tested and refined. It worked. He scaled up the schematics and built a thirty-inch-diameter version, to be tested and put out for construction bid. It was installed as a constant-duty pump in 1912, and remains in service today.

In 1913, Wood upped the designs to a full-sized twelve-foot diameter and had it prototyped and tested. Described as a low-head, high-volume axial-flow screw pump, Wood's invention featured a steel-bladed impeller rotating with a siphon, which drew runoff from a suction chamber at one end, and without any impeding valves or gates, expelled it into the discharge canal. It was primed by a rotatory vacuum pump, which prevented backflow by admitting a cushioning pocket of air into the encasement.[39]

In that regard, Wood's brilliance went beyond engineering excellence and into the sort of beautiful simplicity that the best design schools teach. Maintenance on pump interiors had always been awkward, so Wood, according to writer Sebastian Junger, split the casing "horizontally to allow easy access to the pump's interior." He designed the pumps to hum with so little vibration that they did not need massive foundations. He made them to be self-oiling, and kept their bearings free from contact with grimy water, making them low-maintenance. He put no gates or valves within the turbines, so as not to impede flow capacity or speed. On one of his later inventions, the trash pump, Wood designed the blades to be so smoothly fitted within the casings "that nothing can get caught on them—stockings, rags, two-by-fours, whatever finds its way into a New Orleans sewer," wrote Junger. "In the old days engineers had to open the pumps and clean them out every eight hours of operation. A Wood trash pump has yet to be opened for cleaning."[40]

In January 1914, Wood's superiors at the Sewerage & Water Board sought permission from the Commission Council to put their youthful genius's blueprints into production. But some commissioners balked, as if to say, *who is this kid?* It was indeed a major investment, $15,000 each, in a technology that attorney Edgar Howard Farrar dismissed as "merely an experiment," involving contracts lacking "any penalty in case of the pumps failing to come up to requirements." Farrar incited a "red-hot protest" against the Wood pumps, describing the whole effort as "mere guess work," dismissing Wood as just "an engineer on the board," and suggesting he possibly had financial motivations. "What arrangements the board has with the patentee of this pump, Engineer A. B. Wood, I do not know, and whether he gets a royalty is not the question now." Wood himself testified calmly to technical details, and emphasized he had no financial stake in the contract.

What was needed was the voice of a respected elder to counter Wood's youth. Into that role stepped Charles J. Theard of the Sewerage & Water Board, who stood up and, "looking straight at Mr. Farrar," declared, "I am not posing as a champion of the people[,] but under my oath, gentlemen, under a full appreciation of the gravity of the situation, and with the eyes of all the people whom I am serving upon me, and on my faith in . . . the corps of engineers on the board, I am willing to take upon myself the full responsibility for the action of the board in this matter."[41]

On January 20, the council approved the expenditure. Farrar resigned in disgust, announcing, "I am through the matter."[42] Indeed he was: all along, Farrar had been retained by other pump makers who wanted to bid on the project, and resented that Wood's designs won the day.[43]

On January 26, the S&WB issued a contract to the Nordberg Manufacturing Company in Milwaukee to manufacture eleven of Wood's twelve-foot centrifugal screw pumps and one thirty-six-inch constant-duty pump, plus all affiliated pipes, gates, shafts, bearings, and hardware. The new pumps, along with the already installed constant-duty unit, would increase total system capacity by 250 percent, from 4,400 to 11,200 cubic feet per second, or over 7.15 billion gallons in a single day. That was thirty-thousand times more than New Orleans's circa-1840 system could handle on its best day. Another Milwaukee company, Allis-Chalmers, won the contract to manufacture four six-hundred-horsepower synchronous motors.[44]

Getting the hundred-ton steel behemoths down to New Orleans and into their berths was itself a feat of engineering. Overseeing it was another young wunderkind, A. C. Hoffman, born the same year of the 1895 plan. Hoffman arranged for freight trains to use streetcar tracks to bring the pumps and motors as close as possible to Pumping Station No. 1, on Broad at Melpomene, where timber runways were laid out on streets for the devices to be gingerly rolled into the sheds. It took five months, December 1914 to April 1915, for the first two pumps to be set into place and readied for testing.

Prof. W. H. Creighton, dean of the Department of Technology at Wood's alma mater, conducted the initial tests and came away awestruck. "The pump is larger than any centrifugal pump ever built and among the largest screw pumps," he reported. "While the pump surpasses in efficiency under normal conditions," Creighton wrote, "the superiority is much greater *just when the greatest service is required.* Emergency service is probably the weak point of the old pumps. *It is the forte of the new.* Results show that the pumps easily answer all requirements and that they are the largest and most efficient low-lift pumps in the world."[45] Edgar Farrar had been proven wrong, and Wood, Theard, the board, and the council were all vindicated. Both twelve-foot Wood screw pumps were in operation by mid-1915, and "immediately demonstrated their superiority over the previous equipment."[46] More were on order.

Wood invented additional devices to improve other operations in drainage, as well as in sewage collection and water distribution. Over the next decade, the drained domain of New Orleans expanded to thirty-thousand acres, undergirded by 560 miles of canals and pipelines, capable of removing 13,000 cubic feet of water per second, or 8.3 billion gallons in a day. And all that was before Wood unveiled his latest creation, a fourteen-foot-diameter screw pump, larger than the largest in the world he had previously invented. A photo taken in Pumping Station No. 1 shows four women posing inside the cavernous turbine, which to modern eyes looks like a B-52 Stratofortress jet engine—only much larger.[47]

As the breakthroughs of 1914 rolled forward, a proud Mayor Martin Behrman traveled to Milwaukee in September to address the League of American Municipalities with a rousing paper titled "New Orleans—A History of Three

Great Public Utilities, Sewerage, Water and Drainage, and Their Influences Upon the Health and Progress of a Big City." It was the stuff of which his predecessors could only have dreamed, in this once-wet, once-wan Necropolis of the South. "Cellars, hitherto unknown . . . became an accomplished fact," beamed Behrman of the effects of the drainage system. "Our cemeteries, saturated as they were[,] necessitated the making of nearly all interments in vaults, or tombs above ground"—but now the pumps have rendered that "practice now no longer compulsory." Those were the least of the changes. "Land before worthless, became at once available for agricultural or city development; mosquitoes were perceptibly on the decrease [and] gutters were no longer stagnant."[48] Typhoid cases dropped by half over that same period; malaria by 95 percent; yellow fever by 100 percent; and overall death rates dropped from 27.2 to 19.8 per thousand, leaving 2,664 people alive who otherwise would have perished. Bienville, Pauger, Trudeau, Lafon, Tanesse, Pilié, Dunbar, Barton, Bayley, Brott, Surgi, Bell: how the drainage kings of the past would have marveled!

Then there were the dividends. Assessed value of taxable property, Behrman said, went from $140 million in 1900 to $250 million by 1914. Drained lands blossomed with new automobile subdivisions, while those farther out "are attracting the attention of agriculturalists everywhere," promising "in the future [to] become one of the most thickly settled agricultural regions in the world." All this for a mere $12.5 million investment in drainage, and $30 million for all three systems. "No city in the country can exhibit a more satisfactory bill of health than the New Orleans of the present day," concluded Mayor Behrman, "in which the successful operation of sewerage, water and drainage has been a most potent factor."[49]

Drainage fever spread. Wood's screw pumps were the key booster-shot, and the Broad Street sheds became technology test beds garnering international attention. The Dutch government, embarking on its epic reclamation of the Zuider Zee, caught wind of Wood's inventions and sent a representative to visit the modest savant. In December 1916, Wood "gave blueprints and the exclusive rights for the manufacture and sale of Wood Pumps in continental Europe to the Werkspoor," according to the American Society of Mechanical Engineers. "The pumps were also installed in Egypt, China and India." Wood later consulted for "Chicago, Milwaukee, Baltimore, San Francisco, and Ontario, Canada[;] he redesigned Chicago's entire drainage system and in 1917 was appointed consulting engineer to the Chicago City Water Works." Wood prospered from his patent royalties and consulting fees, but "never at any time did he charge or collect royalties from any of his inventions used by the S&WB."[50]

Before their worldwide diffusion, Wood's screw pumps caught the attention of would-be drainage kings elsewhere in the metropolitan area, whose deltaic hydrology, if not their urban statures, were nearly identical to those of now-drained New Orleans. Among them was the only one of our water-moving monarchs who literally attained the sobriquet "the Drainage King." He had a real name to match: "Hero."

GEORGE ALFRED HERO

George Alfred Hero, born in New Orleans on July 28 during the 1854 yellow fever epidemic, was the son of Andrew Hero of Göteborg, Sweden, a North Sea port city known for its Dutch-style canals. By one account, Andrew and his brother had first arrived to the United States at Providence, Rhode Island, in 1845 before relocating to New Orleans. Other accounts have Andrew arriving at New Orleans in 1837, and later marrying one Caroline Veil Gray of Rhode Island.[51]

The Heros were one of many foreign-born families arriving in this era. For most of the years 1837 to 1860, New Orleans attracted more immigrants than any other southern city, second nationally only to New York; in 1851 alone, over 52,000 immigrants arrived here, equaling those coming into Boston, Philadelphia, and Baltimore combined.[52] Whereas most who landed were laborers from Ireland or Germany, the Heros were educated professionals hailing from a patrician lineage. Their involvement in the Göteborg shipping industry, where English, Dutch, French, and Spanish traders crossed paths, also gave them a useful linguistic advantage in the polyglottic entrepôt that was New Orleans.

Andrew Hero became a prominent notary public, and with his wife, Caroline, they parented ten children, of whom seven survived into adulthood. The eldest, Andrew Jr., born in 1840, became "one of the best-known notaries in New Orleans and distinguished among the Confederate veterans," having fought with the Army of Northern Virginia during the entire Civil War.[53] George, born fourteen years later, came of age during the federal occupation and Reconstruction. It is said he gleaned a fascination with drainage while visiting his brother's plantation in Assumption Parish, where the precocious boy came upon a morass in the rear and, dismayed at seeing it go to waste, figured out rather handily how to dewater it.

That resourcefulness later helped make George Hero "an entrepreneur[,] a speculator," and something of "a gambler"—not on foolish games of chance, but on calculated risks with attainable gains. In the postbellum South, that meant cotton, and New Orleans being its financial nerve center, Hero mastered the art and science of cotton trading and speculating. "It is reported he came close to cornering the cotton market at one time," wrote his grandson.[54] Though his first language was English and he was hardly "old money," George married into the French Creole aristocracy—thrice, his first two wives (sisters, each bearing three children) having died young. Hero eventually became director of the New Orleans Cotton Exchange, placing him at the apex of local society.

By century's end, George Hero sought new investment opportunities and eyed three emerging trends. For one, he understood that New Orleans was running out of developable land. Its population had grown by 45,000 during the 1890s, soon to surpass 300,000, and there just wasn't enough high ground for them. Relatedly, authorities had finally gotten serious about reclaiming new living space through that old interest of Hero's—swamp drainage. Then there was that medical breakthrough confirming that a lowly mosquito caused yellow fever. Following the 1905 epidemic, wrote Hero's grandson in a family memoir, "it was recognized that the prevailing south breezes of our summers were blowing the insects from the swamps behind Algiers and Gretna into the city." Hero's attention thus turned to the West Bank, where he had had some prior investments, and where he deduced that mosquito-control efforts would dovetail with swamp drainage. Unlike the East Bank, where homestead associations were already scooping up soon-to-be-dried lands, the West Bank's bottomlands remained wet, vacant, and available. Now would be the time to revisit that old boyhood curiosity of his, and the West Bank would be the place to do it.[55]

The West Bank's geography, of course, was just as deltaic as across the river. Natural levees ran along the riverbank; backswamps lay behind them; and to the south stretched the marshy, brackish Barataria Basin to the Gulf of Mexico. Whereas the bayous of St. John and Bienvenue naturally drained the East Bank watersheds, the bayous of Barataria, Fatma, and Segnette drained those of the West Bank: same topography, hydrology, pedology, and biota.

But there was a different morphology to the West Bank. Owing to the looping course of the Mississippi River here, this subregion had a shape akin to that of a Greek theater—that is, a semicircular amphitheater. Communities running from Harvey, Gretna, and Algiers to Belle Chasse occupied the curving *theatron,* or seating sections, while Bayou Barataria's juncture with the manmade Harvey Canal formed the orchestra, and the natural Bayou Fatma/Ouatchas River flowed out to the *skene,* or backstage (that is, the Barataria Basin). "The big loop brought the upstream and the downstream levees reasonably close together," creating a semicircular basin that would have delighted a drainage engineer.[56] Why? Because topography made it perfectly clear that the back levees and pumps would best be positioned where the two natural levees came the nearest to each other—that is, between the orchestra and the *skene.* Drainage was never easy, but nature made it easier here on the West Bank, and George Hero knew it.

So Hero gathered up his liquid assets and went shopping for real estate behind the forty-arpent line. "He had the resources to acquire rights to swamp land behind [the] Cedar Grove, Oak Point, and Belle Chasse plantations along the River in Plaquemines Parish, much of it from ex-carpetbagger Governor Warmoth," wrote his grandson. "The price was right since the land was unusable and most of the cypress trees had already been logged out."[57] So as not to tip off speculators of his intentions, Hero made the acquisitions under various names, and ended up with ten thousand acres across Plaquemine and Jefferson parishes.

In 1912, the Louisiana State Legislature, having been lobbied by Hero, enabled police juries to form "drainage districts," a legal entity within which taxes could be raised and bonds issued to install mechanized drainage. Private landowners could include their swamplands in parish-recognized drainage districts, and share the risk and reward of reclaiming them. Alternately, if private owners had sufficient acreage, they could pool their holdings and form their own drainage district.[58] Having been "instrumental in the establishment of the system which the State used to drain many of the lowlands," Hero partook of that very opportunity.[59] His genius was in getting landowners and parish governments to work together on common goals and problems, no easy task in Louisiana. The amassed holdings qualified to become the Jefferson-Plaquemines Drainage District, and Hero became its president.

"The new drainage law will mean a great deal to the business interest of this city," Hero declared in a paid editorial in the *New Orleans Item.* "With proper drainage the mosquito would be forever banished," he wrote, reminding readers that "one of the strongest objections raised to the construction of a navy yard at Algiers was the presence of mosquitoes. This objection will be totally eliminated." Hero predicted his combined landholdings, once drained, would yield $8 million a year, "and it would be a short time before they produce $20,000,000."[60]

That estimate would soon grow, because now the City of New Orleans wanted in on the action to reclaim the Algiers backswamp. Mayor Martin Behrman, champion of improvements and himself an Algerine, got involved both personally and professionally. He became "the director of the West New Orleans Realty Company, which had bought 200 squares [at] $500 per square . . . out of the George A. Hero drainage district and is prepared to market them." Said Mayor Behrman in October 1913, "I did this because I know the work being done by Mr. Hero . . . *is the most monumental task that has ever been attempted by any man in this city in the last generation. . . .* I know the land and the values—and the success of the undertaking is *absolutely assured.*"[61] Behrman also got involved wearing his mayoral hat, as the city itself chipped in $9,000 a year to the effort, forming the Orleans-Jefferson-Plaquemines Drainage District. No one called out a conflict of interest, because swamp eradication was seen as everyone's interest.

The tri-parish West Bank project achieved in one year what the East Bank had been working on since the 1700s. Five thousand owners partook, Hero the largest, and their bottomlands spanned 10,000 acres in Orleans, 12,000 acres in Jefferson, and 17,500 acres in Plaquemines—from "Algiers and Gouldsboro . . . McDonoughville, Mechanicham and Gretna [and] the Harvey Canal [down to] the lower line of the Cedar Grove Plantation," plus the Lower Coast of Algiers's plantations of "Aurora, the Orleans, the Stanton, the Delacroix, and the Beka."[62]

Marrero, hydrologically separated by the Harvey Canal and its guide levees, formed its own Second Jefferson Drainage District in 1909, and funded its drainage through an ad valorem tax.[63] Westwego, on the other hand, determined its backswamp, through which ran the circa-1830 Company Canal, to be economically valuable, and while it did install drains and pumps for street runoff, it never reclaimed its backswamp. To this day, Westwego, as well as Arabi, Chalmette, and Meraux in St. Bernard Parish, mostly remain upon the backslope of their natural levees, going back to the old forty-arpent line. That's Lapalco Boulevard in modern Westwego, behind which is the Bayou Segnette swamp, and in St. Bernard Parish, it's the aptly named Forty-Arpent Levee, behind which is the Bayou Bienvenue wetlands. Into these rear basins each community discharges its runoff.

How to pay for draining nearly 40,000 acres? That's where the 1912 drainage district law came in. Owners within the district voted to tax themselves 75 cents per acre per year for forty years, which enabled state-backed drainage bonds to be issued to front the money. Selling quickly, the bonds raised over $200,000 for construction.[64] Costs were also absorbed by nature of the project's public-private partnership, in which the interests of the state, three parishes, and the landowners intermeshed in ways that could be construed as either wisely collaborative or suspiciously complicit. Minutes of the Jefferson Parish police jury, the elected parish government at the time, record private executives and public officials conversing as if private profit were synonymous with the public interest. For example: "*Mr. Geo. A. Hero addressed the jury and offered the use of the dredge-boat of the Jefferson and Plaquemines Drainage District at cost approximately $4000 to dig a canal . . .*"; "*Request of Geo A. Hero, President of Jefferson-Plaquemine Drainage District, that the Jury order a check for $500 drawn to pay one half cost of dredging drainage canal in rear of Gretna, (second Payment), as per agreement . . .*"; "*Mr. Geo. A. Hero representing property holders in the rear of Gretna petition the Police Jury to make application to the Federal Highway Department for the actual cost of a public road.*"[65] Privatized drainage did not work for New Orleans in the 1800s, due to big-city complexities and inadequate pumps. But now that the technology had caught up to the task, the sort of private-public partnerships spearheaded by George Hero, using the drainage district model, worked quite effectively, particularly in sub-rural areas like the West Bank.

With astonishing speed, Hero's engineers John F. Coleman and Allen S. Hackett teamed with parish counterparts to design and install a system of interconnecting ditches and canals. As happened on the East Bank, they worked within the existing framework of eighteenth-century French long lots and nineteenth-century agricultural ditch-and-gravity systems, as well as those of the village-like enclaves of McDonoghville, Algiers Point, and Gretna, laid out in the 1810s, 1820s, and 1830s, respectively. But unlike the East Bank, which was blanketed with faubourgs each having its own armature of prior drainage attempts, Coleman and Hackett had a veritable *tabula rasa* on the West Bank, with hardly any sunk costs pinning them down. And because of the clarity of the West Bank's topographical "Greek theater," there was no vexing debate about a hydrological outfall, akin to the East Bank's Pontchartrain-versus-Borgne controversy. All physics pointed to the Bayou Barataria's juncture with the Harvey Canal and Bayou Fatma/Ouatchas River; there and only there, on the perimeter of the expanse to be drained, would the pumps be placed. To this day, the West Bank has no major interior pumping apparatus, as does New Orleans on the East Bank. They're all perimeter pumps, along the urbanized fringe.

And of the pumps themselves? Serendipitously, master capitalist George A.

Hero readied his greatest project at precisely the moment—1913—that master engineer A. Baldwin Wood perfected his greatest invention, the low-head, high-volume axial-flow twelve-foot centrifugal screw pump. There's no record of the two men having conversed, but surely they knew each other, being of the same elite circles. Built and installed in 1914, the West Bank's first Wood screw pumps drew water through five gigantic tubes at a speed said to be the fastest in the world, each capable of removing over one million gallons per minute.[66]

Hero knew reclamation itself would not yield profit; the dewatered land had to be primed for development. So he created the South New Orleans Realty and Development Company, and sketched out a "South New Orleans" subdivision in the rear of Harvey. To make the development viable, it had to be accessible to prospective home buyers—from the East Bank, where 90 percent resided. So Hero launched, through an elaborate cartographic illustration in the *Daily Picayune* of March 8, 1914, a proposal to build a bridge across the Mississippi River to connect Jackson Avenue with Gretna. Hero's ad also announced "The Panama Canal of New Orleans," a widened extension of the Harvey Canal elbowed to connect with the Mississippi River south of Belle Chasse, "Showing a Saving of 65% in Distance and Allowing an Additional 72 Miles of Wharfage."[67]

But first things first: reclamation. By early 1915, all components were in place to pull the plug on the West Bank backswamp. Everyone seemed supremely confident that everything would work, and that a civic bonanza awaited. Whereas it took Holland seventy-five years and $3.2 million to drain a similarly sized polder, one journalist pointed out, "construction of the drainage plant opposite New Orleans, under way for less than a year . . . will [cost only] $234,000, and it will require only three days to draw off the water and increase the value of land . . . to a minimum of $300 per acre—[plus], mosquitoes will perish and human health and comfort will increase."[68]

This being New Orleans, merriment would mark the ignition of the pumps—on Carnival weekend, no less. "'Hero Day[,]' regarded by land men as one of the great events in the history of the state," would be Saturday, February 13, 1915, three days before Mardi Gras.[69] George Alfred Hero, the dapper polymath known as the Drainage King, was about to conquer the biggest, oldest problem of the largest city in the South—and who better to hail the king than the president of the United States?[70] Arrangements had been made for President Woodrow Wilson to activate the motors with a button in the White House, connected via telephony to the banks of Bayou Barataria.

That morning, after a sumptuous banquet at the Grunewald Hotel, dignitaries including Governor Luther Hall, Mayor Martin Behrman, Chief Engineer John Coleman, and Hero himself gathered on St. Charles Avenue in the Garden District. There, an automobile parade, fronted by two marching bands, mounted police, and heralds carrying shields spelling out H-E-R-O, proceeded amid cheering crowds to Canal Street and on to the Mississippi River. There they joined three hundred guests aboard the steamboat *Hanover,* among them scores of politicians and real estate agents. The gay entourage steamed over to the Harvey Canal, where they transferred to the yacht *Daisy* and sailed down into the Barataria swamp.[71]

Twelve hundred citizens cheered the arrival of the honored guests to the enormous pump station. Local and national press took their places, and orators gave soaring speeches. Hero, ever the doer, announced his plan to drain a million more acres, from the Mississippi to Bayou Lafourche, and predicted that only four million dollars of drainage investment would make Louisiana a billion dollars richer. More formalities followed. "Then came the President's message, handed up by a hastened messenger. And then the pumps began to whirr."[72]

Whirr they did. Torrents of swamp water levitated from the suction basin, up the five tubes, through Wood's screw pumps, out into Bayou Barataria, and south toward the Gulf of Mexico. The engineers' calculations were spot-on: most of the standing water in the district disappeared within three hours, after which the pumps were slowed to constant duty to remove rainfall and river infiltration.

In only two years—and two hundred minutes—the West Bank had achieved, through a private-public partnership across three jurisdictions, what New Orleans on the East Bank took the better part of a century to achieve. Land valued at $3 per acre increased to $200–$500 almost overnight.[73]

The only thing that didn't work that memorable day was the president of the United States. A miscommunication at the White House led to a delay, giving President Wilson an excuse to step out "for his usual Saturday morning of golf link." Instead, some unremembered aid pressed the button to drain the West Bank of New Orleans.[74]

It's unclear who did all the planning and paying for the festivities of Hero Day and at the "coronation" of the Drainage King. But an advertisement in the real estate section in the next day's *New Orleans Item* made no bones over the fancy leading automobile in the Hero Day parade, all adorned with palms and banners, having been "entered by the South New Orleans Realty and Development Company." The marketers did not wait a day before they parlayed their marketing coup into marketing copy:

> The gigantic pumps, with a capacity of a MILLION GALLONS OF WATER A MINUTE, which STARTED operations yesterday, will drain and make complete HIGH AND DRY 38,000 acres of choice land in Plaquemines, Jefferson and Orleans Parishes.
>
> RIGHT ON THE EDGE OF THE DRAINAGE DISTRICT, and therefore to be drained FIRST is SOUTH NEW ORLEANS SUBDIVISION. . . .
>
> $25 A LOT, $1 down, $1 a month, NO INTEREST, NO TAX. . . . Buy all you can . . . the best opportunity ever offered the small and large investor alike![75]

George Hero was right. The nearly 40,000 acres he reclaimed and drained are now home to over 100,000 people. But his timing was off, and he knew it. Linked only by ferries to the metropolis across the river, the West Bank of the 1910s had barely emerged from the plantation era; it was a sub-rural necklace of village-like communities and industries strung out among truck farms and pastures. What he needed was that bridge he had first broached back in 1914.

Regrouping with his engineer Allen Hackett, Hero reconceived the downtown auto bridge, this time proposing that it cross from Race Street to Gretna. His span plan came to be known as the Hero-Hackett Bridge, and in 1925 it won enthusiastic support from the Jefferson Parish Police Jury. But this would not be a parish or state project; this would be the private asset of the Drainage King.

Two years later, Hero and Hackett applied to the War Department for a bridge permit. Resistance came over issues of height and designs, which led to delays and rising costs. In 1929 they finally secured a permit for a 1,760-foot-long cantilever bridge with curious helical approaches. But then the stock market crashed, and as investment monies dried up, the "structure could not be financed."[76] Meanwhile, a competing proposal for a bridge at Nine Mile Point, backed by powerful city and state interests, gained momentum. When it finally opened as the Huey P. Long Bridge in 1935, the Hero-Hackett Bridge went dormant. Hero's lands went to other productive (though less profitable) uses, and only after the Greater New Orleans Mississippi River Bridge opened in 1958 did "the best opportunity ever offered" get developed.

George A. Hero died of a stroke after having been struck by a car outside his Garden District mansion in December 1932. For a man who was ahead of his times, Hero might have been dismayed that his obituary identified him primarily as "Span Advocate," his one disappointment, rather than "Drainage King," his greatest success.

FRANK HAYNE AND THE LAKE SHORE LAND COMPANY

The East Bank had its own "West Bank," so to speak—that is, a lightly populated prairie-like basin of potential value, only if it could be reclaimed and connected to the urban core. Today we call that space New Orleans East, but historically its nomenclature was as soft as its geography.

Because this area did not adjoin the Mississippi, it had no substantial natural levee. Rather, it was the eastward extension of the main deltaic morass, bordered to the north by the tide-washed Lake Pontchartrain shore and to the south by Bayou Gentilly. Having once been the main channel of the Mississippi, this now-abandoned distributary—known as Bayou Metairie ("small farm") to the west, Bayou Gentilly (from "Chantilly," an estate outside Paris), and Bayou Sauvage ("wild") to the east—had its own relict natural levee, though not nearly as high or wide as that of the current river channel. Travelers formed a *chemin* along this ridge to access these marshlands, where in prehistoric times the Tchefuncte people had created a series of middens (shell heaps), and where French colonials dubbed the scrub forests *Petit Bois*—"Little Woods," still a neighborhood name today.

A French royal concession from 1763 granted the ridge and its flanks to Gilbert Antoine St. Maxent, which in 1796 came into the possession of Borgnier DeClouet. Both men operated plantations along the *chemin,* while the adjacent marshes remained mostly wild, the haunt of hunters and maroons. Among the latter was Jean Saint Malo, who with other escaped slaves in the early 1780s fought a guerrilla war with Spanish Commandant Francisco Bouligny—forefather of engineering prodigy Albert Baldwin Wood.

In the early 1800s, ownership of the eastern marshes came into the hands of that omnipresent engineer-turned-smuggler Barthélemy Lafon, followed by Antoine Michoud in 1827. For a generation to come, most human activity here involved small plantations along the Gentilly Road, fishing and hunting camps along the Lake Pontchartrain shore, pirogues paddled on interbasin bayous, and vessels sailing through two deep channels to the east, guarded by Fort

Wood (Macomb) and Fort Pike and connecting with Lake Borgne, the Missis-sippi Sound, and the Gulf of Mexico. Since the late eighteenth century, the heart of this region had come to be known, for reasons unclear, as *Chef Menteur* ("Big Liar," the name now affixed to one of those channels), while the far eastern end became known as the *Rigolets* (the name of the other channel). Others referred to this general area as *Chantilly,* Gentilly, the St. Maxent or DeClouet plantation, and later as eastern Orleans Parish, eastern New Orleans, and after the 1970s, by the corporate moniker New Orleans East.

What set the stage for eastern New Orleans's reclamation and development were railroads. Trains provided access, but firstly, their track beds, which had to be raised on levee-like embankments, created topography and thus affected hydrology. First came the New Orleans, Mobile & Texas (later Louisville & Nash-ville) Railroad in 1870, whose shored-up track bed and trestles iterated the nat-ural ridge of the Gentilly Road. In the 1880s came the New Orleans and North Eastern Railroad (NO&NE, later the Queen & Crescent and Southern line) to Slidell, whose lakeshore track bed acted like a levee along a previously marshy shore. Together with the tracks by the Gentilly Road, these railroads made the Chef Menteur region into a polder: soft edges became hardened; back-and-forth water became more impounded; and the flat plain became more rimmed like a bowl.

Rail service also put the once-remote eastern region within a short ride from downtown. Stations opened along the routes, and hamlets formed around the stations. The L&N, for example, had given rise to Lee Station, home by 1888 to a few hundred people and a post office. Micheaud (or Michoud) Station became a popular hunting spot; Chef Menteur Station had a nearby fishing fleet; and Rigolets Station was known for its elaborate fishing and hunting camps. Along the NO&NE, stations such as Seabrook, Edge Lake, Little Woods, and South Point became popular picnic and bathing destinations for day-trippers.[77] Scores of family-owned weekend-getaway camps sprouted along the salty littoral with its refreshing breezes, and the soggy prairie off to the south got exposed to ever more eyes contemplating its economic potential. By the early 1900s, regular pas-senger service "carried upwards of 1,500,000 people" through the area annually.[78]

Unlike the West Bank, which had been plantation country and whose back-swamp lay in private hands, the bottomlands of eastern New Orleans were sub-stantially state-owned, the major exception being the chain of title traceable to the 1763 St. Maxent concession. Yet like George Hero's West Bank project, it would be the private sector that would spearhead drainage here, with willing partners in city hall and the state capitol.

In 1908, the newly formed New Orleans Lake Shore Land Company got the state legislature to sell it "certain lands belonging to the State under the water of Lake Pontchartrain, adjacent to said company's riparian lands, on the condition that said lands be reclaimed."[79] The arrangement stemmed from the so-called swamp-buster acts of the 1850s, which had transferred millions of acres of wetlands from federal to state domain, in the hope of steering them into private hands to levee, drain, and develop. Note that the 1908 bill's wordage explicitly described the targeted "land" as being under lake water; what could be more useless? Indeed, maps from the US Topographical Survey of 1892 depict the entire region, minus the Gentilly Ridge, in a blue hachure signifying submergence.

Frank Brevard Hayne was the man behind the Lake Shore Land Company. Born of Revolutionary War stock in South Carolina in 1858, Hayne worked his way up the cotton financing industry, and in 1885, he settled in its fiscal nerve center, New Orleans. Here he rose in prominence, scoring two lucrative cotton-market "coups" in 1903 and 1909, reigning as Rex during Mardi Gras 1904, donating generously to philanthropic causes, and, like George Hero, serving as president of the New Orleans Cotton Exchange. It's hard to imagine Hayne hadn't been influenced by Hero, four years his senior and in the same kingly social and professional circles.[80]

Around the same time Hero had cast his eyes to the West Bank, Hayne gazed eastward to the Chef Menteur region. He organized the Lake Shore Land Company in 1908, hired assistants, including R. H. Downman and M. L. Morrison, spent a few years amassing 7,500 acres "front[ing] seven miles on the shore of Lake Pontchartrain," and in 1911 announced his plans, Hero-style: through a full-page ad in the *New Orleans Item* real estate section.[81] "THREE HUNDRED TEN ACRE FARMS," read the banner headline, "reclaimed prairie land which is admittedly the richest in the world. . . . The company wishes to secure the best class of industrious farmers to place upon this property . . . on the shores of Lake Pontchartrain, the only location which New Orleans has for a beautiful suburb which can be reached without a long and tiresome railroad trip. TEN MILES FROM BUSINESS CENTER OF N.O."[82]

Whereas reclamation and drainage initiatives were historically motivated to improve health and expand living space, Hayne aimed at the lucrative industry of "truck farming," that is, intensive cultivation of vegetables, fruit, dairy, and poultry. Truck farming had boomed in the postbellum South, as emancipation forced many landowners to divide up their once-vast plantations and lease parcels to small growers, mostly former slaves. Families worked their farmsteads to raise delicate edibles, for feeding themselves, for sale at local markets, or for exchange—hence "truck," from the French *troc,* meaning barter.

As early as 1866, farms around New Orleans began shipping "truck" (vegetables) on the river to Memphis and St. Louis. In the 1880s, "when the railroads and express companies furnished adequate facilities for forwarding shipments to distance markets, truck farms sprang up like mushrooms[,] and New Orleans became the center for an intensive agriculture."[83] By 1900, according to the *Daily Picayune,* "about 1000 gardens and truck farmers [operated] in the parish of Orleans, cultivating about four to five city lots up to twenty-five acres each." Most truck sold locally, and another "twenty to thirty carloads daily" shipped nationwide.[84]

Eastern New Orleans seemed to be ideal for truck farming—but for the excess soil water. It had plenty of space, three railroads running regular service, a public road, and a big hungry city close by—but not *too* close. Truck farming upped land value where urbanization had not yet reached; it was a perfect intermediary land use, in that it prepared the terrain for future suburbanization while making a handsome profit in the meanwhile.

But would the agronomy cooperate? To see, Hayne set up a small demonstration plot in Little Woods to lease to the self-proclaimed "Vegetable King," Christopher Reuter, a commission merchant and "largest receiver of Florida and Cuban vegetables on this market." In 1910, Reuter's workers planted endive, escarole, beets, corn, bell peppers, lettuce, and snap beans, and, according to Reuter, found the soil to be "wonderfully productive" and "second to none." Company officials explained that "centuries of decaying vegetation, forming a deep muck deposit," combined with "the silt from the Mississippi River," created a nitrogen-rich "loose arable soil, the fertility of which is greater than any other known land." The balmy waters of Lake Pontchartrain, meanwhile, would put "early truck . . . not in the same danger from frosts as elsewhere."[85]

WARREN REED AND THE NEW ORLEANS DRAINAGE COMPANY

Hayne published Reuter's testimony in an ad for his Lake Shore Land Company. But Reuter's letter had been originally addressed to one Mr. Warren B. Reed, and that was not by happenstance. Reed, an engineer, was Hayne's drainage expert—his colleague and collaborator who had set up a parallel firm to do the actual reclamation.

Reed had devoted his career to the art and science of dewatering, serving as president of the Louisiana Reclamation Club, and, in 1913, assuming the helm of the American Reclamation Association. While Reed did not have Hero's visionary charisma, nor Hayne's *pro bono publico* spirit, he matched them in diligence and shrewdness. Agreeing to do the dewatering for Hayne's land company, to which he had personal ties, Reed incorporated on March 30, 1910, and named his firm the New Orleans Drainage Company—same name as that beleaguered paddle-pump effort of antebellum times.[86]

Just as Hayne would build a demonstration farm, Reed did his homework too, setting up in October 1908 an experimental polder and pump at the eastern end of the Little Woods tract, the first-ever mechanized drainage in what is now New Orleans East. Workers installed a pump with a capacity of 0.48 per inch of rain over twenty-four hours, and kept careful records on precipitation, runoff, canal capacity, and discharge. Results met with Reed's satisfaction, and the findings went into calculations to scale up the system.[87] Reed's 1908 demonstration polder created the drained space that Reuter used for his 1910 demonstration farm: drainage and agriculture working hand-in-hand.

As the field experiments progressed, Reed worked with his lawyers to confirm "absolutely" the legal title of the land his company was about to dewater, tracing it to the St. Maxent royal concession of 1763. His surveyors found that over 91 percent of the 34,057 acres were "free of trees" and "ready for the plow . . . when drainage is completed." Addressing investors in Chicago via a national publication, Reed explained the local political geography. "This property being outside of the present Drainage District of New Orleans," Reed wrote, "it will be necessary to drain it independently," which was to say that this project would *not* be led by the New Orleans Sewerage & Water Board. But he assured investors that "the draining of these lands is not an experiment—the greater part of the City of New Orleans is built on similarly drained land," and because "every acre . . . is within city limits," the company would have to deal with only one government jurisdiction. Hero would have envied that.

In 1910, Reed spelled out his reclamation plan for eastern New Orleans. "The first work will be the construction of the eastern levee, which will connect the embankment of the New Orleans & Northeastern Railroad on the north and the Louisville & Nashville Railroad on the south, completely enclosing 25,000 acres." This levees-first approach would seal off the polder from tidal flows, and help de-submerge its higher rims, the "strip[s] one mile in width" along the two railroads. This would allow for the construction of the pumping stations along the lakefront, and create a staging ground from which workers could dig "drainage canals leading to pumping plants" across the basin, to "dispose of the rainfall." Concluded Reed, "this entire work will be completed well within one year and the cost will not exceed $250,000." Investment dollars were in good hands, Reed reassured, and the whole effort was in good company—that is, with Frank Hayne's land company. "The work of draining these lands is under the direct supervision of engineers resident in New Orleans, of recognized ability, experienced in this particular kind of work, and now engaged in draining the property of the New Orleans Lake Shore Land Company adjoining."[88] Reed, consulting for Hayne, added a transportation element to their dual endeavor—"an excellent speedway . . . from Gentilly Road to Little Woods," along "reclaimed land [once] regarded as fit for only duck hunting. . . . The new road is to be known as the 'Hayne Boulevard,'" the name it retains today.[89]

The eastern New Orleans polder was a drainer's delight—a nice neat rectangle, well accessed, all within one jurisdiction, with zero junk left over from prior reclamation attempts, and no murky old claims or flailing French long lots to complicate matters. What resulted was an orderly lattice of drainage canals, all open as would befit an agrarian landscape. Most were named for company affiliates: the Lamb Canal, the St. Charles Canal, Lawrence Canal, Benson Canal, the Citrus, Farrar, Berg, Jahncke, Gannon, Vincent, and Dwyer canals.[90] Each flowed perpendicularly into the shoreline-parallel Morrison Canal, which, serving as a storage reservoir, redistributed the water evenly to the outflow canals—which flowed gravitationally a half-mile to the pumps, which raised and ejected it into the lake. "The construction of the main drainage channels was carried forward rapidly" during 1913, reported the *Journal of Agricultural Research.* "The spacing of the main channels is ½ mile," designed to create farms big enough for a tractor to ply but small enough for a family to work, and "with the usual field ditches leading into them. A main canal carries the water to the pumping plants," which engineers all concurred would discharge northward to Lake Pontchartrain. Unlike in 1895, this time there was no Pontchartrain-versus-Borgne outfall debate; the latter destination would have been longer and slower, and already loaded with discharge from the city proper. Besides, the runoff ejected from croplands would be minimally polluted, and there were no major recreational facilities here, at least not like West End, Spanish Fort, or Milneburg.

The main canals were each dredged deeper as they neared the lakeside pumps, to draw the runoff into the suction basins. So too the lateral ditches, whose spacing "averaging about 1325 feet, was made to fit the subdivision of land into 5-acre lots," inscribing a farming logic into the future neighborhoods of New Orleans East.[91] From the parallel ditches to the perpendicular main canals, runoff would make its way to the "plants" (stations), where pumps would lift the water over the levee and into the lake, meaning that the whole canal lattice would be below-grade. These edge-of-the-bowl pumps themselves acted to block storm surge from entering the sunken canals, which therefore needed no guide levees or floodwalls.

The perimeter pumps, viewable today along Hayne Boulevard, were an altogether better arrangement than the interior pumps and above-grade canals in Gentilly and Lakeview to the west. More so, Reed's system for the east, designed for agriculture, did not have the main system's impermeability imperative—that love affair with concrete—and did a decent job of storing water on the landscape, in the Morrison Canal reservoir, in the below-grade open canals, and later in ornamental storage ponds. Reed's drainage system, in effect, was green and blue more so than gray, more soft than hard, more open than closed, and perimeter as opposed to interior in its pump placements. In today's parlance, the east had a more sustainable and resilient drainage system. But then again, it wasn't designed for urbanization.

"By the spring of 1912," reported the *Journal of Agricultural Research,* "about 600 acres of the [Little Woods] tract had been drained." By late 1913, about 2,400 acres were drained, of which up to 300 were under cultivation. By late 1914, over 5,400 acres were reclaimed and 1,680 cultivated. By 1915, the entire Little Woods tract had been reclaimed, and 70 percent had truck farms.[92] By 1916, over 6,000 acres had been reclaimed, most of which promptly became farmland. "This land was brought in as fast as it was well drained," stated a federal report. "Usually the land ditched one year was cultivated the next."[93]

Reed's New Orleans Drainage Company lands conflated with those of Hayne's New Orleans Lake Shore Land Company, such that the two partnered efforts may be thought of as one grand scheme. But they did bring different land parcels to the drainage table, and each sub-polder dried at different times via different hardware. According to the *Engineering News,* it was the Lake Shore Land Company that, in 1913, installed what was briefly "the largest pumping plant in the country to be operated by electricity." This was not a Wood Screw Pump, as that invention was not quite available at the time, but rather an enclosed-impeller pump with herringbone gears designed by Tulane engineering Prof. W. B. Gregory. The twin units could handle 76,000 gallons per minute. Had the company waited another year, as Hero did on the West Bank, it could have had a Wood Screw Pump—and 225,000 gallons per minute.[94]

Cultivation came swiftly, as there was money to be made. Within a few months of pumping, workers burned off the natural brush or mowed it with light automobiles refitted with iron-rimmed wheels and cutters. Things went fast because this hydric drawdown aimed mostly to eliminate standing water, and it behooved the goal of agriculture to keep some moisture in the soil. "The land was then plowed with large disk plows drawn by tractors," after which the broken sod was allowed time to settle so that horses and mules shod with "bog shoes" could be used for planting. "The first crop planted usually was corn[,] followed by various truck crops," stated a government report of the 1916 season. "As rapidly as the land came into good condition, it was planted with citrus trees and the cultivation of field crops continued between the rows."[95] Mature orange trees were sent in from Florida and planted, and thousands of acres of orange groves and truck farm went into production. In just a few years, reclamation had converted thousands of acres of *petit bois* into modern mechanized farms, and fresh produce raised here got shipped in refrigerated vehicular trucks and railcars to regional and national markets.

The Hayne-Reed private dewatering effort had been quite successful. But its goal had been to make money, not operate a drainage system. The S&WB, on the other hand, was dedicated to drainage, and though a public utility, it was allowed to consult "for the account of private parties, who advanced the cost." It was under this odd arrangement, of a public agency consulting for private for-profit firms, that key components of Reed's and Hayne's project, such as culverts, canals, and levees, came into place.[96]

In time, the drainage apparatus of eastern New Orleans would be incrementally transferred to the control of the S&WB —though the improved land did not go with it. In 1919, for example, the Lake Shore Land Company agreed to turn over its Citrus Pumping Station to the S&WB, and it's been part of the municipal drainage system ever since. More sub-polders' canals and pumps followed, and eventually Reed's private drainage system had become a public asset—and a public obligation. The reason was pragmatism, not altruism. The companies wanted the endpoint of drained lands, not the operational task of draining. The money was in the leasing of the dried parcels, not the daily removal of runoff. Once the private sector built the system and reaped its fruits, it was happy to turn it over to the public sector to keep it going—because, as private interests had learned in the nineteenth century, there was no profit in drainage per se. As for the city's perspective, this land was, after all, within city limits, and by putting it under the auspices of the city's drainage utility, it was setting the stage for future development, and all the tax revenue that went with it.[97]

To wit, in 1923, Col. R. E. E. De Montluzin, the developer who had launched Gentilly Terrace back in 1909, purchased thirty-five thousand acres in eastern New Orleans, including the reclaimed lands of Reed and Hayne and the farmlands thereupon. Billing himself as "an aristocrat of French ancestry," he named the expanse "Faubourg de Montluzin," said to be the "largest metropolitan land area in the world under one ownership."[98] Dewatering had opened up all sorts of possible futures for eastern New Orleans, starting with agriculture, and this "French aristocrat" had no intention of being a farmer.

L. B. LANGWORTHY AND THE KENNER PROJECT

Kenner and New Orleans East lie at opposite ends of the metropolis, physically and perhaps culturally as well. But their drainage histories and geographies are similar. As with the New Orleans Lake Shore Land Company in the east, investors from Chicago teamed with local real estate interests in 1910 and, capitalized at $700,000, "purchased 5900 acres" to the west, "directly in the rear of Kenner, all semi-submerged lands near Lake Pontchartrain.[99] Up to that time, only hunters prowled the Kenner backswamp, paddling up its main natural outflow, Alligator Bayou, into a series of lagoons, where, as one old-timer remembered, they bagged "a number of black ducks, *poule d'eau* [coots], and some *dos gris* [scaup], and upon our return [made] some jambalaya from the gizzards."[100]

The Chicago men had different plans. Led by L. B. Langworthy, "one of the most enthusiastic believers in the future of the drained wet lands of Louisiana," the Kenner Project aimed to reclaim the Alligator Bayou swamp, "cut [it] up into small farms and truck gardens," and eventually build "attractive suburban homes for people doing business in New Orleans."[101] The aptly named Suburban Realty Company of New Orleans, along with the New First National Bank of Columbus, Ohio, partnered with Langworthy's Kenner Project, and together they purchased $112,000 in drainage bonds from the commissioners of the Fourth Drainage District of Jefferson Parish.[102]

As in eastern New Orleans, a railroad built on berms and trestles had first opened access to this bottomland, starting in 1854. But unlike the east, Kenner also had a fine riverfront perch upon the natural levee, and thus had a colonial history, starting with two French land concessions at Cannes Brûlées; an extensive plantation history, when it was home to the Oakland, Belle Grove, and Pasture holdings; and an early urban history, when in 1855 planter Minor Kenner hired W. T. Thompson to subdivide streets in what is now Old Town Kenner.[103]

While Kenner initially struggled as an urban enclave, it prospered as a truck-farming center, having rich soil, good access by river and rail, and proximity to a big city, yet just enough distance to keep land values cheap. It became an incorporated town in 1866 and a city in 1873, at which time it also started collecting tax revenue for improvements. Irish, German, and especially Sicilian families found Kenner's farming industry to their liking, and "by 1900 the once sluggish and troubled little town had turned into one of the richest and busiest produce-generating areas in Louisiana," wrote historian Craig A. Bauer. "At the height of the growing season over sixty refrigerated carloads of locally grown produce left Kenner every month."[104]

In 1908, Kenner, embroiled in internal disputes, relinquished its city charter and reorganized in 1913 under more favorable arrangements as an incorporated village. During the interim, the community fell under the jurisdiction of the Jefferson Parish Police Jury, and parties in both domains realized they had common interests, namely reclamation of the same backswamp. It was at this time that Langworthy made his move. Instead of "pioneering" a completely new reclamation-for-farming venture, as was necessary in the east, his Kenner Project instead bolstered "an already established condition, [as] Kenner has been for some time noted for the excellence, variety and quality of its vegetable and truck garden products."[105] The dewatering process was just as straightforward here as in the east, having a neat rectangle with simple topography and a clear discharge direction.

First, a levee was built up along the lakefront, followed by a 6.5-foot-high Upper Protection Levee on the western St. Charles Parish line, running straight

north to the lake in parallel with a 45-foot-wide canal, and a smaller interior levee running along the Elmwood Canal to the east. These three embankments,
plus the natural levee to the south, formed the polder. Workers then dug an array of canals, numbered 1 to 20, deepening them such that gravity would draw
runoff and swamp water toward the lake through the main Duncan Canal,
which adjoined Alligator Bayou, where a lakeside pumping station lifted and
discharged to the lake. Another perimeter pump was later added at the Elmwood Canal.

Kenner's canal network did not quite have the rectitude of eastern New Orleans's system, because this part of Jefferson Parish, formerly plantation country, had those splaying French long-lot property lines. This explains Kenner's
modern-day angled street grids and parallelogram superblocks. But its canal network did have the spacing of farm parcels, like in the east, and once the standing
water disappeared and tillage followed, truck farming spread in Kenner from
river to lake.[106]

In the same way eastern New Orleans's private drainage project dovetailed
with public interests and eventually came into the domain of the Sewerage &
Water Board, so too did the private Kenner Project get subsumed into a broader
public parish drainage effort. Of Jefferson Parish's East Bank landmass, only
8,000 or so acres were developed by 1913, and they were entirely on the natural
levee or the Metairie Ridge; the remaining 20,000 acres were "taken up by undergrowth, trees and marsh—useless wasteland."[107] Wrote Justin F. Bordenave in
1938, "several civic-minded gentlemen viewed this spectacle with mingled awe
and horror," and contemplated a mechanism by which it could be drained and
cleared for farming. Their solution was the same tool George Hero had used
on the West Bank: launch a drainage district, tax the landowners, issue bonds
backed by that future tax revenue, sell them to raise capital to install the drainage system, and reap the benefits from the ensuing economic development on
the dewatered land. "Thus, in the year 1913, Sub-Drainage District No. 1 of the
Fourth Jefferson Drainage District was created, comprising about 2400 of those
20,000 idle acres, [all] in the rear of the town of Kenner."[108]

The Kenner subdrainage district issued $120,000 worth of 5-percent forty-
year bonds, backed by an annual tax of $2.60 per acre. These were the bonds that
Langworthy and his partners had acquired in 1913, initiating the reclamation
effort that would convert a thousand acres of swamp into farms over the next
five years. But because so many other concurrent drainage projects had created

a glut of arable land, the Kenner Project defaulted in 1918, and bonds had to be reissued at lower rates two more times before the district stabilized in the 1930s.

So, while Kenner's private reclamation effort thrived only briefly, as happened in eastern New Orleans, it nevertheless succeeded in leveraging the power of state-authorized, parish-organized drainage districts to dewater the polder for later subsurface drainage and full urbanization. Drainage is one of those things that succeeds even when it doesn't.[109]

"NOT HIGH, YET DRY": RECLAIMING METAIRIE

The Kenner Project catalyzed neighboring areas to dewater their swamps, despite the supply of reclaimed land now outpacing demand. In 1915, within months of Wood's pumps and Hero's triumph, "the same civic-minded gentlemen, not satisfied with the thought of reclaiming 2400 acres out of 20,000, branched out further, and created Sub-Drainage District No. 2, a gravity drained district comprising 2000 acres in the Southport area." This was Hoey's Basin, an interior "cypress and gum" bottomland in the unincorporated part of Jefferson Parish known broadly as Metairie. It differed from Kenner's topography in that the economically important Metairie Road ridge precluded digging drainage canals straight north to Lake Pontchartrain.[110] Instead, it made more hydrological sense to steer the water eastward and make a deal with the Sewerage & Water Board in adjacent New Orleans to send the water through its Pumping Station No. 6 on the Seventeenth Street Outfall Canal. This inter-parish arrangement remains in place today, and unfortunately, it has caused more problems than it had resolved. But at the time, the arrangement drained Hoey's Basin quickly and cheaply, and by the 1930s, "every acre in this district is now utilized," and Airline Highway was about to be built across it.[111]

Also in 1915, Jefferson Parish leaders formed Sub-Drainage District No. 4, spanning 1,800 acres bounded by Metairie Road, Bonnabel Place, the Lake Pontchartrain shore, and the Orleans Parish line by Bucktown. Like Kenner, the goal here was truck farming and homesites, and bonds were issued ($60,000) at 5-percent interest and sold to private developers. Like the Kenner Project, the Metairie effort defaulted a couple of times—too much land chasing too few projects—before finally stabilizing. In the meanwhile, it left behind a new Reed-style drainage system, with an orthogonal grid of below-grade gravitation canals intersecting a reservoir canal (now the West Esplanade Canal) that flowed via the Bonnabel Canal to a perimeter pump for lifting and ejection into Lake Pontchartrain. While the hoped-for development initially lagged, the reclamation had succeeded, and the dried land it left behind would have a long shelf life.

Six years later, the state legislature passed an act making most remaining Metairie swamps into Sub-Drainage District No. 3, and gave that district the power to levy taxes, pay off bonds, and get to work. After a 1922 state reorganization of the larger Fourth Jefferson Drainage District, forty-year bonds were issued for $2.5 million at 6 percent to set the stage for "reclaiming the entire 28,000 acres on the east bank of Jefferson Parish." Work began in July 1923 with the construction of a new lakeshore levee, six feet above high-tide level, which, in a cost-sharing arrangement between the drainage district and the State Highway Commission, was topped with the Hammond–Lake Shore Highway. This auto road opened up the area to both New Orleans to the east and Hammond to the north, and the levee it ran upon hardened the polders behind it, readying them for dewatering. There, engineers extended westward the Bonnabel network of gravity ditches and the reservoir canal on what is now West Esplanade Avenue, and dug the Suburban Canal through the levee, joining the Duncan and Elmwood canals to the west and the Bonnabel Canal to the east.[112] At the lakeshore of each outfall, four pumping stations were built "about two miles apart, the first placed in the rear of Metairie and the last in the rear of Kenner."

In all, the three-year effort to drain East Jefferson entailed the excavation of sixty miles of below-grade drainage canals, twenty miles of ditches, and eight pumps in four stations with a total discharge capacity of 640,000 gallons per minute. Jefferson Parish's new East Bank drainage system was "put into full scale operation by 1926," and in ensuing decades, modern subdivisions were nestled into its slanted spatial framework, giving us the map of modern Metairie.[113]

Development indeed ensued—which made permeable surfaces impermeable, which increased runoff, which called for greater pumping capacity. By 1938, the four obsolete pumping stations from 1913 to 1915 had been replaced with "concrete and steel [sheds], with tile roofs, [and] a six-room residence in connection, so that the engineer . . . may be in attendance at all hours." Inside, pumps had been upgraded to "two Worthington Diesel engines, each connected with a lift pump [with] a capacity of 125,000 gallons of water per minute," increasing system discharge capacity by 56 percent over a decade earlier.[114] The pumping apparatus acted as a lock or gate preventing lake water from entering the feeder canals, all of which were below-grade, because, unlike New Orleans, engineers in Jefferson Parish had only one outfall option, Lake Pontchartrain, and thus placed their pumps at the lake perimeter. This was a good move, and to this day, all drainage canals in Jefferson Parish are below-grade, while those in New Orleans proper rise over rooftops.

Even in neat polders with minimal complications, drainage engineering was tough, and nature fought the unnaturalness of it with every bit of physics it could muster. "One of our greatest problems is drainage," acknowledged President John J. Holtgreve of the Jefferson Parish Police Jury in 1953, looking back on forty years of trials and tribulations. "Entire Jefferson is lowland from which every gallon of excess water must be drawn off by pumping[;] it is a constant problem that is both expensive and expansive." By the time of his writing, perseverance had prevailed, and optimism won the day. "All four pumping stations on Lake Pontchartrain are working," Holtgreve wrote, "and *although we can never have a 'high' parish we have a 'dry' parish with every occupied acre ready for a farm, a home, or a factory.*"[115]

It would take many years to reach the point when demand for land caught up to supply, at which point Kenner and Metairie would be fully dried by subsurface pipes and readied for urbanization. But as early as the late 1910s, most Jefferson Parish bottomlands on both banks were reclaimed of standing water and, with the exception of Bayou Segnette and the rural Barataria Basin, were no longer truly swamps. New shell roads favored by drag racers crisscrossed the brushy damp fields of Kenner and Metairie, where just a decade or so earlier, bayous with names like Indian, Tchoupitoulas, Labarre, Laurier, and Alligator flowed naturally into the lake Native peoples once called *Okwa-ta.*[116]

8

Geographies Rearranged,

1910S–1920S

The years 1914–15 marked New Orleans's culminating victory after two centuries battling the backswamp. Those *anni mirabiles* saw the installation of Wood's screw pumps, the final dewatering of the city proper, Hero's stunningly swift reclamation of the West Bank, Reed's bog-to-farm transformation of the eastern marshes, and the steadfast drying of Metairie and Kenner. "The results," reported Weather Bureau forecaster Isaac M. Cline at a scientific conference in late 1916, "may be summed up as follows":

> No water flows in the open gutters on the streets as formerly; storm waters and the millions of gallons of water discharged daily into the sewer wells, to spread out through the soil and maintain a high ground water level, now go out through the drainage and sewerage systems; and the ground water, instead of being level with the surface of the earth, is 6 to 8 feet below that surface. On nearly all sides of New Orleans large areas of marshland which were covered the greater part of the year with tidewater have been drained, thereby reducing the area covered with water much below what it was formerly.[1]

The epic transformation unleashed urban development, increased property values, and augmented tax revenues, while everything from mud to mosquitos to malaria waned. Citizens thrilled to the victory, and drainage kings like Wood and Hero became famous. Reclamation was one of the few things in which everyone, regardless of race and class, agreed was an absolute good—the curing of a geographical disease in a nearly literal sense. In subsequent years, wrote historian John Magill, "the entire institutional structure of the city was complicit" in improving the dewatered wetlands. "Developers promoted expansion, newspapers heralded it, the City Planning Commission encouraged it, the city built streetcars to service it, [and] the banks and insurance companies underwrote the financing."[2] The *New Orleans Item* became a chief oracle for heralding drainage; its coverage on one particular Sunday in May 1915, under the banner headline "*DRAINAGE BOND SALE AT 95 OPENS WAY FOR GREAT THINGS,*" included eight separate drainage stories. They spanned from Hero's heroics on the West Bank, to the 7,000 feet of new levee in Gentilly, to those 7,000 acres of new citrus groves in the east, to the new gravel beds made available for roadbuilding, to a handy local invention for stump-removing, to the visit of "honorary Chinese commissioners to the Pontchartrain Groves of the Lake Shore land company" to discuss if "some 80,000 acres of wet lands in China may be reclaimed by the same methods used in the ninth ward."[3] New Orleans–style drainage was about to go international.

But those same years also brought three disruptions to the victory party. One was a monumental infrastructural decision, another a pedological accident with momentous implications, and the third an inconvenient meteorological shock.

THE INDUSTRIAL CANAL

If New Orleans had drainage fever since the 1890s, it caught Panama Canal fever in the 1910s. So did other world ports, which mused that seaways cut to their harbors would tie them into the new shipping routes emerging from the severed Central American isthmus. Much of it was delirium, but in New Orleans it had dead reckoning on its side: just look at a map! On account of its prograded deltaic geomorphology, with aprons of wetlands on either side of a big protruding river, New Orleans seemed to invite speculative shortcutting to connect foreland and hinterland.

Locals had indulged in such scheming practically since the city's founding. In 1721, for example, an envisioned river-to-lake channel appeared in the *Plan de la Ville de la Nouvelle Orleans,* followed in 1725 by a written call for "a canal communicating between the river and Lake Pontchartrain."[4] Both the Carondelet Canal (1794) and the New Basin Canal (1832) represented variations of that same nexus, as well as concessions to the difficulty of building a lock at the river— which is what sunk the 1807 congressional directive to build a navigation canal on Canal Street. Nevertheless, a *Plan du Canal de jonction du Mississippi au Lac Pontchartrain* surfaced in 1827, this time proposing to extend the Marigny Canal along Elysian Fields Avenue, only to be nixed in 1831 with the opening of the Pontchartrain Railroad.[5]

There were also more ambitious—delirious?—plans for full-scale seaways to be cut from city to Gulf. George Brott's shady Ship Island Canal scheme of 1868 sought to link present-day Uptown with the Mississippi Sound. Ten years later, the Barataria Ship Canal Company won authorization "to construct and operate a ship canal from New Orleans to the Gulf of Mexico," aiming to make the West Bank "the only gate through which the largest ships will reach the port of New Orleans."[6] Both failed, yet the fervor persisted.

Now, in 1914, all the latest lock-building and excavation technology was available. The Panama Canal had opened, and the Port of New Orleans, reorganized in 1896 into a state agency led by a board of commissions, had many motivations to dig deep and wide. A major river-to-lake channel, its commissioners contended, would create a protected inner harbor with a fixed water stage, good for shipbuilding and repair. It would open up ten linear miles of leasable wharf space for cargo handling, storage, and manufacturing. And as a navigation asset with a draft for ten-thousand-ton ships, it would add another segment to the Gulf Intracoastal Waterway, that ongoing effort to dredge a domestic barge canal through the bays and marshes along the entire Gulf Coast and Eastern Seaboard.[7]

So convinced, the state legislature in July 1914 authorized "an industrial and navigation canal . . . to connect Lake Pontchartrain and the Mississippi River in aid of . . . commerce and industry."[8] Advocates extolled the possibilities. Chicago's "Illinois–Lake Michigan Canal and the New Orleans Industrial Canal," wrote the *Item* in 1916, "are complementary links in a new system of waterways connecting the upper Valley through the Mississippi River and New Orleans with the Gulf and the Panama Canal, [giving] the differential to the Valley cities in trade with the markets of the Orient, our own west coast, and South America."[9] With such heady visions, who would be so mundane as to worry about potential impacts on local drainage?

In May 1918, officials identified a 5.3-mile-long swath across the mostly undeveloped Ninth Ward for what was officially called the Inner Harbor Navigation Canal. The slightly dogleg corridor, all within city limits, maximized navigational convenience while minimizing distance, acquisition costs, property expropriations, and residential displacements.[10]

But what it also did was threaten to rework the hydrology of every polder it passed through—not to mention cleave the city in half and introduce tidal water therein. Guide levees would have to parallel the shipping channel, and they would upheave the best-laid plans of drainage engineering, going back to the 1890s.

The problem was that the north/south-oriented shipping channel would obstruct the west-to-east-oriented flow of the main drainage canal. Specifically, the Industrial Canal threatened to undermine the advisory board's carefully made decision in 1895 to discharge most runoff eastward through Bayou Bienvenue to Lake Borgne, and save Lake Pontchartrain only for secondary discharge. In operational reality, more runoff actually ended up in Pontchartrain than originally planned, polluting its waters and curtailing lakefront development, just as the advisory board had predicted. Nevertheless, the Sewerage & Water Board as of 1920 still endeavored to fix this imbalance by augmenting the pumps and capacity of the main canal on Florida Avenue.[11]

Now, the Industrial Canal would plow through that main canal, bifurcate the Bayou Bienvenue basin, and chop off eastern New Orleans from the rest of the East Bank. It would be the single most radical artificial reconfiguration of New Orleans geography to date, and of all the things it would disrupt, drainage topped the list.

To finance construction, the Orleans Parish Levee Board agreed to pay $925,000 annually for fifty years, after which the Port of New Orleans would assume operations and maintenance costs through its board of commissioners ("Dock Board"). It may seem odd for a parish levee agency to foot the navigation bill for a state port authority, but in fact, the Orleans Parish Levee Board was investing in its own mission: the canal would add ten miles of new levees to the OPLB's charge, and along with them would come budgets and staff.[12]

Left out of the collaboration was the city's other water agency, the S&WB, despite being a major stakeholder. Whereas the OPLB stood to gain from the canal, the S&WB stood to lose, because the canal would force the redoing of past drainage work. From this point forward, here and elsewhere, the S&WB and OPLB increasingly found themselves at loggerheads, the former charged to eject interior water, the latter to barricade exterior water.

The Industrial Canal also pitted New Orleans's premier imperatives—maritime commerce and urban drainage—against each other. It was really no contest. The high-profile shipping industry produced enormous profits, and its powerful players advocated loudly for their interests. Drainage, on the other hand, was an unglamorous public utility, and its players had already proved they could work around any obstacle, starting with gravity. They could, in theory, dump the discharge into the new shipping channel. But that would only pollute the lake, which was the main reason to go eastward in the first place, and all that outflow would introduce volatility in the inner harbor, which needed to be stable for the new wharf-side industries.

Surely engineers could pull off another miracle and save the Lake Borgne outfall route. And they did, according to a front-page banner headline in the *Item* of Sunday, May 2, 1920: NEW ORLEANS BUILDS OWN UNDERGROUND RIVER; *GREAT SIPHON UNDER INDUSTRIAL CANAL SOLVES DRAINAGE PROBLEM*.[13] In a triumphalist piece, civic advocate Thomas Ewing Dabney waxed lyrical at a moment (1920) that might be considered the apotheosis of both the Progressive Era and the age of heroic engineering. "Man every day is surpassing Nature," Dabney soared in his opening line. "He flies higher than the eagle, he digs deeper than the earthquake. He turns rivers from their course and mingles oceans.... He has thrown back the giant Mississippi and made it go where it listeth not. He has joined the river and lake" and answered a "terrible drainage problem" with "the world's greatest drainage system." How to reconcile the two clashing undertakings? By creating, Dabney wrote, "a disappearing river, [a] great quadruple passage of steel and concrete that will carry the city's entire drainage underneath the 30-foot deep ship canal now being built."[14]

That is, Dock Board engineers would preemptively install a 378-foot-long "underground river" to send runoff in the main canal beneath the 105-foot-wide shipping canal, and have it reemerge in the lower part of the Ninth Ward at a slightly lower elevation, thus creating a siphon. A longitudinal diagram accompanying the article showed the siphon as a crooked water tunnel—Dabney called it a "bent tube"—40 feet wide and 10 feet high. Inside were four separate chambers: one to eject water from the constant-duty pumps, 4 feet wide; two for storm flow, both 13 feet wide; and a 6-foot-wide duct for cables and other equipment. Each was separated by foot-thick walls made of steel-reinforced concrete. The whole device cost three-quarters of a million dollars.

Engineers had originally planned on a smaller siphon. But they "threw away their blueprints and started over again," initially for 6,000 cubic feet per second capacity but settling for 2,000 cfs, in part to fix that Pontchartrain-Borgne imbalance more toward Borgne, and in part because the Dock Board had suddenly decided to nearly double the size of the shipping channel.[15] Brilliant engineering, it seemed, would once again save the day; indeed, the siphon might even *improve* drainage. "It may mean the reclamation of Lake Pontchartrain shore land," wrote Dabney on the impending corrective, pointing out that because of "drainage contamination, the lake shore . . . has been held back in its development. . . . The highest study of mankind is not Man," Dabney climaxed, "but the works of Man."[16]

With the George W. Goethals Company (same outfit behind the Panama Canal) as consulting engineers, work began on the Industrial Canal on June 6, 1918. It was a tough dig, pitting labor gangs, mechanized excavators, pile drivers, suction dredges, and dynamite against tangled muck, dense silt, and buried cypress logs. Boring in from the Mississippi River levee was too risky, so dredges penetrated in from present-day Seabrook along the Lake Pontchartrain shore as well as from Bayou Bienvenue, with which the Dock Board wanted their inner harbor to connect via a turning basin. Levees had to be built simultaneously, but they did not prevent muddy water from backflowing into the excavation—a problem solved by Baldwin Wood, who adapted his centrifugal pump impeller to the dredges.[17]

Work commenced on the "underground river." First, runoff in the drainage system was rerouted temporarily to Lake Pontchartrain so that the Florida Avenue main canal could be severed, enclosed in two cofferdams, and cut through by the dredges digging the shipping canal. Once the twenty-six-foot-deep pit was dried by Wood's pumps, concrete pilings spaced three to five feet apart were driven thirty to sixty feet into the earth, varying to reflect the "bent tube" that was the siphon. At one point, machines reached sixty-nine feet below grade level, "the deepest excavation that has ever been made in or around New Orleans," according to the *New Orleans Item*.[18] Workers then poured a two-foot-thick sloping floor of concrete, and erected upon it steel and concrete walls separating the internal chambers of the siphon. The great tunnel was sealed with hydraulic-cylinder sluice gates, after which manholes were installed for maintenance access.[19] Just to the north, dredges and steam shovels scoured out the turning basin and connected it to Bayou Bienvenue, to give the inner harbor a second eastward outlet to Lake Borgne and the Gulf of Mexico.

Work wrapped up on the main channel in September 1919, on the Florida Avenue siphon in June 1920, and on the great mechanized lock at the Mississippi River in January 1923. At the May 5, 1923, dedication ceremony, attended by thousands, Governor John M. Parker declared that the Inner Harbor Navigation (Industrial) Canal would "equip New Orleans to be, in the broadest sense, the gateway of the Mississippi Valley for its interchange of products with the markets of the world."[20]

In fact, Dabney's triumphalism, and the supreme confidence of shipping advocates, masked deep discord with the city's drainage establishment as well as with residents, who by now had come to expect reliable services. Three years of construction had stymied drainage, sewerage, and potable water delivery while also provoking stakeholders along Lake Pontchartrain, where polluted runoff had been spewing all along. Statements from the Sewerage & Water Board during these years, according to ecologist Joshua Lewis, show "rather bitterly" that "political dynamics within the Dock Board made it anything but an honest broker." A major sticking point was the Dock Board's insistence that their shipping channel must connect to Bayou Bienvenue, which made for "a strange state of hydrological affairs," in that "the city's main drainage volume was being siphoned underneath a canal, so as not to pollute it, only to flow into it from the opposite side."[21]

The Inner Harbor Navigation Canal, meanwhile, went from trophy to boondoggle in less time than it took to build. Construction costs had far exceeded estimates. Expected shipbuilding contracts disappeared as competing European ports recovered from the recent war. The lock soon proved too narrow. Deepwater vessels had no real need to sail to shallow Lake Pontchartrain, and could not negotiate twisting Bayou Bienvenue, which was dammed off in 1926 and declared unnavigable in 1931. "Horrible example of wasteful extravagance," pondered the Louisiana Engineering Society, or "the future industrial prosperity of New Orleans[?] . . . Some of us may live to see the day when the Industrial Canal lies idle and choked with water hyacinth, a burden to taxpayers and a dangerous liability instead of a most wonderful asset."[22]

The siphon itself had problems. Leaks formed in 1928, filling its chambers and reducing its capacity. Dock Board engineers, having designed the device, went to repair it—but when their barge sank in the process, they sent the bill to the Sewerage & Water Board. The dispute reignited the lingering navigation-versus-drainage tension between the state-backed Dock Board and the city-backed Sewerage & Water Board, and triggered a lawsuit that went to the Louisiana Supreme Court. In 1934, the court ruled that despite that the Dock Board had designed the siphon, and that its navigation project necessitated its installation, it was ultimately a drainage device, ergo the responsibility of the S&WB. Once a heralded solution, the "underground river" now became a costly liability, and a sunk cost to which too much of the city's drainage system had already been committed—yet another example of path dependency steering decisions in bad directions. Matters got worse in 1947 and 1965, when hurricanes pushed surges into the city, filling the underground siphon and causing floodwaters to rise in the now-populated Lower Ninth Ward and St. Bernard Parish.

The 1895 decision by the advisory board to discharge to Lake Borgne had, finally, fully come into focus as a colossal mistake. This outfall was too far away, too prone to disruptions, and too spatially mismatched with the city's main run-off catchment, spanning so many square miles so far to the west. Corrective action came incrementally. From the 1920s forward, more and more runoff was rerouted *away* from Bayou Bienvenue and Lake Borgne, and toward Lake Pont-chartrain, lifted by the great Wood screw pumps in the primary and auxiliary interior pumping stations, and northward through those circa-1870s outfall canals. Only local runoff in the Eighth and Upper Ninth wards went out the siphon to Bayou Bienvenue, joined by that of the Lower Ninth Ward. Eventually the siphon was replaced by a pumping station discharging directly into the Industrial Canal, and in 2004, engineers demolished the concrete chambers of the "underground river" and carted away the riprap. It was an ignominious end to the 1895 Lake Borgne decision, and it went largely unnoticed at the time.[23]

SOIL SUBSIDENCE

Physics created the Mississippi River deltaic plain entirely above sea level. Be they marshes, swamps, relict beaches, barrier islands, distributary ridges, or natural levees, all terrestrial surfaces here originally lay above the mean level of the sea, albeit slightly along the coastal fringe. Geophysical data, maps, photographs, and scientific descriptions all attest to this. French geographer Elisée Réclus observed in 1853 "a difference of four meters between the parts of the city distant from the river and those near the embankment," the former being "only a few centimeters above sea level"—not much, but *above* nonetheless.[24] In an 1887 paper presented to the New Orleans Academy of Sciences, engineer Benjamin Harrod stated plainly that "none of the area of this city is below sea level," its "lowest part" being at "mean high tide level," meaning a foot or so above normal lake or sea level. He contrasted this to certain Dutch polders which had dropped ten to eighteen feet below the North Sea, and iterated his point in 1888, when he described New Orleans's "swampy plain adjacent to the lake [having] about the elevation of high tide."[25] Similarly, a 1918 government document described "the elevation of the surface" of low-lying eastern New Orleans as "a few inches above mean lake level."[26]

Under natural conditions, these soils comprised a wide range of parent materials but only three particle sizes: sand, silt, and clay. Anything coarser, like pebbles or gravel, would have been too heavy to make it this far down the Mississippi. As the river splayed out and deposited that alluvium upon its deltaic plain, the coarsest and heaviest particles (sand) settled first, followed by medium-sized silt particles, and finally fine and light clay particles. The natural levees, forming the highest terrain, attained a loamy texture of mostly sand and silt, whereas the swamps and marshes, the lowest terrain farthest from the river, were mostly silt and clay. The other two components of local soils—water and organic matter (peat, or humus)—were high in quantity everywhere. But the swamps and marshes retained the most water, on account of their topography, and because of the preservative properties of all that water, they also retained the most organic matter.

That was delta soil: sand, silt, and clay particles, intermixed with water and organic matter. Take away the soil, and water immediately occupies its place. "It is impossible to have any subterraneous buildings," wrote a 1770 observer of New Orleans, "as they would be constantly full of water.[27] Conversely, take away the water, and soil occupies its place. That is, desiccation of the soil body leaves behind air pockets, which induce biochemical oxidization of the organic matter, which reduces its volume, thus creating more air pockets. Finely textured particles then settle into the cavities, and the entire soil mass consolidates and subsides. Dewatering alone can cause up to 75 percent compaction of delta soils, resulting in up to twelve feet of sinkage if the water table were lowered by fifteen feet. That is to say, water makes up three-quarters of these soils; remove it, and only mineral constituents and peat comprise the remaining one-quarter.[28]

Subsidence tends to be front-loaded; most sinkage happens shortly after dewatering, and later tapers off. Subsidence is also "unfair." Like a regressive tax, the process sinks lower-elevation areas faster and deeper than higher lands. Because the lowest spots retain the most water, and therefore preserve the most peat, they suffer the most subsidence once drained. Worst yet, the lowest areas generally have the most clay particles, which, on account of their minute size (less than 0.002 millimeters in diameter) integrate most efficiently into the abundant air pockets. As a result of this pedological trifecta, New Orleans's bottomlands, upon being dewatered, sunk the deepest. Once this became understood by the populace, a curious, notorious new phrase entered the local lexicon: *below sea level*—which is where swamps went when they died.

Engineers had long understood that drainage causes subsidence, defined as "the lowering of the elevation of a land area in relation to sea level."[29] The Dutch saw it after they first drained their marshes, and by the 1500s, polders had sunken so deeply that windmills had to be installed to lift and discharge unwanted water to the sea. Today, most anthropogenic subsidence takes the form of aquifer compaction resulting from the over-drafting of groundwater in arid environs, such as Phoenix, Albuquerque, and the Santa Clara Valley in California. Serious as the subsidence is these places—parts of Mexico City have sunk by over thirty feet, damaging buildings, streets, and infrastructure—rarely does it existentially threaten human communities. But in deltaic and coastal environs, subsidence can dangerously lower the terrain below the level of adjacent bays or seas, necessitating levees or dykes to prevent the latter from flooding the former. That's what happened to New Orleans.[30]

Fleeting mentions in historical records reveal some understanding of subsidence in nineteenth-century New Orleans. "The swamps may certainly be drained," wrote the *New Orleans Bee* in 1836, "but as certainly, proprietors . . . will have to elevate the surface of their lots by thick coating [because] drainage cannot give substance to the spongy soil."[31] In 1864, engineer George Bayley insinuated that sinkage would follow drainage when he wrote that "the area to be reclaimed has to be drained several feet below the level of . . . Lake Pontchartrain, consequently[,] the water required to be removed must be elevated [to discharge] into the lake."[32] According to a curious tidbit in *Harper's Weekly* in 1871, "rumor says that New Orleans is slowly sinking: in one locality a batture of about 750 feet long by 120 feet wide has sunk seven feet below the ordinary level." Sounding more like a levee cave-in or a sediment fill project, the piece is nonetheless provocative in intimating that subsidence had become common speculation. Certainly, one could plainly see sinkage of cornerstones and foundations on downtown streets, due to the sheer weight (overburden) on soft soils. An 1860 report to Congress stated that the "St. Charles Hotel [had] sunk 36 inches; St. Patrick's Cathedral even more; the St. Louis Hotel about 24 inches [and the] Custom House [by] 20 inches."[33] The question, *Harper's* concluded, "is—What can be done?"[34]

As for drainage engineers, the closest they came to flagging subsidence was in 1895, when, upon inspecting Linus Brown's contour map produced for the design of the system, colleagues duly noted "a large portion of the basin between the foot of the [natural levee] and Metairie and Gentilly Ridges is below mean gulf level."[35] But they did not speculate as to why, did not extrapolate it, and did not foresee that the sinkage might get much worse.

One day in February 1913, a spiderweb of cracks appeared in a wall of St. Louis Cathedral. It caught the attention of local architects, some of whom had predicted "this outcome for buildings in the downtown district . . . three years ago, when the city's present drainage system was in infancy[.] Others laughed at the idea." Now, visible damage on an iconic building reopened the discussion. "Draining the city has unquestionably caused a lowering of the groundwater level," one architect told a *Picayune* reporter. "It is what we call 'subsardence' [*sic*]." Others thought dewatering had caused the cypress pilings beneath the cathedral to rot, now that air pockets were interacting with the wood. If so, other big buildings would be at risk, as yellow pine and cypress trunks were routinely used for foundations before the advent of concrete pilings and steel frames.[36]

In 1918, US Weather Bureau forecaster Dr. Isaac Cline, an eminent name in the scientific community, issued a pamphlet confirming that "the pumping out of water caused some subsidence or sinking of the soil, with unfavorable effect on some of the older buildings, like the St. Louis Cathedral."[37] Watching the dewatering with a scientist's eye, Cline sensed other changes were afoot—or rather, in the air. "We do know theoretically," he told scientists convening in Washington, DC, in late 1916, "that the water having been removed from a considerable area, leaving the ground exposed to solar radiation, the land . . . would heat twice as rapidly during the day and would cool by terrestrial radiation more rapidly at night than the water surface did." In winter, he surmised, the reverse would be true.

To test his hypothesis, Cline analyzed temperature readings taken at the US Customs House on Canal Street from 1885 to 1899, "the period just prior to the installation of subsurface drainage," and compared them to "1900 to 1914, inclusive, the period during which subsurface drainage has been in operation." Cline found that, before drainage, temperatures in downtown New Orleans reached 95 degrees or higher on thirty-five days, and hit or exceeded 100 degrees on zero days. In the post-drainage period, those figures were seventy-four days and seven days. What about averages? "The average monthly maximum for the period 1900–1914 is above the average monthly maximum for the period 1885–1899 in every month except December, in which there is no difference." He also found that winter months in the post-drainage period tended to be cooler than those of the pre-drainage era.

Like any good scientist, Cline called for more research before anyone could "determine definitely" the relationship between drainage and climate. He probably would have been the first to acknowledge that there had been a fair amount of drainage during and before the years 1885 to 1899, and that street paving and concretization probably also played a warming role (now known as the urban heat island effect). Nevertheless, the empirical data indicated that the turbocharged subsurface drainage that started around 1900 had environmental consequences below *and* above the land surface—neither of which, Cline noted, had been observed in the otherwise comparable city of Mobile, Alabama.[38] "From the foregoing study," Cline told the Pan American Scientific Congress, "we conclude that the changes made in physical conditions at New Orleans have been the cause of higher temperatures."[39] A century later, a study ranked greater New Orleans as number one in the nation in its urban heat island effect, with its (mostly drained) core being on average nearly nine degrees warmer than its (mostly undrained) periphery.[40]

A consensus formed among architects, surveyors, engineers, and scientists that nearly twenty years of dewatering, and five years of accelerated subsurface drainage, had fundamentally altered the New Orleans environment. Warmer highs and colder lows were curious enough, but sinking soils could damage things, flood places, and cost money. The research made its way to the press, and on Sunday, August 3, 1919, reporter W. S. Callender broke the city's geographical story of the century:

NEW ORLEANS IS SINKING SLOWLY BUT STEADILY DOWN TOWARD CHINA
Subsidence is Caused by Drying Out of Earth by Drainage Canals
New Orleans a Foot Nearer to China Lately[41]

"This rising rapidly advancing city is sinking," opened Callender, countering the triumphalism of the times. The evidence was everywhere. Cracks appeared on buildings. Church steeples feel out of plumb. Sidewalks crumbled in new neighborhoods. Track beds had to be shored up. Even geodetic benchmarks were found to have been "entirely out of level," throwing off slope calculations on canals between pumping stations. S&WB Superintendent George Earl had to send crews out to cut the tops off hundreds of manholes, as surfaces had sunk around them. Likewise, on avenues like Napoleon, under whose neutral grounds had been installed big concrete water tunnels on sturdy pilings, "the paving" on the traffic lanes "has sheered off to so steep an angle it is impossible to drive on it."

Most of the city by 1919 had subsided on average by one foot over the past ten years, more than an inch annually, while parts had dropped by three-and-a-half feet in half that time, over eight inches per year. The worst-affected spots were those that had been naturally lowest and wettest, namely the lakeside areas of Lakeview and Gentilly "and sections of the city near the center," present-day Broadmoor and Mid-City, which "are as much as three feet below sea level."[42]

Some subsidence is natural for deltaic soils, Callender wrote after interviewing experts. But this dramatic droppage was anything but. "In the past four or five years New Orleans has become so completely drained that the soil is no longer saturated with water two or three feet below the surface as it used to be, and consequently, the earth has dried and shrunken and settled down."

Callender's article was the first to introduce the phrase "below sea level" to New Orleanians in a detailed and expository manner. In short time, the curiosity became common knowledge, though described in various ways. "One-third of New Orleans is at or below low tide level in the surrounding tidal lakes into which it drains," one national journal explained in 1921; "over two-thirds of it is at or below high tide level in said lakes; and all of it is well below high water level in the Mississippi River."[43] Six years later, in an internal investigation, the Sewerage & Water Board extracted soil samples from Napoleon Avenue at South Johnson Street and photographically documented their dramatic "shrinkage in volume as a result of drying."[44]

Shrinking, compacting, drying, settling, consolidating, subsiding, sinking: soon, "below sea level" would become the best-known topographical factoid about New Orleans, a peculiarity that seemed to suit the city's *sui generis* character. The paradoxical quirk spread so quickly, like juicy gossip, that it dispensed with two important qualifications. For one, about 50 percent of the metropolis south of Lake Pontchartrain remains above the level of the sea.[45] That's the good news. The bad news is that 100 percent of the same area used to be above sea level. Most New Orleanians today believe most of their city is below sea level and always had been, and are astonished to hear that it was humans who inadvertently sunk it—"down toward China," as Callender put it, "or India, or whatever country is on the other end of the New Orleans axis."[46]

Yet New Orleanians of a hundred years ago, convinced that the hydrological demons of their past had been slain by modern engineering, migrated enthusiastically off higher ground and settled into the very areas that were low to begin with and sinking the fastest and deepest. The drainage kings saw no reason to advise them otherwise, because they too bought the canard that topography no longer mattered, that "below sea level" was mere novelty, and that, as Thomas Ewing Dabney put it, "Man every day is surpassing Nature."[47]

THE GREAT STORM OF 1915

On September 22, 1915, sailors reported a tempest brewing over the Lesser Antilles and dubbed it the West Indian Storm. The vortex strengthened over the warm waters of the Caribbean, and curved between Cuba and the Yucatan Peninsula into the Gulf of Mexico. As the system leaned northward toward Louisiana, winds topped 135 miles per hour, a Category 5 hurricane by modern standards. The winds utterly disheveled the barrier islands of Barataria Bay, leading Louisianians to call it the Grand Isle Storm.

Portents appeared over New Orleans on Tuesday, September 28, when the sun rose upon a "cirrus veil" colored like "faint brick dust," followed by sporadic gusts from the system's outermost feeder bands. After the Grand Isle landfall, the track veered to position the city in the dreaded northeastern quadrant—a direct hit. By dawn September 29, winds of 40 miles per hour swept city streets, waters rose in Lake Pontchartrain by five feet, and coastal tides swelled by fifteen to twenty feet.[48]

The surge tested all the engineering interventions of the past generation. Waters overtopped the levees lining the lakeshore and penetrated each of the intruding canals—the four 1870s drainage outfall canals, the 1830s New Basin Canal, and Bayou St. John and its adjoining 1790s Carondelet (Old Basin) Canal. "The overflow from these sources, [plus] about 7¼ inches of rainfall, was a most discouraging feature of this day's development," wrote a Sewerage & Water Board engineer tasked with keeping the pumps operating.[49] Salt water flooded bottomlands from present-day Broadmoor to Lakeview, uninhabitable until a few years ago but now steadily developing. "Over that portion of the city lying between the Old Basin Canal and Broadway and from Claiborne Avenue out to Lake Pontchartrain," wrote forecaster Isaac Cline, "the water depth driven in by the storm ranged from 1 to 8 feet in depth."[50]

Lake Borgne's surge entered the Bayou Bienvenue basin and overtopped the rear protection levee behind Florida Avenue, once the forty-arpent line and now the underground main drainage canal. Flooding ensued in the rear of the Seventh, Eighth, and Ninth wards, which too had few residents until recently, thanks to drainage. Pumps could not relay the heavy rainfall runoff eastward or northward, because far greater volumes of sea surge came into the city westward and southward. Even the indomitable Wood screw pumps were losing the battle.

At 5:10 p.m. winds sustained speeds of 86 and peaked at over 100 miles per hour, after which they abated, leading some to think the hurricane had passed. But around 6:35 p.m., they reversed directions and reaccelerated: the eye had passed twelve miles west of the city, and the trauma was about to repeat. As darkness fell, a *Times-Picayune* reporter described "a peculiar lightening . . . flaring up in sheets not unlike the fire coming out of the mouths of serpents."[51]

Three hundred miles in diameter, the hurricane proceeded northward onto Tangipahoa and St. Tammany parishes, where cooler land surfaces deprived the system of energy. Winds petered out over southeastern Louisiana, and residents in New Orleans surveyed the damage. "CITY CUT OFF FROM REST OF WORLD," read the *Item*'s headline.[52]

By Thursday morning, the sun shined over the Crescent City. Destruction was extensive, mostly from wind: poles and trees were down everywhere, debris blanketed the streets, and over twenty-five thousand structures suffered serious damage, including many prominent landmarks.[53] As for the deluge, it receded on the city's periphery but remained impounded in the levee-encircled and still mostly undeveloped polders, which took four days to pump out via the just-finished drainage system. Damages exceeded $13 million regionally ($340 million today), with roughly half occurring in New Orleans proper. At least 275 Louisianians died, including 43 in the eastern New Orleans community of the Rigolets. The nation's top meteorologist, Isaac Cline, described the storm as "the most intense hurricane of which we have record in history of the Mexican Gulf coast and probably in the United States."[54] No longer named for the West Indies or Grand Isle, the event would go down in history as the Great Storm of 1915.

Yet to city leaders and the press, the marvel was not the extent of damage, but its limit. "STORM OVER, CITY EMERGES VICTOR," crowed the next day's banner headline in the *Item,* followed by an editorial titled "'STORM PROOF!' The Record Shows New Orleans."[55] Flooding occurred mostly in vacant areas, making it seem innocuous. Pumps had removed the water within days, making it forgettable. Politicians extolled how well the system had worked, changing the narrative. In the years following the Great Storm of 1915, those lowlands steadily increased in population, rumors of a recent flood having been converted to evidence of their habitability. To most people, it was as if nothing had happened.

Some authorities, however, saw an emerging vulnerability as more residents moved closer to the lake. All that protected them was an offshore levee built by the Orleans Parish Levee Board to create the polder soon to be drained. Because it used locally available humus rather than pure clay, that meager dyke compacted quickly and afforded little protection. Just as potentially dangerous were the three drainage outfall canals, which connected openly with lake waters. The S&WB came to understand that the same lowlands rising in population were sinking in elevation, so it decided to raise the levees along the outfall canals by three feet. The canal beds themselves did not subside, because their earthen floors allowed water to percolate and keep their underlying soils wet, as delta soils like to be. And so the outfall canals rose higher above sinking neighborhoods, and took their lake-level waters with them—a hydrological peculiarity if ever there was one.[56]

Another disruption like the 1915 storm could reverse the progress drainage had wrought. Something had to be done to reclaim that miracle, and it would have to be along the Lake Pontchartrain shore.

MARCEL GARSAUD AND THE LAKEFRONT IMPROVEMENT PROJECT

New Orleanians had long viewed Lake Pontchartrain's southern shore as both an asset and a liability. Its smooth bight echoed a preexisting barrier island known as the Pine Island Trend, whose mass steered the channel-jumping Mississippi southeastwardly, and in doing so, confined a bay to its north and west, which the French understood to be a *lac* and named for their naval minister Louis Phélypeaux, Count de Pontchartrain.[57] Later, Lake Pontchartrain's brackish waters and balmy breezes offered respite to denizens of the hot and crowded city, while its abundant natural resources lured commerce to its coast. As a result, starting in the 1790s and particularly the 1830s, three coastal communities formed where three transportation routes—the Carondelet Canal/Bayou St. John, the Pontchartrain Railroad, and the New Basin Canal—met the lake. By the late 1800s, those three enclaves became known as Spanish Fort, Milneburg, and West End, the last of which adjoined a Jefferson Parish fishing enclave known as Bucktown. New Orleanians flocked every weekend to these recreational getaways, which catered to every whim—bathing, dining, amusements, entertainment, sports—while local workforces handled the cargo coming in and out of the Pontchartrain Basin.

With these assets came liabilities. Peaty soils were too soft and low for urbanization, requiring all structures to be raised up on stilts, while swollen lake waters often intruded into the marshes and swamps, and occasionally to the city proper. In June 1871, for example, a crevasse in the levee at Bonnet Carré sent high river water into western Lake Pontchartrain, where easterly winds prevented it from flowing eastward to the Gulf of Mexico. Lake levels rose, inundating the lakeshore and backflowing the New Basin Canal up the Hagan Street Canal, past its "drainage machine" and into the rear streets of the Second, Third, and Fourth wards. It was among the first major floods in New Orleans to have been caused (that is, communicated inland) by manmade navigation and drainage canals.[58] City Surveyor William H. Bell is said to have warned city officials of the dangers of such canals, advising officials to place pumps along the lakeshore, else "heavy storms would result in water backup within the canals, culminating in overflow into the city."[59] The flooding of 1871 was the city's worst between Sauve's Crevasse of 1849 and the Great Storm of 1915, and it framed Lake Pontchartrain as the source of the problem.[60]

The Bonnet Carré Crevasse Flood occurred at the same time that maverick speculator George Brott got authorization to tear up New Orleans for his preposterous Ship Island seaway, the planning of which had embroiled the diligent surveyor, William Bell. It was under these circumstances—of designing against future floods and humoring Brott's flood-prone folly—that Bell released his *Plan of Proposed Improvements for the Lake Shore Front* (1873). Bell's dazzling vision for shoring up the lakefront, an illustrated answer to George Bayley's spoken suggestion from 1864, included scenic promenades and inner harbors within a high, broad breakwater straddling the natural shore. The concept never made it off paper, but neither did it disappear.

After the 1890 formation of the Orleans Parish Levee Board, the state legislature authorized the board to make piecemeal improvements along the lakeshore, the largest of which was at West End. This project occasioned building a new concrete seawall five hundred feet out into the water and creating a cofferdam (enclosure) into which dredged sediment was siphoned to create new land. Planners then landscaped the resulting oval-shaped landmass with live oaks, lagoons, and a scenic bridge, becoming today's West End Park. The project took years, delayed in part by the Great Storm of 1915—the very impetus that got planners thinking about scaling up the West End effort for the entire lakeshore.[61]

Mayor Martin Behrman, the political embodiment of progressive vim, had long made lakeshore improvements a campaign platform. After his sixteen years in office ended in late 1920, he and other city advocates prevailed upon state legislators to revise the constitution to empower the Orleans Parish Levee Board to acquire private lakefront land, from the Jefferson Parish line to Little Woods, be it by donation, purchase, or expropriation. This sweeping power spelled doom for the many family camps and recreational enclaves which had dotted the lakefront since the 1820s. But the 1921 constitutional revision went beyond moving people for leveeing and landscaping; it also called for building "a Lakefront development . . . designed as a flood control project." That is, the law granted the levee board the right to dredge the lake bottom, use the fill to create new upraised land, and develop and sell it for a profit, so long as at least 30 percent of the acreage became public parks and roads. Most importantly, the landmass would block lake surge from entering subsiding neighborhoods to its south. The more they sank, the more the perimeter had to be raised, and that's what lakeshore improvement aimed to do.[62]

Enter Col. Marcel Garsaud, a determined leader and skilled engineer with a knack for turning water into land. Born in Bordeaux in 1881, Garsaud immigrated with his family in 1893 and studied engineering at Tulane University, where he surely brushed shoulders with Baldwin Wood, two years his senior. Graduating cum laude in 1901, Garsaud worked as an engineer for a number of railroad companies as well as the Sewerage & Water Board, where he developed expertise in waterfront reclamation. He fought in the Great War and attained the rank of colonel in the Engineering Corps, gaining more experience in his areas of specialization. In 1924, the French immigrant became chief engineer of the Orleans Parish Levee Board, whereupon he eagerly delved into the lakefront initiative lagging since the 1921 constitutional revision.[63]

On August 19, 1925, Garsaud unveiled his favorite of three proposals. "HERE'S COLONEL GARSAUD'S PLAN FOR LAKEFRONT DEVELOPMENT," reported the *Times-Picayune* the next day. Garsaud's vision called for some 117 million square feet to be reclaimed from the lake, using a process akin to what had been done at West End except a hundred times larger.[64] No tax dollars would be used for the construction; instead the $27 million cost would be paid as the OPLB sold off some of the improved land, a legal power that required a second revision to the Louisiana Constitution in 1928 which "solidified the [Levee] Board's role as an agent of residential and commercial development."[65]

The design called for a curving five-mile-long levee, built three thousand to four thousand feet offshore, forming a cofferdam from the mouth of the New Basin Canal to the mouth of the Industrial Canal. Engineers would then fill that enclosure with 36,000,000 cubic feet of sediment, until two thousand acres of new land rose five feet above the level of the lake. The original plan called for sandy beaches and five ornamental lagoons with islands. The lagoon plan was abandoned, and all but one of the beaches were to be replaced with a stepped concrete seawall.[66]

The Lakefront Improvement Project commenced in 1926. Workers installed a temporary wooden bulkhead offshore to create the cofferdam, and began filling it with the dredged hydraulic fill. Next, the bulkhead was strengthened, raised, refilled with more mud, drained, and repeated. Setbacks came when "large sections of the new bulkhead were washed away by storm waters, and banged out of position by roaring waves, [and] at times as much as 3,000 feet of wooden wall had to be replaced."[67] Gradually, water became mud, mud became land, and the land rose higher, four to six feet above the lake. In 1930, workers built the earthen levee on the landside and finished off the stepped concrete seawall, modeled after similar structures in Florida. Indeed, when the project was completed in 1934, the new seaside landscape looked more like the breezy Florida coast than the steamy old port city of New Orleans.

"It is some measure of the project's scale," wrote the late geographer Peirce F. Lewis, "that a municipal airport was added to the Lakefront scheme almost as an afterthought."[68] The reclaimed terrain was ideal for that purpose, requiring no land acquisition, providing for unobstructed take-offs and landings, and allowing for future expansion without interfering with existing infrastructure. The runway was built in the same sequence as the Lakefront, starting with a cofferdam and dredged sediment slurry, and ending with a seawall, only in this case it was vertical rather than stepped. Shushan Airport, named for OPLB President Abraham Shushan, spanned over 470 acres and rose seven feet in elevation; when it opened in 1934, it was billed as the "the Air Hub of the Americas." Today the facility is Lakefront Airport, and it's second only to the Lakefront itself as the largest water-to-land reclamation project in the city's history.

The OPLB decided on two main land uses for the Lakefront: public-access recreational facilities between the levee and the lake, where ran a gravel road that later became Lakeshore Drive, and residential subdivisions inside the levee, up to where the old shoreline used to be (today's Leon C. Simon and Allen Toussaint boulevards). This being during the Depression, lots did not sell at the initial asking prices, as high as $1,875, but after the OPLB built a dozen sample houses and sold them at cost, interest picked up, and so did prices. Eventually, half the acreage was sold to private developers, and new neighborhoods opened, including Lake Vista in 1936 and 1948, Lake Terrace in 1953, Lakeshore in 1955, and Lake Oaks in 1961.[69]

The Lakefront Improvement Project came with costs. It hardened the ebb and flow of diurnal tides across ecologically rich seagrass beds, and replaced it with a concretized shore and a scoured lake bottom, to this day the lake's deepest trench. It eradicated scores of picturesque family fishing camps, and landlocked the century-old resort community of Milneburg. It turned West End's eastern flank from water to land, and what had once been the "Coney Island of New Orleans" became instead a small park and yacht harbor. The Lakefront Project also cost Black New Orleanians their recreational space. Whereas all the old resorts had segregated sections, and Spanish Fort was Black-only from 1896 to 1909, the artificial Pontchartrain Beach at Milneburg (1939) was designated for whites only. Not until 1955 did Blacks get their own lakefront recreational facility, Lincoln Beach, and it was located well to the east, at Little Woods.[70]

From a broader urbanistic perspective, however, the Lakefront Project was a success, and New Orleanians thrilled to the scenic new amenity. "Lakefront was and is an ornament to the city," wrote Peirce Lewis, a scholar of the American landscape and a tough critic to please. The geographer called it "one of the very few places where twentieth century city planning has truly improved a large area of an American city."[71] It also reconfigured the map of New Orleans, adding a "knuckles-and-thumb" profile to the lakeshore bight and inserting a high, broad brim along the sinking topographic bowls of Lakeview, City Park, and Gentilly. In this regard, the Lakefront Project achieved its original purpose, to protect the new lakeside neighborhoods from 1915-like storm surges. It took thousands of workers, millions of dollars, and eight years to build, but behind it all was really just one man, Col. Marcel Garsaud, and New Orleans would look very different today without him.

Garsaud died in 1958 after an illustrious career that included work on the Bonnet Carré Spillway, years on the Dock Board, service in both World War I and II, consulting on the Bayonne Tunnel in New Jersey, and many local civic activities—an illustrious citizen, like other drainage kings going back two centuries.

But also like them, Garsaud, an assertive and prickly personality, wielded his prowess to the tune of the powerful. He was the chief engineering voice pushing to dynamite the Poydras levee during the Great Mississippi River Flood of 1927, triggering deep flooding in rural Plaquemines and St. Bernard parishes. The draconian move, intended to relieve pressure on city levees, proved unnecessary, and victims were never fully recompensed. Garsaud also chaired the engineering committee for digging a seaway that eventually became the Mississippi River–Gulf Outlet Canal, a main cause of Katrina flooding in 2005.[72] And while Garsaud's Lakefront Improvement design was superb, it left open a major problem. He originally designed the new land to steer the transecting Orleans and London Avenue outfall canals to flow through a system of ornamental lagoons before pouring into the lake.[73] When the lagoon idea was dropped, the two outfall canals were extended straight to the lake—ungated, as they had been before, and as had been the Seventeenth Street Outfall Canal by West End.

Why? Because a gate or lock to prevent storm-surge intrusion (which was the charge of the Orleans Parish Levee Board and later the US Army Corps of Engineers) would have impeded the ability to pump out rainfall accompanying the storm (which was the charge of the S&WB).

Why not take this opportunity to relocate the pumping stations to the lake end of the outfall canals, rectifying the 1895 mistake while arranging the pumps to also serve as gates? "At the time that the seawall was built," explained two investigators many years later, "it was presumed that there was little economic incentive to move the pumping stations to the lakefront, as existing channel capacity was adequate to contain all pumped water, and the new seawall was expected to provide protection from hurricane-driven surges that might be pushed into the canals."[74] And so the pumps remained far inland, where they raised the hydraulic head too early, and the outfall canals remained ungated, even as they rose higher relative to the sinking landscape.

Catastrophe would come through those outfall canals in the form of Hurricane Katrina's surge in 2005. But one man saw it coming 132 years earlier. In his "Lake Shore Front" plan of 1873, City Surveyor William H. Bell had carefully drawn in locks on the recently dug canals, modern corollaries of which would have prevented most of the lakeside flooding during Katrina.

RISK, REAL ESTATE, AND RACE

Upon the drained swamplands now buffered by the reclaimed Lakefront, thousands of pretty new houses arose—California bungalows, Spanish Revival villas, English cottages, and midwestern ranch houses, among other nonlocal styles and types. They appeared in places like Lakeview, Gentilly, Broadmoor, Fontainebleau, Metairie, and Kenner, and across the river in Algiers, Gretna, Harvey, and Marrero. Into them moved thousands of families, along with modern electrical appliances and an automobile in the driveway. Between 1920 and 1930, nearly every census tract lakeside of the Metairie-Gentilly Ridge at least doubled in population; low-lying Lakeview grew by about 350 percent; Gentilly, by 636 percent; and Gerttown, by 1,512 percent. Older neighborhoods on the higher natural levee, meanwhile, lost residents: the historic faubourgs of Tremé and Marigny dropped by 10 to 15 percent; the French Quarter declined by one-quarter; and the Lee (now Harmony) Circle area lost 43 percent of residents. People were effectively migrating vertically as well as horizontally, and in doing so, they were reshuffling the demographic composition of the city.[75]

Drainage had made greater New Orleans bigger, wealthier, drier, warmer, lower, and riskier. It also made it more segregated. That is, power players in the real estate industry, backed up by a racist political and judicial establishment, deemed that more wealth could be made from drained spaces if they were designated white-only. What resulted was a great irony of human geography: the enthusiastic shift of middle-class white families into the geophysically riskier environs of sinking former swamplands, and a concurrent residential accrual of poorer Black families upon the relatively safer higher ground of the natural levee.

Racial geographies in antebellum New Orleans were spatially heterogeneous. Free and enslaved people, despite their polarized life experiences, dwelled in residential propinquity, on adjacent lots or abutting structures. Unlike plantation slavery, where the slaveholding family lived in the "big house" apart from slave cabins ("about a hundred yards," by one 1834 estimate), urban slavery took place in pricey high-density neighborhoods and prioritized for convenience and control.[76] Proximity enabled whites to have their enslaved servants at their beck and call, and to monitor their actions and inactions. No threat to social status came from this co-residency, because white supremacy was so powerfully inscribed into the social hierarchy that no one would have confused that spatial proximity with social equality. For this same reason, *gens de couleur libre* (free people of color), who occupied an intermediary caste, also tended to live inter-mingled among whites and the enslaved.

But free people of color were not evenly distributed citywide. Being mostly francophone, Catholic, and Creole, they predominated in the older, mostly Cre-ole First and Third municipalities in the downriver half of the city. Being less moneyed than whites, most members of this mixed-race caste tended to settle in the lower-cost, lower-elevation streets farther back from the river. Joining them in this "back of town" were recently manumitted Blacks who had attained freedom and, while not necessarily accepted as *gens de couleur libre,* also ended up around these same backstreets.

A look at the 1820 census illustrates the relationship of race and elevation. That year, the population of the rear-most (lowest) street within the French Quarter, Burgundy, was 37.6 percent free people of color (FPC). That percentage steadily decreases as we move closer to the river, and rise higher in elevation: Dauphine was 29.3 percent FPC; Bourbon 21.7 percent; Royal 8.6 percent; and Chartres and Levee (now Decatur) only 5.8 and 5.6 percent. Most whites, con-versely, lived in the higher blocks closer to the river, and most enslaved persons were evenly distributed, regardless of elevation.[77] This pattern was replicated citywide: more white in the higher front-of-town, more FPC in the lower back-of-town and in the downriver neighborhoods, and the enslaved intermingled throughout. As a reporter for the *Daily Picayune* put it in 1843, "The Negroes are scattered through the city promiscuously; those of mixed blood," implying free people of color, "showing a preference for the *back streets* of the First and part of the Third Municipality," meaning the Creole-dominant downriver neighbor-hoods of the French Quarter, Faubourg Tremé, and Marigny.[78]

After the Civil War, thousands of emancipated people moved to the hope of the city. The Black population in New Orleans rose 110 percent between 1860 and 1870, and by another 54 percent by 1900. Scorned and destitute, most poor Black families had little choice but to settle in the muddy shantytowns of the back-of-town, near cutover swamps.[79]

By the early twentieth century, a racial topographical trend had become in-grained. Most Blacks resided in the lower-elevation rear of the city; most whites were in the higher front; and working-class neighborhoods throughout were fairly integrated. In general, the rear backslope of the natural levee was about two to three times more Black than the higher front of town.

Numbers from the 1910 Census help illustrate. When analyzed at the enumeration-district level (a forerunner of census tracts), we see that 9,244 Black families lived on the back (swamp) side of Dryades Street, from Howard Avenue to Louisiana Avenue, making up 66 percent of that area's 14,071 total households. Conversely, on the front (river) side of Dryades Street, also between Howard and Louisiana, there were 3,995 Black families, or only 19 percent of that area's 20,764 total households. Nearly all of the other 81 percent were white.[80] *This means that in this large section of Uptown in 1910, a drop in 5 to 10 feet in elevation equated to more than a tripling of Black occupancy, from 19 to 66 percent.* Looking downtown, at the more-integrated Seventh Ward, a similar drop in elevation equated to nearly a doubling of Black occupancy. Households riverside of St. Claude Avenue were 25 percent Black, whereas the lakeside blocks were 46 percent Black.[81] (New Orleans as a whole was 74 percent white in 1910, and had been majority-white since the late 1830s.)

These patterns matter because urban space is rarely equitable, particularly in New Orleans. People who lived "back" got left back, because, in a tragic synergy of racism, poverty, and delta hydrology, the back-of-town was more flood-prone, more divested of services and amenities, and more plagued by urban nuisances. Literally and figuratively, it constituted the margins, and to be there was to be marginalized. Floods in the 1700s and 1800s were floods of the back-of-town, and Black residents suffered them disproportionately—deeper, longer, and more frequently. Then came the Progressive Era, the drainage system, and the *anni mirabiles* of 1914–15. What followed was an astonishing reversal of racial topography.

"LEAPFROGGING" TO LOWER GROUND

Since the end of Reconstruction, white supremacists throughout the South had disenfranchised Black voters, backed their threats with violence, passed segregationist legislation, and won an 1896 Supreme Court ruling in *Plessy v. Ferguson* sanctioning a "separate but equal" society. In New Orleans and throughout the South, WHITE ONLY and COLORED ONLY signs went up at nearly every facility in the public domain, such as parks and schools, as well as private spaces in the public domain, such as department stores and diners. Even the dead were racially segregated. For decades in McDonoghville Cemetery, for example, "both white and colored were indiscriminately buried[;] there was no separation of the graves of negroes and whites. [But] in 1891 Mayor Shakespeare of New Orleans ordered their separation," and by the 1930s, "the negroes all lie at the southern end of the cemetery, [and] whites and negroes are separated by a fence."[82] This being on the West Bank, the southern end of McDonoghville Cemetery meant the lower section closer to the swamp—the back-of-town, from cradle to grave.

Yet in residential patterns, Blacks and whites often remained close neighbors, especially in working-class areas. This might seem paradoxical, given that we have already established that the back-of-town was more Black and the front-of-town more white. But those same statistical tendencies left room for exceptions, and there were plenty of exceptions. The obverse side of the same 1910 statistics cited above show that, contrary to prevailing trends, fully 34 percent of households in the back of Uptown were white, as were 54 percent of the back of downtown. "Segregation always existed," recalled an elderly African American woman from Algiers, "but most of the whites and blacks got along. We lived side by side, we were in and out of each other's houses, and the children played together."[83] One part of Algiers, Freetown, the oldest Black neighborhood on the West Bank, was and remains on high ground right by the river—further evidence of the importance of exceptions among prevailing trends.[84] Racialization of the older parts of town was the result of *de facto* racism and historical inequities, not explicit *de jure* zoning ordinances or housing policies. There were certainly Jim Crow signs in the streetcar, on the schoolhouse door, and at the lunch counter, but not on the shotgun house or the Creole cottage.

All this changed when drainage opened up the swamplands, and Jim Crow found a comfortable new home in modern subdivisions. Those on the production side of the real estate industry—developers, agents, financiers, home-builders, mortgage-lenders, underwriters, tax assessors, the city itself—had a vested interest in making them white-only, because in their minds whites brought money to the table, whereas Blacks brought down property values. Those on the consumption side—middle-class white families, accustomed to prerogative in a segregated society—were not about to protest the preclusion of Black neighbors. As for Black families, they had limited means to protest, being denied both a voice and a vote, and were now about to be denied their place in the new New Orleans.

Starting in the 1910s, white families began "leapfrogging" over the mostly Black back-of-town on the backslope of the natural levee, and settling in the *even lower and faster-subsiding terrain of the former swamps.* Having been convinced that pumps and levees had solved an old problem, and oblivious to the new problem of subsidence, white families gleefully settled at the bottoms of topographic bowls, where, for the first time in New Orleans history, their neighbors were not just more likely to be white, or mostly white, but 100 percent white.

By what legal mechanism? Authorities eyed new zoning ordinances as one possible segregationist tool. Adopted by cities nationwide, zoning aimed to protect property values by spatially separating residential land use from commerce and industry. Extending that reasoning to the social realm, segregationists argued that Black neighbors also affected property values, and thus ought to be zoned. Certain cities passed racial zoning ordinances, in which selected neighborhoods would be predesignated for whites or for Blacks only, and deviations could only occur with the explicit approval of neighbors within a certain distance of the "transgression." Civil rights groups took action, and fought Louisville's racial zoning ordinance all the way to the US Supreme Court, in the 1917 case of *Buchanan v. Warley*. Despite the openly racist sentiment of the times, all nine justices found Louisville's ordinance to be unconstitutional.

Undaunted, the City of New Orleans passed its own racial zoning ordinance in 1924, whereby, in accordance with a recent state law, neighborhoods would be racially designated, with mostly white blocks off-limits to Black residents. Jefferson Parish contemplated the same in 1926, when whites filed a request to the district attorney "for zoning of white and colored residents."[85]

Once again, a legal challenge arose, this time involving two neighbors on Audubon Street in Uptown, and the case of *Harmon v. Tyler* again went to the US Supreme Court. Assistant City Attorney Francis T. Burns traveled at taxpayers' expense to Washington to make the city's case for discriminatory zoning, knowing that not only the fate of the drained swamps was at stake, but the interests of "many of the larger cities of the country, especially in the North [where] the increasing migration of negroes . . . has made segregation a live issue, [and] where the negro problem formerly was considered exclusively a Southern problem."[86] Once again, the Supreme Court ruled unanimously that racial zoning ordinances were unconstitutional. There had to be another way.

RACIST DEED COVENANTS

Indeed there was. Instead of relying on government to segregate neighborhoods, real estate operatives did it themselves. They found they could circumvent the Supreme Court rulings by having white buyers commit, through deeds and cov-

enants, not to sell or rent to non-white families. State courts viewed such restrictions as private contractual agreements not in their purview, and agents
eagerly stewarded the paperwork. Two real estate agents explained what would
happen in the event of a transgression: "Individuals and associations [would]
pledge not to participate in any transaction in which either white or colored
would attempt to obtain residence in any section reserved for the opposite race.
The real estate man would decline to sell or lease, the lawyer to examine title,
the notary to pass the act, the insurance man to protect, the architect to design
or remodel, and the homesteads to grant loans, where any such invasion would
be intended. *Co-operation along those lines would be more effective than formal
law,* [and] the outcome would guarantee that both races in New Orleans would
continue to reside here in peace and tranquility."[87]

Racist deed covenants became an industry norm in new subdivisions nationwide, from the drained swamps of New Orleans to the prairies of Minnesota, from Long Island to Los Angeles, from Miami to Seattle.[88] Locally, they
proliferated in Lakeview and Gentilly, in Aurora on the West Bank, and in the
subdivisions of Jefferson and Metairie. When the Gentilly Terrace Company first
subdivided a neighborhood in now-drained Gentilly to become "Little California," it assured prospective buyers, in a section subtitled "THE COLOR LINE," that
"each purchaser binds himself and his heirs and assigns to never lease of sell to
a negro and negroes . . . nor can a Chinaman build a laundry shack on the corner next door to your home."[89] Racist covenants are what enabled Broadmoor,
freshly drained and also billed as "Little California," to "boast . . . that it is exclusively Caucasian, and that all the homes are owned, not rented."[90] In Lakeview,
restrictions for properties at Canal Boulevard and Louis XIV Street in 1934 read:
"The buyer agrees that there shall not be erected on this property any residence
to cost less than three thousand dollars; that no business establishment shall be
erected or operated without the consent of the seller and of the owners of two-
thirds of the property within a radius of three hundred feet . . .; that the front
of no house improvements or extension thereof shall be built or set nearer than
fifteen feet from the property line; that lots shall be sold only to people of the
white race."[91]

By the late 1930s, every single one of the 206 populated blocks of Lakeview—
that is, the former swamp between the Metairie Ridge (City Park Avenue) and

Lake Pontchartrain—were 100 percent white. Those same blocks had also subsided two to four feet below sea level. Similarly, 100 percent of the 195 populated blocks in Gentilly Terrace and adjacent subdivisions were 100 percent white, and similarly subsided, while three small enclaves elsewhere in Gentilly were nearly 100 percent Black. In what is now considered Old Metairie, 96 percent of populated blocks were 100 percent white. What racially balkanized these spaces were discriminatory deed covenants enacted *en masse* on the blank social space that was the drained backswamp.[92]

Meanwhile, in the historical neighborhoods of the natural levee, despite their housing stock generally not having racist deeds, the old back-of-town got incrementally Blacker, and the front whiter. That is to say, whites and Blacks were pulling apart—a sort of racial spatial disassociation, abetted by government-backed segregation in schools, facilities, and public housing. Revisiting the aforementioned 1910 Census data by comparing it to 1939 block-level racial demographics, we see that the back of Uptown that was 66 percent Black in 1910, became 83 percent Black in 1939. The front of Uptown that was 81 percent white in 1910, became 85 percent white in 1939. In downtown, the back of the Seventh Ward that was 46 percent Black in 1910, became 48 percent Black in 1939, while the front section of that same ward went from 75 percent white in 1910 to 88 percent in 1939.[93] "The Wood Pump," wrote scholar Daphne Spain of the social effects of drainage, "was an unwitting agent of residential segregation in New Orleans."[94]

Drainage paved the way for the positioning of the more privileged segment of society on the more vulnerable parts of the city—quite the opposite of what usually happens in geographies of race and class. Vulnerability was increasing because, all along, those drained bottomlands continued to subside, while new subdivisions continued to expand closer to surge-prone waters. By 1935, roughly 30 percent of the urbanized land surface had sunken below sea level,[95] and the "ground water or [the] permanent line of saturation" at one spot downtown had dropped to "about 18 to 20 feet below ground surface."[96] A 1937 WPA report stated that, "since 1897, when the first installation of sanitary and storm sewers was started, the subsidence has been very marked and in some sections amounts to three to six feet or more[.] The surface of the ground over the entire [city] has changed very materially."[97] Worse, the lowest areas, which were sinking the fastest and deepest, had the highest population growth rates. One Lakeview family living in a house built in 1941 saw, with the help of two scientists who analyzed

fifty-one years' worth of family photographs, fully thirty-two inches of absolute droppage of their property, not to mention greater subsidence (invisible to
them) relative to the rising sea.[98]

Yet few people saw the imminent problem. To the contrary, the best and
brightest minds seemed almost willfully blinded by their own triumphalist man-
tras. Listen to the well-regarded public scholar Thomas Ewing Dabney as he cast
his eyes over "the Jefferson Parish frontier" in the 1940s, savoring the dividends
forthcoming once new levees are erected along the lake:

> The immediate prize will be 13,000 acres of land which are *below sea level,* nearly
> half the total area which has already produced a spectacular development, in-
> dustrial and residential. . . .
>
> Take Metairie, for instance. Already it is nearly in the fabulous class. Its popula-
> tion has more than trebled in the past decade, [to] 26,000 residents. . . .
>
> Metairie first built on the ridges [and later] in the lower ground. It believed that
> its drainage system was sufficient to cope with the tropical rainfall of this section
> *and so it was.* It went *joyfully ahead* in its beautification and increase, *suffering oc-*
> *casionally from wet feet, but catching no great disaster.* . . .
>
> If you doubt that [Metairie is] the most beautiful part of this general area, choose
> well the company in which you utter the heresy.[99]

One might pardon Dabney's cheer because no major hurricane had struck since
1915, before which Metairie was mostly swamp. But in fact, Dabney wrote these
passages in *response* to a major hurricane, which had just struck in September
1947. By his own account in the very same article, the storm "drove so much water from the Gulf into Lake Pontchartrain . . . that the waves overtopped the concrete seawall [and] rampaged through Metairie and the rest of Jefferson Parish.
. . . Vast areas were under a four-foot flood, and the water in some Metairie
houses was 5 feet deep."[100]

How did Dabney reconcile this recent disaster with the prize of his below-
sea-level frontier? By devoting the rest of his article to all the latest plans for even
better drainage and ever higher levees.

9

Drainage Becomes a Utility,
1920s–1950s

Col. Marcel Garsaud, the genius behind the Lakefront Project, was perhaps the last of the old-style drainage kings—those poised personalities who, through prowess and power, managed to push forward reclamation and drainage agendas. They held their sway for nearly two hundred years, but the modern metropolis had become too massive, and laws and regulations too complicated, for any one maverick to master. Drainage infrastructure by this time was now largely in place, and what was needed to run it (and upkeep it, and upgrade it) were not flamboyant kings with speculative investors, but diligent civil servants backed by tax revenue. Drainage was no longer a dream; it was now an expectation, a millage, a utility.

It was also complex. Greater New Orleans's drainage infrastructure now comprised eight semiautonomous subsystems, established mostly during the 1890s to 1910s, to drain fifteen polders.[1] They differed in their history, motivations, funding, and management, but all had the same fundamental concept: make a polder by building a back levee; let gravity draw runoff through gutters and ditches; and install pumps to push, lift, and discharge the water to an outfall. Suburban jurisdictions learned the tricks—and the mistakes—from New Orleans proper, which had been working on the problem since the 1720s. Now, the city had finally solved "the world's toughest drainage problem," and while it took two centuries to perfect, it did so on its own. This was a home-grown world-class engineering accomplishment, something of which New Orleanians could truly be proud.[2] "'Pumps' are as symbolic of New Orleans as 'skyscrapers' are of New York City," wrote a federal writer in 1940, "because from the standpoint of volume New Orleans has the largest pumping system in the world."[3]

Geographer Peirce Lewis analogized pumps in New Orleans to elevators in New York, "one of those potent inventions that people in later years would take for granted," as each radically changed the geography of their respective cities.[4]

With a total investment of $21,809,077 as of late 1934, roughly half a billion dollars today, the system in New Orleans proper comprised ten pumping stations powered by 35,000 horsepower electrical generators capable of moving 27,000 cubic feet per second, through 73 miles of open and closed canals and 870 miles of pipes, into three outfalls. The equipment had outgrown its own electrical power supply: the Central Power Station down on Florida Avenue proved inconvenient and antiquated, such that by 1923, it supplied only 6 percent of the power needed for drainage. The rest came from the new Main Power Station built in 1908 adjacent to the Water Purification Plant in Carrollton, and in time, these Uptown generators would power most S&WB equipment.[5]

Hourly removal rates improved ten-fold in forty years, from 0.0625 inches of rainfall in 1900, to 0.125 inches in 1916, to 0.3 inches in 1930, to over 0.6 inches in 1940, enough to "make a river 52 feet wide and 10 feet deep running at rapid speed."[6] The system reliably drained thirty-nine thousand acres on the East Bank and another eleven thousand on the West Bank, making home for 495,000 people, still the largest city in the South. Neighboring systems served another 80,000 people, with commensurate equipment. And then there were the drinking-water and sewerage systems, into which the S&WB had invested another $22.3 million and $14.9 million, respectively.[7]

By this time, the framework for draining the modern metropolis was largely inscribed into the landscape. Most future changes would constitute capacity upgrades, component replacements, or system extensions, not wholesale redesigns. Although the metropolis would not reach its full spatial extent until century's end, the early-1900s drainage framework foretold exactly where it would go—because, whereas development *preceded* drainage in the urban core, it *followed* it in the suburban periphery. The West Bank would not become the Terrytowns and Timberlanes of today until the 1960s–70s, but Hero's drainage network of the 1910s predetermined its shape. Eastern New Orleans would not become today's New Orleans East until the 1970s, but Reed's drainage network of the 1910s prefigured its morphology. So too Lakeview, Gentilly, Kenner, Metairie, Marrero, Westwego, Arabi, Chalmette, and beyond: *pipes preceded streets, and dewatering preceded peopling.*

THE 1927 FLASH FLOOD

Starting 11:45 p.m. on Good Friday, April 15, 1927, a thunderstorm dumped nearly fifteen inches of rain at a pace faster than the pumps could remove. Winds knocked down electrical transmission wires of the New Orleans Public Service, Inc. (NOPSI), which tangled with those of the S&WB's own generators, "cutting off the power from all of the pumps."[8]

It had been an unusually rainy season, but this was the worst in years. Had such a downpour occurred decades earlier, it would have made a muddy mess, but the runoff would have steadily flowed back to the cypress swamp. Not anymore: modern drainage had turned New Orleans into topographic bowls, increasing in population, decreasing in elevation, and dependent on complex machinery working perfectly to keep dry. When it didn't, the backswamp came roaring back to life—not around cypress knees, but people's knees.

Soon, four feet of water had accumulated in Broadmoor, Fontainebleau, the back of Carrollton, and the rear of the Lower Ninth Ward, some areas of which had experienced street flooding five times since February 1. "Stalled automobiles dotted every street Saturday morning," reported the *States* of scenes all too familiar today. "Many establishments operated with only a handful of employees, the others being unable to leave their homes."[9] The pumps came back on later Saturday afternoon, and by Easter Sunday, the sinking city got back to abnormal.

The 1927 Good Friday flash flood has a curious place in history, starting out as a footnote and ending up driving the narrative. By pure coincidence, this local incident occurred while an immense volume of valley-wide rainfall and snowmelt made its way down the Mississippi River, breaching levees at various spots and eventually inundating twenty-six thousand square miles from Cairo, Illinois, to lower Louisiana. As the crest made its way downriver, all eyes were on New Orleans: would her levees hold?

In covering the story with datelines from New Orleans, news editors published photographs of the minor local pluvial flooding adjacently to stories of the major region-wide fluvial flooding. Readers conflated the two completely unrelated events, and bankers, investors, and other out-of-state interests called their counterparts in the Crescent City for assurances that New Orleans was safe. Officials with the US Army Corps of Engineers wistfully knew that it was, because they understood that levees farther upriver would likely breach first, thus relieving pressure on those downriver.

But for power brokers in New Orleans, chiefly bankers, that wasn't enough. They won over state officials in strong-arming Army Corps officials into taking radical action: destroying the levee at Poydras (Caernarvon) to divert millions of cubic feet of river water across Plaquemines and St. Bernard parishes, with the intention of reducing pressure on levees at New Orleans. The chief engineer advocating for that decision was the same Col. Marcel Garsaud who led the ongoing Lakefront reclamation project. The dynamiting took many attempts—it proved to be a well-built levee—until a rupture opened, releasing millions of cubic feet of water over thousands of acres, making refugees of hundreds of farming, trapping, and fishing families.

The Great Mississippi River Flood of 1927 put a decisive end to the Army Corps's ill-advised "levees only" policy and, by means of the Flood Control Act of 1928, replaced it with a more accommodating strategy incorporating levees to restrain the river, but also new spillways and floodways to divert it, and reservoirs to store its excess waters. The congressional act put the federal government permanently in the business of flood control, mandating financial and engineering responsibility for controlling the Mississippi and its tributaries, while also immunizing itself from liability should these systems fail.[10]

The Good Friday pluvial flooding in New Orleans did not cause any of this, of course. But it did inadvertently help justify the dynamiting of the Poydras levee, which itself evidenced, under outrageously unfair circumstances, that controlled spillways were indeed a better way to manage a high river. To this day, photographs of those backstreet floodwaters from that weekend in April 1927 confusingly circulate with stories of the Great Mississippi River Flood of 1927, the two separate events forever linked by happenstance.

The Good Friday flash flood had another, more positive consequence. It pushed long-time S&WB General Superintendent George G. Earl to seek desperately needed upgrades to the Achilles' heel of the drainage system, its electrical power. Also advocating for this fix were "business men and representatives of the commercial exchanges at the Association of Commerce," the same forces that backed the dynamiting of the Poydras levee. A dry New Orleans made money, and they were willing to do anything to prevent it from getting wet. One of them even offered a personal loan of $500,000 "to relieve the city of the flood condition to which it has been subjected five times during the past year. I trust the city of New Orleans."[11]

Superintendent Earl needed $7 to $9 million to make the needed upgrades.

The most critical was a better connection from each pumping station to the main S&WB generators at the water-purification plant in Carrollton, and to NOPSI electrical generating capacity, "so that if our current fails they could supply current, and if theirs should fail we could let them have some."[12]

It would take nearly a decade to carry out the upgrades. In 1936, the S&WB laid new stormproof underground cables connecting the pumps with the Carrollton generating plant, itself upgraded, and to "a change-over facility by which [electricity] may be obtained from the New Orleans Public Service."[13] The impetus for change, wrote the *Times-Picayune/New Orleans States,* was "the incident of 1927, when public opinion rose to a high peak for modernization of the system to prevent recurrence of street flooding."

Once again, the versatile engineering brilliance of Albert Baldwin Wood played a key role. Wood figured out how to lay the cable efficiently in segments of eighteen hundred feet, rather than pulling three hundred feet at a time through cumbersome conduits. He also found a simple way to detect a break in the underground cable: not with some costly contraption, but just a "dollar compass" passed along its path, and when its wavering needle froze steady, there's your break. Said Wood, "We can now locate the trouble to a gnat's eyebrow."[14] But Wood's engineering acumen had less sway in resolving complex ministerial questions such as power sources, and to this day, nearly a century later, devising a sustainable strategy of municipal versus in-house power generation, with all the equipment and electricity on the same frequency, remains a vexing problem.

DEPRESSION, WAR, AND A TROPICAL LULL

As dewatering preceded peopling, funding had to precede dewatering. In New Orleans proper, drainage funds came mostly from the two-mill tax on assessed property values won in that spirited landslide referendum of 1899. That millage gave the S&WB a reliable stream of revenue, and the city an incentive to improve conditions, so that properties would gain value and reap more revenue. But the tax also pegged the S&WB's revenue to a robust economy, and if property values should slip, so too would its budget—even though people would still need just as much sewerage, water, and drainage.

When the stock market crashed in 1929 and the economy followed, down went the S&WB's revenue stream, and at a most inopportune time. The utility had just launched $8 million of improvements triggered by the 1927 flash flood, including electrical enhancements, pump upgrades, and building a new Pumping Station No. 9 in Algiers, among other things.[15] The city's population, meanwhile, continued to grow by over 3,500 residents annually, to 494,537 by 1940, as rural denizens migrated in search of city jobs. The S&WB was in trouble, as were utilities elsewhere. In Jefferson Parish, for example, many landowners who had formed drainage districts now could not afford the taxes, nor could they find buyers for their land. Fully nineteen thousand acres of drained swamp temporarily reverted to state ownership, and dense vegetation enveloped soggy fields.[16]

To fill the gap between means and needs, the S&WB submitted proposals to two federal New Deal agencies created expressly for this purpose, the Public Works Administration and Civil Works Administration. The proposals were awarded, and after years of political wrangling with Sen. Huey P. Long, work finally got underway in the late 1930s, by which time laborers hired through the Works Progress Administration had joined the effort.

Federally backed improvement projects popped up everywhere in New Orleans, from landscaping at City Park and Audubon Zoo, to beautifying the Lakefront and Bayou St. John, to restoring Jackson Barracks and renovating the French Market. For drainage, the Public Works Administration effort "involved the redredging, widening, and enlarging of outfall canals; the lining and covering of drainage ditches; and the laying of concrete pipe—not the $7 million project originally hoped for, but it provided needed improvements that would not have been possible otherwise during the Depression." By one estimate, the drainage upgrades made it possible to remove up to three inches of rainfall per hour in the downtown area, triple the system at large.[17]

Needs grew that much more during World War II, when New Orleans became an official Port of Embarkation for hundreds of thousands of troops, and a destination for millions more in transit or taking leave from southern boot camps. Wartime worker migration swelled the resident population to 559,000 in the city proper, and 630,000 in the metro area. Housing was now at a premium, and property values soared. The S&WB's three critical systems kept the city humming, and enabled nearly 30,000 workers to manufacture over twenty thousand landing craft and other naval vessels at seven Higgins Industries plants, most of them built on or near former swamplands.

Wartime bustle breathed new economic life into the Industrial Canal, whose inner harbor was well suited for boatbuilding and testing. It also motivated the excavation of the adjoining Gulf Intracoastal Waterway (GIWW), a federal effort ongoing since 1907 to piece together a free inland navigation channel from Texas to Massachusetts. This particular segment, inaugurated on July 15, 1943, enabled domestic barge traffic to evade German submarines by avoiding open seas. Forking off from a turning basin in the Industrial Canal, the GIWW gave the Port of New Orleans what it had always wanted—an eastward outlet to the Mississippi Sound, something it once hoped to wrangle out of Bayou Bienvenue, only bigger, better, and federally funded.

But the GIWW also created a new ingress for storm surge, at the same time that just-drained eastern New Orleans began to sink, and as new neighborhoods formed along the Industrial Canal, itself a surge conduit.[18] The risk did not go unnoticed. Three times in 1941, the eastern Highway 90 and Highway 11 "were closed to traffic by reason of over-flow caused by high winds and storm conditions." Three years later, according to historian Nicolle Youngerman, the director of the Division of Public Health Engineering, John H. O'Neill, wrote to S&WB General Superintendent Baldwin Wood that "a considerable part of the area east of the Industrial Canal, due to its low elevation and the character of the soil, is presently not suited by residential development," and that the continued installation of water and sewer lines would only "accelerate the subdivision of certain areas in this section."[19]

Mercifully, the weather behaved during the war years, part of a lull in tropical activity ongoing since 1915 which saw only minor storms in 1922, 1926, and 1940. A whole generation came of age in New Orleans without having experienced a major hurricane, and much had changed in the interim. The city had shored up its defenses, with better federal levees and "safety valves" on the river, a broad breakwater along the lake, and the most sophisticated drainage system in the world.

It also had new vulnerabilities. Metro-area population had increased by nearly 70 percent since 1915, from under 400,000 to over 650,000, and expansion had occurred almost entirely on drained, sinking backswamps. Two new shipping canals now penetrated eastern New Orleans, bringing in tidal waters from the north and east. So too did three outfall canals to Lake Pontchartrain, all ungated, each rising relative to subsiding residential surroundings. And all along the coast, a new industry had dredged hundreds of linear miles of canals to extract oil and gas, with "few if any restrictions," wrote historian Jason Theriot, because "most people viewed wetlands as wastelands." Up these channels intruded salt water, turning forested swamps into grassy marshes and thence open water. Lafitte, a freshwater community in the middle of the Barataria Basin, had become brackish by mid-century, by which time two to five square miles of surrounding marshlands had been disappearing annually.[20] The loss of that terrestrial impedance, plus the canal ingresses, could send storm-induced surges into populated areas from the rear, protected only by back levees designed for drainage and not hurricane protection. The US Army Corps of Engineers, charged in 1928 to take responsibility for Mississippi River flooding, "was just beginning in the 1940s to examine the possibility of a larger federal role in municipal storm surge protection" for coastal communities in the hurricane zone.[21] Greater New Orleans, with all its river and lake levees and interior drainage systems, remained largely unprotected on its lateral flanks.

THE 1947 HURRICANE

In early September 1947, a storm formed off the West African coast and rode the North Equatorial Current across the warm South Atlantic, where it gained energy. Big, strong, and slow, the system stalled over the Bahamas, its winds peaking at 160-plus miles per hour—a Category 5 hurricane on the contemporary Saffir-Simpson scale—and made landfall at south Florida. After killing eleven people and causing over $31 million in damages, "the Fort Lauderdale Hurricane" proceeded northwestwardly across the Gulf of Mexico, toward Louisiana.

On September 18, as the system's track jogged to target New Orleans, authorities alerted citizens with a mix of admonition and aplomb. Mayor deLesseps "Chep" Morrison, the young reformer whose sunny disposition could shine through the darkest clouds, closed schools "only as a precautionary measure," not to imply an "impending danger," and expressed confidence in the drainage system. "All our wires are underground," concurred S&WB General Superintendent Baldwin Wood. "I don't anticipate any great danger," he told a reporter, "if the winds stay below 88 miles per hour."[22] Most residents stayed home; others took shelter in schools or other public buildings. Only coastal denizens evacuated, and they came *to* New Orleans—quite contrary to modern evacuation planning, which steers evacuees *from* New Orleans.

The hurricane arrived at New Orleans 7:30 a.m. on Friday, September 19, with winds approaching 100 miles per hour, the equivalent of a Category 2. Between 9:32 and 10:45 a.m., the eye of the twenty-five-mile-wide vortex passed directly over downtown, giving huddled residents a moment of deceptive tranquility. "My dad took me down Marais Street to Canal, to survey [while] the eye passed over," recalled a resident of the Iberville Housing Projects who was twelve at the time. "The sky was perfectly clear, [calm] like a church, [and] there was very little damage to the Projects."[23] Authorities appeared to be right, at least over the inner city.

But in the suburban periphery, conditions worsened. Overtopping along the drainage outfall canals flooded the mostly reclaimed but still-vacant lowlands in East Jefferson Parish, and put two feet of water on the newly opened Moisant Airport in Kenner. Surge intruded the Seventeenth Street Outfall Canal and flooded Bucktown and Jefferson Parish to its west, putting "vast areas . . . under a four-foot flood, and the water in some Metairie houses was 5 feet deep."[24] Fully thirty-three of East Jefferson's forty-eight square miles went under water, nearly all of it former swampland, and the deluge became impounded when seven of the eight pumps became inoperable.[25]

In New Orleans proper, lake water had overtopped the levees along the London Avenue Outfall Canal and the Industrial Canal, flooding parts of Gentilly deep enough to stall military trucks and require surplus Higgins boats to carry out rescues. Surges from lakes Borgne and Pontchartrain completely covered the Rigolets land bridge and most of eastern New Orleans, including the Higgins Industries plant at Michoud (now NASA), as well as parts of St. Bernard Parish. The S&WB got to work with its pumps, with some success. "The Gentilly and lakefront area, which appeared to be the worst flooded section in town, is rapidly being drained of the heavy floodwaters," Wood told the *States* on September 19. Pumps "were working satisfactorily, although taxed to capacity in all sections."[26]

That outlook changed as unexpectedly high water came in via the Florida Avenue Canal behind the Ninth Ward, the same route selected in 1895 as the city's main drainage discharge. Now, it worked in reverse, sending swollen waters of Lake Borgne and Bayou Bienvenue westward into the city and making Pumping Station No. 5 inoperable. Worse yet, the incoming seawater breached multiple sections of levees behind in the Lower Ninth Ward, some up to fifty feet long, many of which were not really levees but raised track beds of Southern Railways. Similar backflowing occurred on the Upper Ninth Ward's section of the Florida Canal, putting the S&WB in the Sisyphean position of trying to take seawater coming westward from Lake Borgne and pump it northwardly into Lake Pontchartrain, two connected bays of the same sea. It took four thousand workers (including soldiers and prisoners), tens of thousands of sandbags, and three full days to plug the breaches, after which the pumps got back to their "normal" business of pushing water uphill.[27]

Unlike in 1915, when wind was the main destructive force, the Hurricane of 1947 was mostly a water event. Ninety percent of damages to Louisiana came from flooding—sixteen hundred houses destroyed; twenty-five thousand damaged; $100 million in costs. Better forecasting and media warnings had reduced the death toll compared to 1915, yet still fifty-one people perished. Why did this weaker storm on a more favorable track cause so much water damage? Because hazard and exposure had both increased. The Industrial Canal, the Gulf Intracoastal Waterway, oil and gas canals, and wetland loss had allowed the hazard of storm surge to come inland, while new subdivisions in sinking swamp basins had increased human exposure.[28]

As in 1915, civic optimism became the prevailing post-storm narrative. The historian Nicolle Youngerman noted how Mayor Morrison had "repeatedly insist[ed] that 'any flooding is purely local conditions'"; how newspapers had declared "Toll of Hurricane Surprisingly Low" and promised "soon living will return to normal"; and how the mayor beamed that "New Orleans is not letting a storm hold of civic progress."[29] The business-advocacy group Greater New Orleans, Inc., took the message nationwide, spending $10,000 on papers from New York to Chicago to run "a dozen half-page advertisements across the top of which are emblazoned the words 'HURRICANE? NEW ORLEANS BREEZED THROUGH IT.'"[30]

The 1947 hurricane is one of the more multi-interpretational disasters in New Orleans history; one can look back on records and find fodder for both triumphalists and declensionists. Boosted by postwar optimism, the former assessment prevailed at the time, whereas hindsight now casts 1947 more problematically: had New Orleans never dug the Industrial Canal nor the Gulf Intracoastal Waterway, and had it never drained and developed its backswamp, then September 19, 1947, would have been mostly a windy day. Morrison's response to the event—"downplaying its effects on New Orleans's outlying areas and recently developed suburbs and focusing instead on older parts of the city that were not so badly damaged"[31]—seemed to ascertain that, despite decades of drainage fever and development hoopla, the old metropolitan geography had been safer all along. The historic inner core did fine; it was the drained suburbanizing periphery—low, sinking, and minimally protected—that flooded.

To be fair, imagining New Orleans without the navigation canals and drained swamps obligates us to deduct all the economic activity and wealth-generation that came from those interventions. Might that maritime activity have gone elsewhere? Would Higgins have landed those lucrative boat-building contracts and hired tens of thousands of people? Would greater New Orleans have fallen behind other major cities and ports, putting it more on par with Mobile and Galveston than with New York and San Francisco? At what cost, wealth? At what cost, safety?

THE CONFLICTED FRANK CLANCY

The year 1947's ten storms inaugurated a period of heightened tropical activity that would endure through 1969, over which time eight hurricanes struck coastal Louisiana, three seriously. The direct hit of 1947, plus a smaller one in September 1948, renewed levee-improvement efforts along the lake, and spurred discussion of lateral levees across the marshes. The need was greatest in Jefferson Parish, whose boosters seemed genuinely shocked by the 1947 surges, and again by an intense ten-inch downfall over Metairie in March 1948.

Parish officials got serious in the aftermath, and at least one got frank. "Eastern Jefferson Parish must live like a walled town," admonished Sheriff Frank J. Clancy in 1949, "with high earthworks thrown up on all its borders, because over 26,000 acres are below the 5 foot contour, 13,000 of which are below sea level. . . . [Our] drainage scheme had worked so well the water table had been reduced and in some places the surface of the land had sunk lower still."[32] For once, a public official had understood subsidence as a flood risk and communicated it in plain language. According to congressional testimony, "once the waters had blown into this area" during the 1947 storm, "a very long period of time was required to pump it out to the lake again." Drainage had not only triggered subsidence; it lured people into subsided spaces. As a direct result, thirty-three square miles went underwater, two thousand homes flooded, a thousand people had to be evacuated, eight people died, two thousand head of livestock drowned, $4 million of damage had been incurred, and fears arose of "a grave danger of epidemic"—the very thing swamp drainage was supposed to eradicate.[33]

Sheriff Clancy went further. Sounding like a modern-day sustainability advocate, the lawman lectured constituents on Urban Planning 101: "Each new home, each improved street," he wrote, "adds to drainage problems, and large developed areas create large drainage issues." But the forces erecting those houses and paving those streets, he added, lagged when it came to managing water. "The Government has provided facilities for financing new homes, but has not provided for . . . the enlarged demands on drainage and other public utilities that new home construction makes." By some measures, Jefferson Parish had a tougher drainage job than New Orleans, putatively the world's toughest, because Jefferson had more space to dewater and only 20 percent of its neighbor's revenue and facilities. "Jefferson Parish has fought a hard drainage and flood control battle, uphill most of the way," sighed the sheriff. "The tropical rains, almost 60 inches a year, fall on the land without regard for parish lines, and angry waves on storm- and flood-swollen Lake Pontchartrain . . . necessitate a bulwark just as high and as strong as the New Orleans seawall, with a population under one-sixth the size footing the bill."[34]

Yet Clancy, in the end a loyal parish official, framed the risk not as deterrence for development, but as incentive for protection. What was needed, he and other authorities agreed, was that bulwark—that is, higher and stronger lakeshore reinforcement, not unlike New Orleans's Lakefront Project of twenty years prior.

In fact, authorization for just such an effort had initially occurred in July 1946, but the 1947 hurricane forced supporters to update their request. Working with their congressional delegation in Washington, parish leaders secured $5.1 million in support from the US Army Corps of Engineers to build a major storm-protection project along the lakeshore. It would mark the inception of a new federal/local partnership that would formalize in 1955, mobilize in 1965, and completely dominate flood protection thereafter.

The aim was not technically land reclamation, because, unlike the Lakefront, this project would not turn water to land. Rather, it aimed to shore up the marshy littoral into a rigid barricade, which required the same sort of dredging and siphoning that went into the Lakefront Project twenty years prior. In 1948, dredges positioned twenty-two hundred feet offshore tore up the mucky bottom of Lake Pontchartrain, pumped mud through conduits supported by pontoon boats, poured the slurry between a two-hundred-foot-apart pair of concrete retaining walls inside the shore, dewatered the accumulation, and repeated the process until a massive embankment rose ten feet above lake level. The operation then moved up and down ten miles of Jefferson Parish's lakeshore, from Orleans to the St. Charles parish line, building the foundations for today's lakefront levee. More federal support came in 1950, and the project was completed in 1953.[35]

"Though we cannot prevent torrential rains from falling," wrote Clancy, tacitly acknowledging continued drainage deficiencies, "we will have the facilities to dispose of the water." But at least now, with the lakeshore reinforcement complete, "our earthen bulwarks will keep out all but cataclysmic tidal waves."[36]

Across the parish line, New Orleans also got some of the federal support. The Orleans Parish Levee Board raised the levees on the three drainage outfall canals in Lakeview and Gentilly, to the tune of $800,000, and the Port of New Orleans installed sheet-piling walls along the Industrial Canal during 1947–48.[37] Walls, in sum, were going up all around greater New Orleans. It was becoming a fortified city, sinking in the soft middle and rising along the hardened perimeter.

The walls got a test on September 23–24, 1956, when Hurricane Flossy passed well south and east of the city, yet still pushed enough surge across Lake Pontchartrain and dumped enough rain to flood certain lakefront neighborhoods, particularly Gentilly, nearly as badly as 1947. But Jefferson Parish fared well, Metairie resumed its strident growth, and hardly anyone remembers Flossy today.[38]

10

The Feds Wade In,
1960s–1990s

Up to now, storm-surge protection had mostly been the problem of local coastal jurisdictions. Controlling floods on the Mississippi River, on the other hand, had been the responsibility of the US Army Corps of Engineers, starting in 1879 and particularly after the Flood Control Act of 1928. Toward this charge, Army Corps engineers computed a theoretical maximum probable river deluge, called the Project Flood, and engineered their levees, spillways, and other devices to protect from that threat, known as Standard Flood Protection. All this was for the Mississippi River; only when the Office of Emergency Planning or local politicos were able to arrange it did the Army Corps assist in coastal surge protection, as it did for the Jefferson Parish bulwark in the late 1940s.[1]

But now, as the nation suburbanized and subdivisions sprouted along sunbelt bays and beaches, local jurisdictions struggled to mitigate the growing sea-surge risk, and appealed to Washington for help. So long as Congress appropriated funds to pay for its authorization of new duties, federal entities were perfectly pleased to expand their fiefdoms. In the case of the Army Corps, a department in the military with a civilian and uniformed workforce, that meant adding a new skill set to their 150-year legacy of engineering rivers, dams, locks, harbors, canals, roads, and forts.

Accordingly, the Army Corps in 1952 adopted a new internal policy "to provide no less than Standard Project Flood (SPF) protection for river areas where catastrophic storms may result in social disruption and loss of life."[2] This enabled Army Corps officials to engage with local beneficiaries on storm-surge protection, at least where rivers discharged into the sea. It was a small but significant uptick of jurisdiction.

Two years later, as hurricanes battered coasts from New England to the Carolinas three times in six weeks, constituents pressed their representatives to do something. On June 15, 1955, Congress authorized the secretary of the Army and other federal agencies to study "the eastern and southern seaboard of the United States with respect to hurricanes," and determine "possible means of preventing loss of human lives and damages to property."[3] Covering barely half a page in the *Congressional Register,* the act set the stage for a major federal commitment to coastal communities.

More laws and policies followed. In 1958, Congress stipulated that fully 30 percent of the costs of hurricane-protection projects had to be borne by local beneficiaries. Considered very high at that time, the local cost-share took a substantial burden off federal budgets. But it also made the Army Corps beholden to the wants and limitations of local partners, and turned both sides into compromising negotiators. Locals now had a bill to pay, but that meant they also now had leverage.

In 1959, Army Corps engineers extended their Project Flood and Standard Flood Protection concepts to coastal environments, resulting in a sequence of Standard Project Hurricane and Probably Maximum Hurricane declarations.[4] With each additional federal commitment to "protect" flood-prone areas from storms, coastal communities attained that much higher assessed value in the affected real estate—which justified more protection, which yielded higher land values. That feedback loop would spiral upwardly in the years to come, as would a parallel conundrum. The more the feds agreed to protect the coasts, the more costs they would bear. And the more the feds required local governments to share those costs, the more power they relinquished to local politicking.

OUTFALL FALLOUT

A flashpoint for these changing dynamics involved New Orleans's troublesome outfall canals, dug in the 1870s, incorporated into the modern system in the 1890s, and made into its primary discharges after the 1920s. Because these channels handled internal runoff, they were ostensibly a local rather than federal concern. But lacking gates at their mouths, and with erodible soil bottoms without concrete lining, the canals also gave external water three ingresses into the levee-rimmed metropolis, and thus became the problem of the Army Corps of Engineers.

In fact, the ungated outfall canals effectively negated the much-higher levees fronting the Lakefront project. Lake water had intruded exactly in this manner during the 1915 and 1947 hurricanes, flooding places like Gentilly and Bucktown. After each incident, the Orleans Parish Levee Board and Jefferson Parish Levee Board (which shared jurisdiction over the Seventeenth Street Canal) raised the outfall canals' levees—"simple berms," really—by an additional three feet. But subsequent subsidence, of both the levees and the neighborhood, nullified much of this gain.[5]

When the Army Corps's New Orleans District got involved with the outfall canals in the late 1950s, it had to reconcile its surge-protection focus first with the S&WB's drainage priority, and also with the two parish levee boards' canal-maintenance duties. No one solution pleased all four parties, much less the public. Installing gates would block storm surge (if they worked), but would also impede the S&WB's ability to remove runoff (heavy during tropical storms). Increasing pumping capacity would help the S&WB's job, but the augmented flow could erode the soft earthen canal bottoms, flummoxing the levee boards' charge of channel integrity. Diminishing pumping capacity would have an opposite but equally pernicious effect, allowing sediment to accumulate and reducing capacity to the pumps. Widening the three channels would increase that capacity and reduce the need to heighten the levees, but would have necessitated costly expropriations, mere mention of which incurred public wrath and political resistance.[6]

In 1960, the Army Corps recommended a compromise: install gates at the canal mouths, and equip them with bypass pumps to remove runoff when the gates are closed. Even though this strategy obviated home expropriations, it still met with resistance. "The S&WB and local residents feared that the tidal gates would malfunction," wrote the authors of a later investigative report, "inhibiting outflow of pumped storm water, which would, in turn, allegedly cause flooding." Discord also arose among local agencies. The S&WB, tasked with pumping out rainfall, was concerned that the gates would be under the control of its rival, the Orleans Parish Levee Board. Along with the Army Corps, this board was tasked with maintaining levee integrity and keeping out surge. Closing the gates would block surge and absolve the Levee Board and the Army Corps, but it would also set the S&WB up for failure in regard to rainwater flooding. The corps "argued (correctly as it turned out)," wrote investigators in 2006, "that the preferred solution would be to place storm gates at the north ends of the three canals which could be closed in the event of a Hurricane. [But] this proposal was bitterly contested by the local Sewerage and Water Board, who were concerned that the gates would be under the control of the local Levee Board." As a result of this "internecine distrust," federal authorities could not proceed with the gates, and "the canals would remain open to hurricane-induced storm surges[,] essentially 'allowing the enemy (storm surges) right into the backyard' of metropolitan New Orleans."[7]

It's as if stakeholders were more concerned with frequent minor pluvial flooding than with rare but disastrous surge flooding. Perhaps it is human nature to view quotidian problems as a higher priority than a theoretical outlier; after all, humans evolved to survive knowns more so than known unknowns, much less unknown unknowns. Whatever its psychological origins, the impasse would go on for decades. "The Corps soon found itself embroiled in a clash of cultures and goals with the levee districts, the S&WB, and the local citizenry, who flatly opposed the Corps' proposal." And with that 30-percent cost-sharing formula, the Army Corps could only leverage 70 percent of its authority.[8]

THE BARRIER PLAN

With the gating plan in a standoff, engineers with the New Orleans District of the Army Corps contemplated another way to keep surge out of the outfall canals, one that didn't cause internal drainage problems. They did so by rescoping the problem to a much larger hydrological realm. In 1961, Army Corps engineers proposed a "Barrier Plan" in which surge would be fought not at the mouths of the outfall canals, but at the mouth of Lake Pontchartrain. That is, along the eastern New Orleans-St. Tammany Parish land bridge, a massive 1,047-foot-long barricade would be installed across 22-foot-deep Rigolets Pass, and an 827-foot gate across 27-foot-deep Chef Menteur Pass. The plan also called for a gate at the Seabrook mouth of the Industrial Canal, as well as new lateral levees across the marshes and along the Gulf Intracoastal Waterway.[9] Such a strategy, the corps contended, "would keep out 90 percent of the water which would otherwise enter [the lake] with hurricane-boosted tides."[10]

On September 1, 1965, the US Congress approved the Barrier Plan, with an $84.8 million budget and the standard 70/30 federal/local funding formula. Eight days later, Hurricane Betsy struck, and the ensuing flooding, some of which came through the eastern land bridge, seemed to assure the envisioned barrier would soon be built.[11]

But the nation was changing, and new social movements voiced new concerns. Complications ensued, and the Barrier Plan remained on paper for years. All the while, the very possibility of its construction—indeed, the likelihood—had the effect of mooting further discussion of gating the outfall canals. Why revisit that ill-spirited cross-agency urban impasse when the best minds had already moved their focus out to the rural eastern fringe? And so the three canals remained wide open to the tempestuous lake.

<h2 style="text-align:center">"THE SEAWAY MOVEMENT"</h2>

In a confounding dissonance, the Army Corps planned to barricade one major eastern ingress for storm surge even as it worked to create another one nearby. The project was called the Mississippi River–Gulf Outlet, and it was the latest incarnation of a bedeviling temptation as old as the city.

This particular idea, like earlier ones, promised to make a call at the Port of New Orleans faster, better, and cheaper. Port cities are founded for their maritime convenience, and it behooves them to hone their forte, lest competitors lure away their customers. The favored tactic to achieve this advantage was as simple cartographically as it was difficult hydrologically: you dig a canal to shorten a shipping route. As early as 1725, colonials envisioned such a canal connecting the Mississippi River and Lake Pontchartrain; in 1794 and 1832, two canals were dug partway across the backswamp to connect city and lake; in 1904, St. Bernard Parish cut what is now the Violet Canal to connect the river with Lake Borgne; and in 1918, work began to finally connect the Mississippi with Lake Pontchartrain via the Industrial Canal.

Then there were *seaways*—that is, bigger deeper tidewater channels designed for oceangoing vessels, not just regional schooners and barges. In 1868, entrepreneur George Brott devised the cartoonishly ambitious Ship Island Canal, which he proposed to cut clear across New Orleans to the Mississippi Sound. Thankfully it failed, but not before triggering an affiliated drainage component, including those three problematic outfall canals. Ten years later, in 1878, Congress authorized the Barataria Ship Canal Company across the West Bank to circumvent shoaling at the river's mouth. But Eads's jetties solved the shoaling problem, and the Barataria Ship Canal joined the Ship Island Canal in the wet grave of bad ideas.[12]

A half-century later, Jefferson Parish refloated the idea of a tidewater canal cut through the Barataria Basin. With the blessing of the Army Corps of Engineers, the idea grew in the late 1940s into a proposed $75 million Mississippi Valley–West Bank Seaway, with a lock and inland harbor at Westwego and a fifty-five-mile-long channel to the Gulf by Grand Isle.

Shipping interests on the East Bank had their own seaway dream: cut a channel eastward from the Industrial Canal along the northern shore of Lake Borgne to reach the Mississippi Sound. The main proponent of this route was former Dock Board president Col. Lester F. Alexander, who had advocated for it since the early 1900s. The "Alexander Seaway" route had momentum on its side: the Gulf Intracoastal Waterway had just been dug along this very path in 1943. Converting it to a seaway would require widening, deepening, and extending it out to the Chandeleur Sound.[13]

After a decade of political wrangling, the East Bank proposal prevailed. Its New Orleans–based backers had more political clout than those in West Jefferson, and their argument benefited from Robert Moses's *Arterial Plan for New Orleans,* a 1946 diagnosis of metro-area transportation needs which called for a gargantuan Tide Water Ship Canal and harbor on the East Bank—and nothing for the West Bank.[14] A bill to fund the seaway was signed into law by President Dwight Eisenhower on March 29, 1956, fulfilling what historian Gary A. Bolding, author of an article titled "The New Orleans Seaway Movement," called a "century-old dream for a shorter and safer water passage from New Orleans to the open sea."[15]

The approved seaway would not quite follow Colonel Alexander's route, but rather cut southeastwardly through forty miles of St. Bernard Parish marshland into the Breton Sound. Its western terminus would dovetail with the Gulf Intracoastal Waterway and merge with the Industrial Canal—itself now fully paid off, by the Orleans Parish Levee Board in 1959, to the tune of $36,266,040.84 over the prior forty years.[16] The new project would not bear Colonel Alexander's name, nor be called a seaway. Instead it would be officially named the "Mississippi River–Gulf Outlet Canal," with the chipper sobriquet "MR-GO."

Digging the MR-GO was reclamation in reverse, converting 8,000 acres of freshwater swamp and saline marshlands into deep open seawater. Phase one began in 1958 by removing 20 million cubic yards of soil to enlarge the Gulf Intracoastal Waterway between the Industrial Canal and Paris Road. The second phase commenced in 1959 and tore up an additional 27 million cubic yards to cut a narrow access channel across St. Bernard Parish to the Breton Sound. Phase three took five years, 1960 to 1965, as barge-mounted dredges enlarged the access channel to a width of five hundred feet and depth of thirty-six feet. The final phase, concluding in 1968, deepened the draft to the minus-thirty-eight-foot bathymetric contour, where the Breton Sound opened into the Gulf of Mexico. Nearly a quarter of a billion cubic yards of peaty marsh soils had been torn up, mostly at the expense of St. Bernard Parish, some of which went into forming guide levees or shoring up land for what was to become the massive Centroport, USA inner-harbor terminal.[17] Upon completion, the MR-GO ran forty-five miles in inland length, plus another thirty miles offshore; it saved shipping traffic scores of miles of twisting, contra-current river navigation, and created ample space for new centralized wharves and containerized facilities. Reported the Army Corps of Engineers, "sailing time, ship turnaround time, navigation hazards, and congestion all tend to be reduced" by the MR-GO.[18]

West Bank advocates were disappointed to see their proposed seaway go to the East Bank. But having their own network of smaller navigation canals, they now pressed to augment them into something bigger—their own little "seaway movement." That inland shipping system traced back to 1830, when the Barataria and Lafourche Canal Company used slave labor to dig the Company Canal to connect what is now Westwego with Lockport on Bayou Lafourche and points west. Ten years later, Nicolas Noël Destréhan hired Irish contractors to dig a channel connecting his plantation in present-day Harvey with Bayou Barataria and points south, becoming today's Harvey Canal. In the early 1900s, after oil wells were discovered throughout Texas and Louisiana, both private and public interests surmised that some sort of free inland public channel, dredged through marshes and bays, would allow commerce to travel safely along the Gulf Coast and eastern seaboard. The idea, known variously as the interstate inland waterway or intracoastal canal, gained momentum and eventually came under the charge of the Army Corps of Engineers as the Gulf Intracoastal Waterway (GIWW).

Throughout the 1910s and 1920s, Army Corps officials bought out private properties and secured public water bodies to extend the GIWW all along the Gulf Coast. New Orleans being the nexus, Army Corps officials contemplated how best to thread the channel through the metropolis. In 1917, they contacted the Barrow and Harvey families, respective owners of the Company Canal and the Harvey Canal, to gauge their interest in a federal buyout. After years of negotiations, the corps selected the Harvey Canal to become the linchpin of America's Gulf Intracoastal Waterway. Soon, barges and other coastal vessels were passing through its upgraded lock at the Mississippi, down the river and through the lock at the Industrial Canal, and, after its 1943 opening, eastward on the latest leg of the GIWW dug through eastern New Orleans and on through the Mississippi Sound.[19]

By the late 1940s, the GIWW had become so popular that traffic constantly bottlenecked at the two narrow river locks, costing time and money. Barge companies pressed the Army Corps for solutions. For the East Bank, officials proposed widening the lock on the Industrial Canal or relocating it inland, only to find that dense Ninth Ward urbanization complicated the reengineering, making it extremely costly and disruptive to neighbors—a dispute that is ongoing to this day, one lifetime later.

The area around the West Bank's Harvey lock was also quite congested. But the reclaimed backswamp to its rear was still mostly rural, and it invited another proposal to solve the bottleneck problem: dig an alternative route for the GIWW and build a second lock. In 1948, funds totaling $14 million were authorized to excavate a trench from the main channel of the GIWW (which here shared the natural channel of Bayou Barataria), follow along the Orleans-Plaquemines parish line toward the Mississippi, and build a lock at Cut Off along the Lower Coast of Algiers.[20] The West Bank would never get its seaway, but two federally funded portals to the Father of Waters were a nice consolation prize.

In the early 1950s, backhoes and bulldozers gnawed away at 9.4 miles of West Bank earth, making a channel 12 feet deep and 125 feet wide. Engineers encountered unstable soils and had to dredge down to 25 feet, pump out water, and drive six thousand pilings into a layer of sand to support the 760-by-75-foot concrete lock chamber. The GIWW Alternative Route, known popularly as the Algiers Canal, opened in April 1956.[21]

The GIWW Alternative Route represented a radical alteration to West Bank geography. It effectively severed the Algiers and English Turn promontories from the delta landmass. It forced some infrastructural reshuffling, requiring a new tunnel and bridges, while offering a convenient new drainage outfall for subdivisions in all three abutting parishes, whose pumping stations were relocated to its banks. The canal created new economic opportunities for Plaquemines Parish along job-rich Engineers Road, a haven for the oil-and-gas services sector. But it also created new risk for Plaquemines, Jefferson, and Orleans parishes as a potential conduit for storm surges. Together with the Harvey Canal, the Algiers Canal formed a sort of inverted counterpart to the GIWW/MR-GO "funnel" on the East Bank, this one forming a narrow "waist," through which two artificial waterways could allow surge into developing, sinking West Bank neighborhoods.[22]

HURRICANE BETSY

On August 26, 1965, a tropical depression formed off Surinam and amassed over the Windward and Leeward islands. Officially named Betsy, the season's second storm stalled for two days north of Puerto Rico, leaned northwestwardly on September 1, and paused off central Florida, keeping millions on edge for days. Late September 5, the now-hurricane dithered and darted west, striking between Miami and Key West early on September 8. Once in the Gulf of Mexico, Hurricane Betsy accelerated to a twenty-knot clip northwestwardly toward New Orleans.[23]

Most city dwellers felt safe within what they understood to be a robust levee-protection and drainage system. They either sheltered at home or decamped for sturdy neighborhood buildings, namely public schools, as they had done in 1956, 1947, and prior.[24]

Betsy approached southern Louisiana on what appeared to be a worst-possible track. Radar systems tracked the eye offshore at 11:03 a.m. September 9, and eleven hours later, Betsy made landfall at Grand Isle, sending 100-mile-per-hour winds into Venice, Buras, and Port Sulphur.

Then came the surge. Low barometric pressure lifted a dome of Gulf water, and counterclockwise winds pushed it inland. The Mississippi River, unable to discharge up a hill of water, flowed backward, swelling and pouring into lower Plaquemines Parish. To the east, off the coast of Mississippi, the sea rose to 6.4 feet above normal levels at Pascagoula, 8.6 feet at Biloxi, and 10.7 feet in Gulfport, while to the south, waters hit 9 feet in the lower Barataria Basin. Betsy's surge ranked as the highest recorded in the region to date, due largely to the system's powerful forward momentum.

The most vulnerable flank of the metropolis lay to the east, where the Army Corps of Engineers had proposed the Barrier Plan yet also commenced digging the MR-GO. Water in Rigolets Pass rose to 10.6 feet above normal levels, and within the not-yet-completed MR-GO, the 9.3-foot surge flowed westward into the metropolis, courtesy of the ungated juncture it made with the GIWW. Because of the funneling effect, water in the merged GIWW/MR-GO channel rose to over 10 feet, while lake surge entering the Industrial Canal at Seabrook came in at 6 feet above normal sea level. All around were populated neighborhoods in former swamps which had sunk to 4 to 6 feet below sea level.

Hurricane Betsy struck New Orleans around midnight of Thursday-Friday, September 9–10, with sustained winds of 75–85 miles per hour and gusts up to 125 miles per hour. Over the next thirty hours, it dumped 5.13 inches of rainfall, after which the system drifted up the Mississippi Valley and petered out over Ohio.

As dawn broke on Friday, September 10, most of southeastern Louisiana was under water; only the natural levees of the river and distributaries, plus the levee-rimmed western half of metro New Orleans, remained dry. Worst hit were regions to the east: 96 percent of homes in Plaquemines Parish and 61 percent of those in St. Bernard Parishes flooded, while Orleans Parish had the most flood victims, 141,600 out of over 600,000, nearly all of them in the eastern half of the city.[25] That was the area perforated by the three navigation canals dug through the swamps and marshes over the prior forty-seven years: the Industrial Canal, the GIWW, and the MR-GO. Through this hydrological trifecta entered Betsy's surge, which overtopped or breached levees bordering five polders, inundating the former backswamps of the Seventh, Eighth, and Ninth wards with up to seven feet of seawater and damaging or destroying 6,350 homes and 400 businesses. Hardest hit of all was the Lower Ninth Ward, where overtopping and breaches along the Industrial Canal and the Southern Railroad tracks filled the mostly African American neighborhood with three to five feet of water up to St. Claude Avenue, and nine feet along Florida Avenue. Only the streets closest to the Mississippi River (present-day Holy Cross), a working-class, mostly white area at the time, evaded the deluge.[26]

To the north, the floodwaters did not surpass Gentilly Boulevard, but they did work their way through the drainage canals traversing that natural ridge. What resulted was two to four feet of flooding in an all-white portion of Gentilly Terrace as well as the nearby Pontchartrain Park, built ten years earlier as the city's first modern subdivision available to African Americans. Over 200 homes and a dozen businesses were swamped in these two neighborhoods. Farther to the east, Betsy's surge overtopped the Industrial Canal and GIWW levees and sent water into the former citrus orchards of the sinking polder drained fifty years earlier by Frank Hayne and Warren Reed. Residential development had since extended along Chef Menteur Highway and into the lakeside Citrus and Little Woods enclaves along Hayne Boulevard. Now, 1,330 homes and 140 businesses in those areas endured an average of three feet of water.

The only good news was to the west and south. Jefferson Parish on both banks had largely evaded serious flooding, because that bulwark built along the lakefront in the late 1940s had worked well. Levees along the three outfall canals had also managed to hold, sparing lakefront neighborhoods of significant flooding except for those parts of Gentilly affected by the Industrial Canal breach.

In nearly all other coastal environs, storm surge drains off gravitationally, once the storm passes and the sea returns to normal levels. That also happened historically in New Orleans—until drainage sunk half the landscape, after which gravity needed a lift from pumps. Removing Betsy's surge required the coordinated effort of the New Orleans District of the Army Corps of Engineers, the Orleans Parish Levee Board, and the Sewerage & Water Board. Engineers devised three approaches: (1) cut levees at certain spots to let gravity drain out the highest of the impounded waters back out through the very canals that sent the water in; else (2) bring in portable pumps on barges or trucks, lay pipes over the levees, and pump the water out, to the point that (3) S&WB pumps may be put back on line, and the municipal drainage system can take care of the rest.

All three approaches were deployed in various combinations, depending on local circumstances. In the Citrus community, for example, truck-mounted pumps were brought in to eject water into Lake Pontchartrain, after which S&WB pumps along Hayne Boulevard were restarted to drain the rest. In the nearby Pines Village and Donna Villa subdivisions, barge-mounted pumps ejected water into the Industrial Canal. South of Chef Menteur Highway, incisions were made in the northern levee of the GIWW to let trapped water flow out gravitationally. Similarly, three cuts were made in the Industrial Canal levee to dewater the Ninth Ward. The imported pumps provided 19 million gallons per hour of drainage capacity, and together with the levee cuts, most floodwaters were drawn down within two weeks.[27]

In one little-known incident, brought to light recently by historian Andy Horowitz, the S&WB made a wrenching decision regarding the famed Florida Avenue siphon, the so-called "underground river" built beneath the Industrial Canal in 1920. Still used to drain runoff from the Upper Ninth Ward into Bayou Bienvenue, the siphon now bore the brunt of Betsy's surge coming in the opposite direction. Worried that the east-to-west surge might overpower the west-to-east runoff discharge and thus worsen flooding in the main part of the city, engineers made a decision on the morning of September 10 to close the floodgates within the siphon. "At the height of the flood, in other words," wrote Horowitz, "officials directed water to gather higher in the Lower Ninth Ward," sacrificing residents there and in adjacent Arabi and Chalmette "in order to protect the rest of the city."[28]

Hurricane Betsy killed 81 Louisianians, injured 17,600, and caused $372 million in damages within the state (over $3 billion today), about one-third in New Orleans proper. "We have suffered the greatest catastrophe in our State since the Civil War," Governor John McKeithen told Congress later that month. "Nothing has approached it in the way of a natural disaster."[29] When McKeithen testified on September 25, pumps were still at work removing Betsy's surge, nearly all of which came in on artificial navigation canals, spilled over or through artificial levees, and collected within artificially drained swamps developed into anthropogenically sunken neighborhoods. *Natural* disaster?

Hurricane Betsy showed the MR-GO to be a terrible new risk to human safety. Its completion in 1968 would be followed by a comparable share of ecological damage, as well as economic disappointment. Bankside erosion necessitated costly dredging, else the shallow draft would force vessels to lighten their loads and lose money. Shipping firms hesitated to build containerized wharves at the planned Centroport along France Road, which had the effect of limiting vessels' calls, which further deterred new port investments. Intended to replace the Mississippi River as a shipping route, the MR-GO instead carried only 11 percent of port tonnage by 1990, and barely 5 percent by 1998. Shippers instead reinvested in riverside containerization facilities in Uptown New Orleans, and Centroport became secondary. Old Man River beat out "Mister Go," and the young upstart seaway became an expensive liability and enormous environmental problem, as a source of saltwater intrusion, coastal erosion, and storm-surge entry.[30]

STRENGTHEN THE FORT OR CLEAR THE BATTLEFIELD?

As Congress responded to the 1840s floods with the "swamp-buster" acts of the 1850s, and to the 1927 deluge with the Flood Control Act of 1928, it also reacted to Hurricane Betsy's inundations. Passed six weeks after the storm, the Flood Control Act of 1965 authorized dozens of water initiatives nationwide, one of which was "the project for hurricane-flood protection on Lake Pontchartrain, Louisiana." That effort, estimated at $56,235,000, came to be known as the Lake Pontchartrain and Vicinity Hurricane Protection Project (LP&VHPP), and it represented the next step in the federalization of storm-surge protection, ongoing since 1955.[31]

Deemed to have a benefit-cost ratio of 11.6 to 1.0—that is, every dollar spent would yield nearly twelve in direct or indirect benefits—the LP&VHPP put the

Army Corps of Engineers in charge of design and construction of storm-surge protection in the Lake Pontchartrain Basin. Local agencies in Orleans, Jefferson, St. Bernard, and St. Charles parishes would have roles of maintenance and operation, and together, feds and locals would share costs along the standard 70/30 ratio.

The LP&VHPP was not one mega-project but a roster of dozens of smaller tactical interventions throughout the targeted parishes, such as new or raised levees, floodwalls, channels, canals, diversions, gates, and a lock at Seabrook.[32] For example, the Orleans Levee Board used LP&VHPP funds to drive new steel sheet–piling walls along the three drainage outfall canal in Lakeview and Gentilly, reducing risk by raising their effective height. This and dozens of other comparable projects evidence that the LP&VHPP was not really a bold new vision; rather, it was a partial mobilization of piecemeal plans and proposals deliberated since the 1950s.[33]

The most fundamental of those ongoing tactical debates pitted the so-called High-Level Plan against the Barrier Plan. The High-Level Plan took the philosophy of strengthening the fort—that is, raising lakefront levees from 16 feet to 18.5 feet, to prevent the "enemy" (surge) from flooding the city. The Barrier Plan, on the other hand, aimed to clear the enemy from the battlefield, so to speak—that is, barricading the Rigolets to prevent the surge from entering Lake Pontchartrain in the first place, such that you wouldn't have to strengthen the fort.

Nested within the High-Level-versus-Barrier debate was a subsidiary dispute focused on the drainage outfall canals. Should they get "Frontage Protection"—that is, gates—or "Parallel Protection," meaning higher levees or floodwalls? According to investigators Douglas Woolley and Leonard Shabman, "the District [of the Army Corps of Engineers] favored placing gates at the canal mouths [which] would close automatically when there was a threatening storm surge," viewing that approach as "the most cost-effective plan." The Orleans Levee Board and the Sewerage & Water Board, on the other hand, "adamantly preferred higher walls along the canals, termed 'parallel protection,' as the best means to protect against hurricane surges . . . while still allowing the canals to be used to pump storm water from the city."[34]

An impasse developed, and it would endure for decades. Abetting it was the pending resolution of the High Level-versus-Barrier Plan debate regarding surge in the lake, which of course would deeply affect the parameters of the Frontage-versus-Parallel Protection dispute about the outfall canals.

Impasses tend to favor the status quo, and that's exactly what happened here. The Parallel Protection Plan prevailed: levees would be raised through Lakeview and Gentilly, as they had been previously, and frontages (mouths) would remain ungated, so that runoff could be discharged as has always been done. *Status quo.*

As for the lake-surge debate, the Barrier Plan was "believed to be less expensive and quicker to construct," and in regard to the new LP&VHPP, it had momentum on its side. Congress had approved it eight days before Hurricane Betsy struck, and eight weeks before it passed the Flood Control Act of 1965.[35] So by the time the LP&VHPP launched, the Barrier Plan became its flagship project, and by its very nature, it tended to nullify any further discussion on the Frontage Protection Plan to gate the outfall canals. Why seal off every hole in your fortress when you've already committed to keep the enemy off the battlefield?

At first, the Barrier Plan had plenty of support. But opposition gradually fomented among navigation interests, landowners, agencies concerned with operations and maintenance costs, and communities worried about getting flooded by waters trapped by the barriers. Mississippians fretted that barricades to protect Louisiana would send surge reverberating in their direction, particularly after Hurricane Camille ravaged their historic seaside towns in August 1969. The federal/local cost-sharing arrangement had the effect of amplifying these local voices, and depending on one's political philosophies, those empowered voices could be interpreted as either courageous resistance to dangerous proposals, or selfish obstruction of necessary interventions.

Political philosophies had indeed changed since 1961, when the Army Corps first broached the Barrier Plan. The national ethos had gone from trusting government to questioning authority, and from heralding progress to redefining it. One new social movement hardly had a name when the Barrier Plan was first announced; now it was called environmentalism.

Legally empowered by the 1969 passage of the National Environmental Protection Act (NEPA), environmentalists and other engaged citizens accused the Army Corps's New Orleans District of inadequately scrutinizing ecological impacts of the Rigolet barrier, something now required by NEPA through the mechanism of an Environmental Impact Statement. The new law also gave citizens the right to review and lodge concerns through public hearings, which brought civic rigor to governance but also lengthened the list of issues to which the corps had to respond, all of which took time and money. Modern democracy was proving to be complicated, in more ways than one. Had these same measures been in place just ten years earlier, they might have prevented a major navigation project that ended up letting surge *in.* Now, they were obstructing a major flood-control project that aimed to keep surge *out.*

Delays in the Barrier Plan hindered progress on the larger LP&VHPP effort, because the fate of the former affected all the variables of the latter. "We continue to be greatly concerned about protecting coastal Louisiana residents against storms," reported Maj. Gen. A. P. Rollins Jr. during a 1971 Senate hearings on the delays in New Orleans. "I must point out, Mr. Chairman, the urgency of proceeding as fast as possible with the barrier complex that extends from east of Rigolets to the tie-in with the New Orleans east levee. This barrier complex will provide the earliest and most complete protection to the entire area by preventing hurricane surges from entering Lake Pontchartrain." In the meantime, the corps had made progress on other aspects of the LP&VHPP, including levee work on the Industrial Canal and lateral hurricane-protection levees around the new real estate development known as New Orleans East. But Rollins made it clear that these were subsidiary devices all designed in the expectation that the Rigolets barrier would drastically curtail lake surges. "The construction of the barrier complex, including the levees and structures, is the most important single item necessary to provide the greatest amount of hurricane protection. We feel the authorized plan is adequate."[36]

Others dissented. In 1975, environmentalists organized as Save Our Wetlands, Inc., teamed with conservationists in filing a lawsuit against the Army Corps's Barrier Plan on grounds that it had not properly considered all environmental impacts on lake ecosystems. In December 1977, the US Circuit Court of Appeals agreed with their argument, finding that the corps failed to consider alternatives to the Barrier Plan, and that its Environmental Impact Statement was inadequate. After sixteen years of planning and twelve years of delays, the key component of the LP&VHPP had become an unknown variable.[37]

American culture had fundamentally changed over those years. The last gasps of the age of heroic engineering had given way to opening arguments of the age of environmental lawsuits. A man-against-nature mindset had yielded to a man-in-nature awareness, one that challenged the very notion that water was an "enemy" forcing humans to live in a "fort." That new cognizance went more along the lines of, *Why risk damaging nature when we don't understand all the impacts, and haven't considered all the alternatives?* In the years ahead, more and more Americans would agree.

The 1977 injunction prompted the Army Corps to reevaluate the entire LP&VHPP, a process that would take an additional eight years.[38] That period happened to see little tropical activity, and it let inertia set in. That which had been problematized became normalized. The Rigolets remained unbarricaded. The MR-GO grew wider. The coastal marshes further eroded. The outfall canals remained ungated, and their channels stayed narrow so their human neighbors did not have to move. Their waters rose higher above the sinking land, and their levees had to be raised again, and later their floodwalls. It was water management by path of least resistance. "The threat of litigation by environmentalists delayed the project," concluded a later Government Accounting Office investigation regarding the relationship of the LP&VHPP to the Barrier Plan. "[L]ocal opposition to building the control complexes at Rigolets and Chef Menteur had the potential to seriously reduce the overall protection provided by the project."[39]

Environmentalists would have scoffed at such an assertion, dismissing it as feds defending feds. Why didn't government officials seem nearly as concerned over that *other* Army Corps project, the MR-GO navigation canal, which really did "seriously reduce the overall protection," flooding the city twice in forty years? During that same 1971 Senate hearing in which General Rollins stressed the importance of the Barrier Plan, he also spoke approvingly of "the new facilities [of] Centroport, U.S.A. being built by the New Orleans Dock Board" astride "the Mississippi River-gulf outlet," and of the need "to produce a study satisfactory to navigation interests."[40] It's as if corps officials wore an eyepatch while inspecting their maps, the good eye seeing the Rigolets, the bad eye missing the MR-GO.

For eight years following the 1977 court decision, the New Orleans District of the Army Corps proceeded with the pre-injunction LP&VHPP plan, minus the legally moribund Barrier Plan. It built new steel and concrete floodwalls along the Industrial Canal, lateral earthen levees around new suburbs, ring levees for St. Bernard Parish, guide levees along the MR-GO and GIWW, and floodgates at Bayou Dupre and Bayou Bienvenue.[41] By the early 1980s, it became clear that the Barrier Plan was dead, and in a revised Environmental Impact Statement in February 1985, the corps reversed its stance by finally abandoning the Barrier Plan and approving the High-Level Plan.[42]

No longer would the enemy be kept off the battlefield; instead the fort would be strengthened—and that meant a full focus on levees around the metropolitan perimeter, 125 miles of them: hurricane-protection levees for Orleans and Jefferson; rear levees for St. Bernard; a better seawall for Mandeville; a new mainline levee for St. Charles Parish; a parish-line levee with Jefferson, and more. The only major gate envisioned was a lock for the Seabrook mouth of the Industrial Canal—and that project was deferred until navigation interests had their say on fate of the (ungated) MR-GO.[43]

Levees, levees, levees: the days of the drainage kings were over, and no longer would New Orleanians dread the swamps, and throw parades for the heroes who drained them. Now, storm surges had become the new dread; the salvation was levees; and the engineers responsible for them were more likely to be berated as bureaucrats than heralded as kings.

Supposedly the levees of the High-Level Plan would deliver the same protection promised by the Barrier Plan, namely against a twelve-foot surge and up to 130 miles per hour, or a Category 3 storm. But that could only be ascertained if and when the LP&VHPP was completed.[44] It never was. "The project, when designed, was expected to take about 13 years to complete and cost about $85 million," wrote the General Accounting Office of the LP&VHPP. Instead, the Army Corps "encountered project delays and cost increases due to design changes caused by technical issues, environmental concerns, legal challenges, and local opposition[;] costs had grown to $757 million and the expected completion date had slipped to 2008."[45] Four decades after Betsy, the LP&VHPP remained up to 40 percent incomplete, and some of what had been built turned out to be dangerously under-engineered—due in part to underfunding, confusion, resistance, and incompetence, and the rest to continued soil subsidence, coastal erosion, and sea-level rise.

And what of drainage? The S&WB and its counterparts in adjacent parishes had an odd seat at the table of surge protection, where sat Army Corps officials and levee boards at one end, and environmentalists and special interests at the other. That crowd had the luxury of thinking about the future. Utility operators, on the contrary, had to think about the present—about delivering clean water to a million residents' faucets, getting sewage from their toilets, and removing rainwater from their streets, today, tomorrow, and constantly. Whereas corps engineers contemplated hypothetical storms and checked their maps, drainage engineers eyed dark clouds out the window and checked their watches.

What people at both ends of the table were starting to understand was that external and internal water sources were increasingly converging to become one and the same flood threat. Subsidence and sea-level rise had conspired to merge the threat of storm surge trying to enter the bowl, with heavy rainfall trying to get removed from the bowl. Unlike fifty and a hundred years ago, S&WB engineers could no longer only focus on pumping out rainfall; now they were also expected to pump out seawater, should a surge make its way in. Planning against surge flooding had, in effect, become the S&WB's fourth charge; perhaps it ought to have been renamed the Sewerage, Water, Drainage and Surge Board. And if surge managed to get in, the deluge might knock out the S&WB pumps and generators, in which case neither the rainwater nor seawater could be removed, and New Orleans would drown in a bowl of its own making. The bowl, meanwhile, was sinking deeper, and in the mid-1970s, it made front-page news.

"SMILE: YOUR HOUSE IS SINKING"

At 8:45 a.m. Monday, September 1, 1975, in the quiet Country Club Homes subdivision in Metairie, a ranch house at 3501 Henican Place suddenly exploded in a fireball so powerful it shattered windows a mile away. The structure itself plus four adjacent houses were reduced to rubble, and fifteen others were damaged. Eleven people were seriously injured, including the homeowner, who suffered severe burns. "There was immediate panic," said an eyewitness. "People came running outside. It was like nothing you've ever seen."[46]

In fact, it had happened before. The first was in April 1972, at 3905 Henican, three blocks to the north. Two years later, in May 1974, a similar house exploded three blocks to the east, at 3405 Haring. Five months later, family members were severely injured when their house at 1705 Airline Park Boulevard erupted. At least eight times between 1972 and 1977, well-maintained homes in modern subdivisions, all within a mile radius, exploded without warning. "Scores of Metairie residents," reported the *Times-Picayune,* "wondered whether they are living in what amounts to time bombs."[47] At one point the Jefferson Parish government sought federal disaster aid to figure out the problem, even if it put the community in a bad light. "It scares me to ask the governor to declare East Jefferson a disaster area," said Councilman Lawrence W. Heaslip Jr., "but it scares me worse to do nothing."[48]

Unnerved, Jefferson Parish authorities and Louisiana Gas Service Company technicians investigated the smoldering ruins and determined the proximate cause to be a broken gas line. But the clustering suggested an ultimate cause, one that went beyond shoddy workmanship or tragic happenstance.

The culprit, it turned out, was severe soil subsidence. This part of Metairie had been the very nadir of the swamp basin in today's East Jefferson Parish, lying farthest from both the Metairie Ridge and the natural levee of the Mississippi. As such, its earthen floor was particularly low, and its standing water correspondingly deep, which preserved an unusually thick layer of subterranean peat—eight to twelve feet of ancient marsh grass, leaves, wood and other humus integrated into the mud like coffee grinds.

When reclamation came in the 1910s, out went the standing water. When subsurface drainage came in the 1950s to pave the way for subdivisions, out went the soil water. The dryness allowed the peat to oxidize, and on account of its thickness, the subsequent subsidence tended to be fast and severe, dropping one to two feet since the houses were built, and nine to ten feet since the swamp was first reclaimed.[49] Uneven settlement buckled concrete slab foundations, which had been poured directly on the soil, and buried utility connections stretched and twisted with the cracks and shifts. In some cases, brittle gas lines ruptured and leaked gas into the cave-like cavities beneath slabs, or wafted up into attics. All that was needed to detonate the house, as Heaslip put it, was "for some citizen to plug in a toaster."[50]

Those who suffered injuries might be considered victims of improper construction. But ultimately, they were victims of drainage. Drainage enabled them in reside in that geographical space, and it triggered the subsidence that destroyed their residence—just as it had also trapped the floodwaters that poured into that sunken space in 1947.

But of course drainage is a process and not an agent. The agent behind drainage is us; it is three centuries' worth of human decisions made to wring water out of wet dirt to render a city dry. Decades of trying were followed by apparent triumphs, and now, increasingly, came the tragedies.

The explosions made for frightening news throughout the 1970s, and probably did more to educate residents about deltaic precarity that anything outside of flooding. Headlines reveal how a community grapples with an unsettling new understanding of itself, and how it proceeds from there.[51] First, there is the realization that humans are part of the environment, and subject to its risks:

- "Marshlands in Trouble—Homeowners Are Too" (*Times-Picayune*, November 1, 1972)
- "Wetlands in Trouble; Drained Marshland Poses Hazards" (*Times-Picayune*, May 29, 1974)

- "Rats, Nutria, Snakes, and Mosquitoes—Not to Mention Sinking Backyards" (*Times-Picayune*, July 28, 1974)

Then comes the trauma:

- "Eleven Persons Are Injured as Metairie House Blows Apart" (*Times-Picayune*, September 2, 1975)
- "Jeff Goal: Prevent More Explosions" (*Times-Picayune*, September 2, 1975)
- "Gas Line Link Discovered at Littles'" (*Times-Picayune*, September 3, 1975)
- "Gas Explosion Destroys House" (*Times-Picayune*, January 20, 1977)
- "State, Federal Help Sought to Avoid Jeff Explosion" (*Times-Picayune*, January 21, 1977)

People next grapple, reckon, and adapt to the newly learned set of circumstances:

- "Soil Survey Needed" (*Times-Picayune*, February 13, 1977)
- "How Can You Cope With Sinking Soil?" (*Jefferson Parish Times*, February 2, 1977)
- "Soil Sinkage News Not Good for Kenner" (*States-Item*, February 10, 1977)
- "Warning: Hazardous Soils" (*Times-Picayune*, March 28, 1977)
- "East Bank's Soil Sinkage 'Severe'" (*Times-Picayune*, March 30, 1977)
- "Soil Testing Starts in East Jeff" (*Jefferson Parish Times*, March 26, 1977)
- "Soil Survey Finds Muck in Metairie" (*States-Item*, July 15, 1977)
- "Seeking Solutions to Kenner's Soil Subsidence" (*States-Item*, September 1, 1977)
- "Soil Map of Jeff Tells 'Hole' Story" (*States-Item*, September 29, 1977)
- "What Are Ways to Cope with Subsidence?" (*States-Item*, October 6, 1977)
- "West Bank Is on Shaky Ground" (*Times-Picayune*, December 10, 1978)
- "Soil Sinkage Plagues 84% of West Jeff" (*States-Item*, December 11, 1978)
- "They're Losing Ground" (*States-Item*, December 16, 1978)

At the same time, political bodies react:

- "Administration Reacts to Explosions" (*West Bank Guide*, February 26, 1977)
- "Steps Urged to Prevent Jeff Gas Blasts" (*States-Item*, June 9, 1977)
- "E. Jeff Soil Not Best for Building" (*Times-Picayune*, October 3, 1977)

- "Study Shows Jeff Land Unsuitable for Urban Use" (*States-Item,* October 6, 1977)
- "Subsidence Panel Established" (*States-Item,* October 6, 1977)
- "Development Slowdown Recommended in Jeff" (*East Bank Guide,* October 25, 1977)
- "Parish Seeks New Soil Study" (*East Bank Guide,* November 2, 1977)
- "Halt Marsh Development for Soil Study, Team Urges" (*States-Item,* December 10, 1977)

As does the private sector:

- "Gas Firm Halts Service Expansion, Kenner May Fight (*States-Item,* April 19, 1977)
- "Flexible Gas Lines Urged for Jefferson" (*States-Item,* September 1, 1977)
- "Land Developers Asked to Provide Soil Subsidence, Foundation Data" (*West Bank Guide,* October 19, 1977)
- "Builders Seek Help in Kenner" (*Times-Picayune,* October 26, 1977)
- "Soil Reports by Developers Urged in Jeff" (*States-Item,* November 9, 1977)

And the lawyers:

- "Louisiana Gas Sues Jeff, Says Parish Partly at Fault for House Blasts" (*Times-Picayune,* July 6, 1977)
- "House Blast Victims Sue NOPSI" (*Times-Picayune,* July 15, 1977)
- "Jeff Soil Sinkage: Seller Can Be Mum" (*States-Item,* November 30, 1978)
- "West Bank Couple to File 'Sinkage' Suit" (*States-Item,* November 9, 1977)
- "Some Subsidence Blamed on Builders" (*Times-Picayune,* 1978)
- "Warning: House May Sink" (*States-Item,* February 16, 1978)
- "Couple Wins, Loses in Sinking-Home Suit" (*Times-Picayune,* February 16, 1978)

Finally, new ordinances, regulations, and building codes are passed:

- "Resolutions Offered to Curtail Soil Subsidence" (*East Bank Guide,* October 26, 1977)
- "Soil Study Requirements Are Proposed" (*East Bank Guide,* November 16, 1977)

- "Builders Soon May Need Soil Test Prior to Permit" (*East Bank Guide,* December 21, 1977)
- "Jeff Requires Report for New Subdivisions" (*Times-Picayune,* February 16, 1978)
- "Homes North of Jeff Line May Pay for Gas Hoses" (*States-Item,* August 16, 1978)
- "Ordinance Would Require Jeff Soil Sinkage Data" (*Times-Picayune,* October 25, 1978)
- "Sinkage Help Urged for W. Jeff" (*States-Item,* December 21, 1978)
- "Jeff Council Declares New Gas Connectors Necessary" (*East Bank Guide,* 1978)
- "Building Code Is Target of New Jeff Panel" (*States-Item,* January 25, 1979)
- "Jeff to Consider Requiring Pilings Over Sinking Soil" (*Times-Picayune,* March 7, 1979)
- "New Law Requires Pilings" (*Times-Picayune,* March 8, 1979)

After a flurry of finger-pointing and court cases, parish officials agreed to new building codes. No longer could concrete slabs be poured directly on soils; first an array of pilings had to be driven fifteen to twenty feet down, where lateral friction would stabilize them, after which a concrete slab may be poured upon their protruding tops. Load would therefore no longer be borne by the superficial soils most prone to sinkage, but by the more consolidated layers deeper down. Codes also called for goose-neck utility hook-ups designed to bend with the geological dynamism.

The new codes, now standard throughout the region, put an end to the exploding houses. Yet everything else that constitutes a neighborhood—driveways, streets, parks, older houses—continued to sink, and flood risk only got worse. "*New Orleans is sinking*" became another local curiosity, along with "*New Orleans is below sea level,*" as if they were expectable conditions that came with the territory, like Santa Ana winds or Minnesota winters, and perhaps even made light of. As the *States-Item* put it in a 1977 headline, "Smile: Your House Is Sinking."[52]

NEW ORLEANS EAST

Had hurricanes struck or the economy sank in the 1970s, greater New Orleans might have a smaller footprint today. Instead, the tropics cooled and the economy heated up, namely the oil and gas sector. Skyscrapers arose in New Orleans's Central Business District, rivaling Houston, while subdivisions sprouted farther out in the suburbs, accommodating incoming workers and fleeing urbanites. Now there were better-than-ever ways to commute from periphery to core, with the new Interstate 10 and connecting expressways accessing bridges in every direction.

No longer were the departing city-dwellers all white, as they were in the 1960s following school integration; now families of color also savored suburban living. The 1964 Civil Rights Act had nullified explicitly racist restrictions in housing, and the 1965 Voting Rights Acts enabled the rise of Black representation in politics, while African Americans made up an ever-growing percentage of the New Orleans population, surpassing 50 percent in the late 1970s for the first time since the 1830s. Racism retreated from its *de jure* form to its *de facto,* systemic, and/or institutional forms, becoming less overt but still pernicious. What remained in the inner city, meanwhile, were divested communities with diminishing tax bases, and residents who were more likely to be impoverished and uneducated. Housing projects in particular became overwhelmingly poor and African American, isolated in every way—spatially, economically, and socially. The resulting crime and decay, centripetally intensifying, triggered more divestment, while the resources and assets that help communities prosper all seemed to spin outwardly to the suburbs, as if in a centrifuge. Social and economic decay intertwined with environmental ills, chief among them the growing flood risk of drainage-induced soil sinkage. "The decline of New Orleans as a major American city," observed geographer Craig Colten, "paralleled its physical subsidence."[53]

Because a suburban house takes up more space than its inner-city equivalent, outlying parishes saw their concretized surfaces spread spatially even faster than their residents did numerically. Jefferson Parish's population more than doubled between 1960 and 1980, after having doubled in the 1950s; St. Bernard Parish's population nearly doubled between 1960 and 1980, after having nearly tripled in the 1950s. The City of New Orleans, meanwhile, lost 70,000 people, and many of those who remained had in fact internally migrated out of the urban core to low, sinking subdivisions to the east.

Were they safe? New LP&VHPP hurricane-protection levees purportedly guarded those bowls from surges; their name said as much. Flood-insurances rates went down, and real estate interest perked up. Eastern New Orleans got marketed as a "suburb within the city," trademarked as New Orleans East, and developed with thousands of new houses, nearly all at grade level. Each subdivision was nestled within the framework of drainage ditches first installed by Hayne and Reed sixty years earlier for the purpose of farming, at which time under 1,100 people lived in the vicinity. That figure rose into the low thousands after the opening of the Industrial Canal, and over 10,000 with the paving of highways and building of bridges. By 1960, fully 23,562 New Orleanians lived east of the Industrial Canal, of whom 80 percent were white.

Then came Hurricane Betsy, NASA and its Saturn rocket jobs at Michoud, the LP&VHPP hurricane-protection levees, and Interstate 10. In a remarkable (and equally counterintuitive) replication of the early 1900s "leapfrogging" of white families into lower-elevation, higher-risk Lakeview and Gentilly, now a similar demographic wave was moving eastward into comparably risky spaces. By 1970, eastern New Orleans's population was 44,526, over 82 percent of whom were white.

In the next ten years, eastern populations would grow even more. But their racial composition would flip, the white component plummeting from 82 percent to 55 percent, and by century's end, a scant 11 percent. Why?

The proximate cause was a series of multifamily housing complexes built along the I-10 corridor in the 1970s, many of them with Section 8 housing vouchers. With an affordable version of the suburban ideal beckoning, the new housing opportunities attracted nearby working-class Black families eastward out of the inner city, while public bus lines connected them with inner-city jobs and resources. The ultimate cause of this demographic shift can be traced back to historic settlement patterns in the "Creole faubourgs." Many middle-class and upper-class Black New Orleanians descend from Creoles of color, formerly *gens de couleur libre,* who lived on the lower (eastern) side of nineteenth-century New Orleans—what are now the Sixth, Seventh, Eighth, and Ninth wards. When the swamps of Gentilly were drained, members of this predominantly Franco-African-American ethnicity moved steadily lakeward, as new blocks were laid out deeper into the former swamp. Nudging them in this direction were a number of Black enclaves in greater Gentilly, including Dillard University, opened on Gentilly Boulevard in 1931; the adjacent Sugar Hill neighborhood; and the St. Bernard Public Housing Development, opened in 1940. All other Gentilly-area subdivisions were strictly white-only—until philanthropists in the early 1950s funded the first modern suburban-style subdivision for Black home-buyers, complete with curvilinear streets and a golf course. Completed in 1955 and home to the Black-only Southern University in 1959, Pontchartrain Park had water on two sides and unfriendly neighbors on the others, but it was a hit among up-wardly mobile African Americans, and it had the effect of steering Black atten-tion eastward, as new housing options opened.

With the banning of racist deed covenants in the 1960s, Black families who moved were much more likely to settle in the east–of–City Park neighborhoods of greater Gentilly, including Pontchartrain Park. Lakeview, on the other hand, did not adjoin a historically Black settlement area, nor did it have the equivalents of Dillard, Sugar Hill, the St. Bernard Projects, or Pontchartrain Park. As a result, by the 1980s and afterwards, Gentilly and areas to the east of City Park became racially integrated or majority-Black, while Lakeview and areas to the west of City Park remained mostly white.[54]

That's what brought African Americans to the cusp of eastern New Orleans. What got them over the Industrial Canal was the new I-10 high-rise bridge, be-yond which were new subdivisions and multifamily complexes, many of which later qualified for federally subsidized rental vouchers. The integration of the city's civil service also meant that Blacks could now work in local government, reside within city limits, and yet enjoy a suburban milieu—a suburb within a city!

The rise of the Black population in the east corresponded to a decline in the white population. In 1980, when 76,712 people lived in the east, the white percentage had dropped to 55 percent, while a new Vietnamese contingent had grown to 5 percent since their first arrival as refugees in 1975. By century's end, the population of eastern New Orleans had swelled to 96,363, of whom 81 per-cent were black, only 11 percent white, 6 percent Asian, and 2 percent of His-panic ethnicity. What had been swamps and farms had, within two generations, become home to nearly 100,000 people—and within a single generation, flipped from 82 percent white to 81 percent black.[55]

The dramatic racial turnover had the effect of positioning the city's Black population disproportionately in the eastern half of the metropolis, by many measures the riskiest place in town. That risk was a function of its natural prox-imity to open saltwater bodies, to its lack of substantial natural levees, to the three navigation canals, and to extreme subsidence. LIDAR elevation data shows the Eastover neighborhood—a wealthy gated community opened in 1986 and home predominantly to affluent African Americans—to be the single lowest subdivision in the metropolis. Its severe droppage had been foreseen by a 1918 government report pointing out eastern New Orleans's thick layer of organic matter prone to shrinkage. "The elevation of the surface was a few inches above mean lake level," it stated, noting "a considerable depth of turfy humus or muck covered the entire area[,] to as much as 10 feet in the portion one mile back from the lake. The average depth was perhaps 5 feet."[56] The LIDAR data showed most of the larger New Orleans East polder to be five to eight feet below sea level, and Eastover specifically to be ten to twelve feet below sea level. That is, the thickness of the peat estimated in the early 1900s turned out to be the amount of subsidence by the early 2000s.

In the early 1980s, the world oil market collapsed, taking with it the New Orleans economy and the fortunes of New Orleans East. The city did not recover until the mid-1990s, and some say eastern New Orleans never recovered. But while the recession caused great pain, it also alleviated future suffering. The twenty thousand acres of marshes slated by the corporation known as New Orleans East, Inc., to be leveed, drained, and developed, were instead transferred to the federal government in 1986 to become Bayou Sauvage National Wildlife Refuge. For once, wetlands were valued for their water and wildlife, and dewatering and peopling them came to be viewed as bad policy. It took an oil bust, a corporate bankruptcy, and a federal buyout to do what centuries of flooding could not: convince people that maybe they shouldn't live in certain places. Years later, when Hurricane Katrina's sea surge flooded nearly all of eastern New Orleans, only Bayou Sauvage National Wildlife Refuge saw no human suffering.[57]

THE 1991 DEFEAT

With a stagnant economy, rising crime, and a civic malaise, the last thing many citizens wanted was higher taxes for what appeared to be failing services. Long gone was the spirit of 1899, when "the men and the women of New Orleans unite[d]" and voted sixteen-to-one to tax themselves for "sewerage . . . drainage [and] plenty of pure water."[58] Now was an age of what some might call overdue skepticism, and others, undue cynicism.

The latter-day version of that 1899 millage, up for a fifty-year renewal for the first time since 1941, came up for voters' approval in March 1991. S&WB leaders were nervous about the anti-tax sentiment in the air, and the little time they had to make their case. "Flood Protection? It's Your Call March 23," read a full-age S&WB ad in the Times-Picayune on March 14. "Better Drainage . . . No New Taxes." The copy invoked the prideful history of drainage in New Orleans, putting it on par with cherished aspects of the city's culture:

Around the world, New Orleans is known for its food, music, and ability to throw a party—and for its amazing success in protecting people from flooding. Engineers from all nations . . . marvel at our 172 miles of drainage canals, 18 massive pumping stations, [and] ability to pump billions of gallons of rain water out of the city within a few hours. . . . These international experts have tremendous respect for the Sewerage and Water Board, [and] the secret of this success, of course, is that the people of New Orleans have been willing to pay for this system.[59]

Whereas the water-treatment and sewerage systems got their funding from monthly user fees, drainage depended on real estate taxes, specifically a voter-approved four-mill *ad valorem* tax on land and buildings. Those calculations were pared down to exempt the first $75,000 of equity of qualifying homeowners—the politically untouchable "homeowners' exemption"—as well as all nonprofits, such as universities, churches, and civic institutions, some of which were big landholders and thus big drainage users.

What arrived into S&WB coffers was less than the cost of drainage, and not enough for needed improvements. Over the 1980s the agency had spent a quarter of a billion dollars on upgrades, and for the 1990s it needed an additional $450 million, of which two-thirds it hoped would come from the four-mill tax up for renewal. It was tenuous fiscal footing for such critical infrastructure, but because the public tended to take drainage for granted, few citizens were inclined to continue to pay for it, if asked. Research indicated many citizens thought the S&WB already had enough funds, and that additional tax dollars would not result in improvements in their neighborhoods.[60] "We have had at least seven 'super' rainstorms in recent years and more are likely," the S&WB countered. "It's your future, your home, your city. Vote on March 23rd."[61]

Most didn't. Competing with a beautiful spring Saturday and with no big races on the ballot, the election drew "one of the lowest turnouts in the city's history." It yielded a 52-to-48 percent defeat of the drainage millage renewal, driven by "unexpected opposition and voter apathy."[62]

The defeat spoke volumes about the state of New Orleans society and its relationship to infrastructure. Whereas in 1899, public enthusiasm to self-tax for drainage yielded front-page headlines ("How the Glorious Victory Was Nobly Won"), now, in 1991, an ignominious defeat ended up on page B-2 of the Metro section under the somnolent headline, "Drainage Tax Dies in Sparse Voting."[63]

The outcome sent S&WB members scrambling to make up for the loss. Their reluctant solution: split off part of the drainage system and put it under the auspices of the Department of Public Works. No one was happy with the arrangement; the DPW was itself hard-pressed for budget, and the S&WB pained to see its vital charges stray from its control, like struggling parents sending off a child to live with relatives. For the first time since 1902, the S&WB was no longer in charge of the entire drainage system, but rather only its heart and veins—that is, the electrical generation plants, the pumping stations, the open and underground drainage canals, and pipes measuring thirty-six inches in diameter or wider. The blood vessels of the system—twelve hundred miles of smaller subterranean pipes as well as seventy-two thousand catch basins—now went to the Department of Public Works.[64]

To some, the cross-agency arrangement proved that, if pushed by exasperated citizens, government could make do with less. To others, the awkward split would only breed further inefficiency, while still leaving unaddressed the critical upgrades now unaffordable by two strapped agencies. "Of the three divisions, drainage is chronically underfunded," stated a later internal investigative report. "It has not gained a new revenue source since 1982, and in 1991 it lost a dedicated millage. Twice during the last 30 years, in 1985 and 1998, S&WB unsuccessfully proposed supplementing its funding by implementing drainage fees."[65]

Leaders could not convince the public that something seemingly so routine, invisible, and unglamorous actually required resources. Budgetary starvation was the latest manifestation of a replicating theme in the centuries-long dewatering of New Orleans: drainage became a victim of its own success.

RISE OF THE "DATE STORMS"

The year 1995 marked the point at which the decade-long oil bust gave way to a rise in the tourism industry, buoyed by a rigorous national economy. Violent crime began to diminish, civic spirit rose, and something of a cultural renaissance ensued into the new millennium.

August of 1995 also brought a startling sight to the nightly weather reports: four storms swirling westward across the Atlantic simultaneously, ending a decades-long lull and initiating a period of high tropical activity ongoing to this day. The quiescence prevailing since 1969's Hurricane Camille, of course, had not been entirely quiet; Hurricane Juan had flooded the West Bank in 1985, and

Hurricane Andrew disheveled the region in 1992. In hindsight, the summer of 1995 seemed to bring something different in weather patterns, something more anthropogenic. "Global warming" had by now entered the popular lexicon, later understood as climate change, and with it came a realization that sea levels were rising at increasing rates. That meant the drained polders of greater New Orleans, and the fraying coast of Louisiana, were that much lower. It also meant more irregularity and intensity in precipitation, because warmer air holds more moisture. Neither bode well for the drainage system of New Orleans, on this, the hundredth anniversary of its modern inception.

No hurricane struck New Orleans in 1995, but the year did bring "a rain of biblical proportions" which caused nearly as much damage. On Monday, May 8, as thousands of music fans departed after Jazz Fest and people everywhere marked the fiftieth anniversary of V-E Day, a fast-moving front made its way across the Great Plains toward the South. The ensuing storms had been brutal on Texas, killing nineteen people in tornadoes, and now, at around 5 p.m., the system reached southeastern Louisiana. Fifty-mile-per-hour winds buffeted the metropolis, from St. Charles Parish through St. Bernard and on both sides of the lake. Twisters struck Arabi and Slidell, and falling branches pulled down power lines and darkened fifty thousand homes. Most of all there was rain, intense rain, falling far faster than any pump could remove it, which was putatively one inch the first hour and a half-inch every hour thereafter. All pumps were working, the S&WB made clear; they were just "overwhelmed by the amount of water."[66]

Totals measurements the next day confirmed that "the May 8 Flood" was truly exceptional. Mandeville got 10 inches; Audubon Park 11 inches; Covington 12 inches; Reserve 13 inches; Broadmoor and Mid-City, both in the bottom of their respective bowls, 14 inches; River Ridge ended up with 17.5 inches, and Ormond Estates in St. Charles Parish broke records with 18.5 inches. In a region that gets 60 to 70 inches of rainfall in a year, May 8, 1995 brought 15 to 30 percent of that annual total in a single evening.

What the pumps could not handle accumulated in streets and lawns, to one, two, sometimes three feet of depth. Any house built at grade level took on that much water, and there were tens of thousands of them. "Virtually all streets and highways [were] impassable," and underpasses filled so deeply that stranded motorists had to swim to safety. Each of the fifteen polders throughout the metropolis filled with their share of what historically would have been harmless swamp water. "Everything is flooded in Jefferson Parish," despaired one policeman. "Water, water, water everywhere."[67]

Flash flooding like that of May 8 had happened before, such as the one on Good Friday 1927. But in recent years, the downpours seemed to be increasing in frequency or intensity, as climate patterns changed, urbanization spread wider, as polders sunk deeper, as soils became concretized, and as more runoff raced to the same number of pumps. Because most were unaffiliated with named tropical systems, people tended to identify these flash floods by their dates. Thus was born the "date storm," hours-long intensive rainfalls that overwhelm local drainage capacity, cause sudden inundations, and leave behind ruined cars and soaked sofas, though rarely death or major destruction.

There had been the May 3 Flood in 1978 (next day's front page: "*pumps were overloaded ... official rain gauge 'drowned' ... [worst] since flooding in April 1927 ... insurers brace for sea of claims*"); the April 13 Flood on the West Bank in 1980 ("*May 3 Yardstick in Jeff Replaced by April 13, 1980*"); the June 10 rain of 6.7 inches in 1981; the April 23 downpour of 5.6 inches in 1982; the April 7 Flood of 1983 ("*torrential overnight rain, the most widespread in years ... disrupting the lives of a million people and leaving them cut off from the outside world*"); the December 28 rain of 6.7 inches in 1983; the November 7 flood of 1989 ("*10-Inch Rain Covers St. Bernard ... downpour caught weather forecasters by surprise ... 'It just happened'*"); the Mother's Day storm of 1990 (5.7 inches); the December 3 flood later that year ("*Swamped*" ... *[like] freak storm in November 1989*"); and the May 9 Flood of 1994 ("*Storm dumps 7 inches ... five times the capacity of the city's drainage pumps ... downpour ensnarled traffic and caused extensive street flooding*").[68] The "biblical" rain that came one year later topped them all, with double the precipitation in half the time, and the effects were commensurate with a moderate hurricane: six dead and $761.4 million in damages throughout twelve parishes, including $563 in National Flood Insurance Program payouts, $123 million in FEMA payments, and $75 million in auto damages, to a total of over 113,000 total claimants.[69]

After years of focus on storm surge and hurricane-protection levees, the May 8 Flood got people talking about drainage again. What they realized was that their marvelous but aging system, homemade and home-paid, was no longer up to the task of solving the world's toughest drainage problem.

THE SOUTHEASTERN LOUISIANA URBAN
FLOOD CONTROL PROJECT

The May 8 Flood—"one of the largest non-tropical rainfall events in New Orleans history"—did for drainage what the Great Mississippi River Flood of 1927 did for river control, what the 1947 hurricane did for lakefront levees, and what Hurricane Betsy in 1965 did for lateral surge protection: it brought the federal government into what had previously been a local problem.[70] Passed with the help of Louisiana's powerful congressional delegation, the Southeastern Louisiana Urban Flood Control Project, or SELA, won authorization through the Energy and Water Appropriations Act and the Water Resources Development Act of 1996. Costing $455 million with a 75/25 federal/local funding-contribution formula—"the largest influx of federal dollars for drainage dollars ever"—SELA aimed to augment canal and pumping capacity primarily in New Orleans and Jefferson Parish, for protection from a theoretical "ten-year storm," calculated to be 9 inches of rain over twenty-four hours. (The May 8 Flood was 10 to 18.5 inches in about six hours.) The process entailed local jurisdictions deciding on specific drainage improvements to submit for federal funding, which Army Corps engineers would then subject to "reconnaissance, 'a study that determines if there should be another study.'"[71] If the project makes it past reconnaissance, the corps would proceed with a benefit/cost analysis, and if positive, the project would be authorized. With the help of fast-tracking, the SELA roster over the next five years comprised fifty-four separate projects across the metro area, ranging from lining ditches to widening canals to adding pumps. Contracts were first let in 1997 for what was expected to take fifteen years to complete. Construction started in Jefferson Parish, worst hit on May 8, and later proceeded to New Orleans.[72]

The slate of upgrades may be read as a century-belated addendum to the 1895 advisory board plan, and one can only imagine how Benjamin Harrod, Rudolph Hering, and Linus Brown might have reacted to how their visions had played out. Twelve million dollars went to two new 1,200-cfs pumps at the Melpomene Pumping Station No. 1 on Broad, the linchpin positioned on what had been projected to be the Main Drainage Canal toward Lake Borgne. Over $19 million went to adding two new canals beneath Napoleon Avenue, for all that additional runoff flowing into the biggest, deepest polder of the city's heart. Over $23 million went into more and larger box culverts along South Claiborne Avenue, which once marked the edge of the backswamp, and had an uncanny way of remembering it. Another $25 million went to a whole new drainage subsystem in Hollygrove, which fell within that hydric stepchild named Hoey's Basin, straddling the parish line. The Dwyer Road area in eastern New Orleans would also get a new subsystem, priced $25 million. To its south on Florida Boulevard at the Industrial Canal, where the great "underground river" siphon once functioned until it malfunctioned, a new $4 million generator would be installed in Pumping Station No. 19. That's nineteen out of a total of twenty-two pump stations in the system, plus ten underpass pump stations, compared to five main pumps and three auxiliary pumps planned in 1895.[73]

"The world's toughest drainage problem" was tougher than the graybeards had thought—or rather, the solution they designed had unforeseen consequences more vexing than the original problem. In the prior two centuries, locals had feared swamps; local tenacity prioritized for drainage; local brilliance devised how to drain; and local funding made it happen. Then hydrological processes sunk those drained swamps, and economic drivers turned them into valuable real estate, whereupon developers and citizens proceeded to create homes and communities. Now the topographic bowls were trapping rainwater, their concretized surfaces steered too much runoff too swiftly to the pumps, the pumps could not discharge it fast enough—and everyone surmised that the problem lay with the *hardware* of drainage rather than the process itself. And so the "date storms" increased, expenditures rose, local coffers emptied, frustration mounted, and flooding became a constant, costly burden to living in New Orleans. "Our philosophy is pretty simple," declared Jefferson Parish president Tim Coulon in justifying SELA's federal taxpayer support. "You pay me now or you pay me later. That's why we went to the federal government for money, so hopefully we won't have so many flood claims on the back end."[74]

In the ten years between the May 8 Flood and Hurricane Katrina, a total of $823 million in federal dollars flowed into storm protection and drainage in Louisiana, of which LP&VHPP accounted for only 17 percent, compared to 53 percent for SELA.[75] Surge flood was the true existential threat to greater New Orleans; pluvial flooding was an expensive nuisance. But in a world of limited resources, the struggle to abate that internal nuisance crowded out spending on stronger floodwalls and levees against external threats. *Pay me now or pay me later:* the dewatering and the peopling of the New Orleans backswamp was turning into a downward-spiraling feedback loop of sunk costs justifying further expenditures—a risk-production machine that made soils sink, infrastructure crack, houses explode, and residents flood. But how do you tell that to a half-million people who innocently moved into those spaces?

11

The Only Thing That Can Mess Us Up, Early 2000s

By 2005, the S&WB drained over ninety-five square miles of Orleans Parish, on both sides of the river plus Hoey's Basin in Jefferson Parish, with an annual discharge capacity of nearly thirteen billion cubic feet. That's what the entire Mississippi River sends to the sea in six hours—enough, as writer Sebastian Junger put it, to "suck the Thames dry at London."[1] The system used natural topography to collect runoff into thousands of stormwater drains and hundreds of miles of underground pipes, through ninety miles of covered canals and eighty-two miles of open canals, to ninety-nine pumps in twenty-two stations plus ten underpass pump stations, which lifted and expelled the runoff into four interconnected water bodies: Lake Pontchartrain, the Industrial Canal, and Bayou Bienvenue on the East Bank, and the Algiers Canal on the West Bank. Power for the pumps came from the S&WB's electrical plant at the Carrollton Water Treatment Plant on Eagle Street, consisting of diesel turbines generating up to sixty-one thousand kilowatts. That in-house power was supplemented by another eighteen thousand kilowatts produced by Entergy, as well as a number of emergency generators hooked up to individual pumps. The S&WB could also tap power across the Mississippi. "A lot of people don't know [that] the Sewerage and Water Board's power system [has] two power lines between the West Bank and East Bank," noted the executive director at the time, Marcia St. Martin, "so that [we] can move power from the east to the west."[2]

Half of the ninety-nine drainage pumps were Baldwin Wood Screw Pumps, including all of the biggest, and they remained in excellent condition. The newer pumps were designed by S&WB engineer Joseph Sullivan, who matched Wood in acumen and dedication. "Everything that was built from 1972 forward was a Joe Sullivan design and a Joe Sullivan implementation," St. Martin said of her late colleague, who died in 2011 at age eighty-five. "He was just a brilliant engineer."[3] What was no longer up to par, however, were the veins of the system—the water lines, the sewer lines, the pipes, the connections—as well as the juice of the system, the electrical power.

THE ELECTRICAL FREQUENCY PROBLEM

S&WB did have impressive electrical generating capacity, enough to power a city of eighty-thousand people. The problem was the frequency. As radios need to be on the same frequency to communicate, generators and end users need to be wired for the same number of oscillations of alternating current flowing on their grid. In the late 1880s, pioneers in electrical power generation chose their system frequencies individually, almost arbitrarily, within the range of $16\,^2/_3$ to $133\,^1/_3$ cycles per second, or hertz (Hz). As those systems grew and merged into grids, frequencies had to be standardized—but only after a higher-level decision was made regarding the use of an alternating current (AC), in which the electrical charge reverses directions in the form of oscillations, versus a direct current (DC), where the charge flows in one direction and does not oscillate. Each side had its arguments and its lobbies in the "War of the Currents," with Thomas Edison backing DC (his invention) and Nikola Tesla advocating for AC. Telsa's winning argument was that AC could be more easily converted to different voltages through a transformer, and therefore was more conductive to widespread electrification for various uses, from lighting homes to running streetcars to powering drainage pumps.

As America's nascent DC electrical power grid shifted to AC, it got a major boost from Westinghouse's adoption of AC for its Niagara Falls Power Company generation plant in upstate New York, built in 1895—same year as the advisory board's drainage plan for New Orleans. Westinghouse selected a frequency of 25 Hz for its system, due to reasons of technical happenstance (namely the preset speed of their turbines), and when its generators successfully illuminated the city of Buffalo in 1896, the industry was sold. General Electric adopted AC, as did other power companies, and soon it became standard—along with the 25 Hz frequency coming out of the Niagara turbines (though some entities went with 40, 42, 50, 60 Hz or other frequencies, depending on their end users).[4] Engineers in New Orleans also adopted the emerging national standard of AC electricity with the option of 25 Hz frequency. S&WB generators, transformers, transmission wires, and pumps were all rigged up accordingly, and because it was no easy thing to change, subsequent upgrades reinforced the commitment to 25 Hz.

The rest of the electrical world, however, did change. In the interest of standardizing the manufacture and trade of electrical equipment and to allow power grids to be interconnected, power companies in North America switched over to 60 Hz, and 50 Hz in most other areas.

For a relatively small city with complicated infrastructure and insufficient funds, New Orleans hesitated to adopt the new convention, and each passing year made it that much more difficult to do so. In the same way the S&WB ended up getting stuck with the circa-1879 Cairo Datum for its topographical mapping, its electrical infrastructure got stuck with a circa-1895 electrical frequency. "Welcome to the oldest operating technology museum in the southern United States," joked one S&WB superintendent to a journalist visiting Pumping Station No. 5—and that was thirty years ago.[5]

It all worked until it didn't—that is, until parts had to be replaced, new components installed, or when the in-house 25 Hz generators could not supply enough power and needed NOPSI's (later Entergy's) 60 Hz power to fill the gap. Today, the S&WB has a split system, with 63 percent of its pumps getting their power from—and only from—its own 20-megawatt generators using the obsolete 25 Hz frequency, and the other 37 percent on the grid-compatible 60 Hz frequency. If an in-house turbine breaks down or a generator needs replacements, parts have to be custom-made in the S&WB's machine shop, or obtained within the dwindling community of other century-old systems.[6]

And where were these generators positioned? That was the *other* power problem. "Unfortunately," wrote later investigators, "all of these generating stations are located below mean Gulf level and subject to shut-down by flooding."[7] The main plant was on Eagle Street in Carrollton, by South Claiborne Avenue, which, citywide, hugs the sea-level (zero) contour. The auxiliary plant on Florida Avenue in the Upper Ninth Ward sits four feet below sea level. The five diesel-powered steam turbines in the Carrollton Plant, wrote journalist Richard F. Snow in 1992, are "about the size of a three-story house," inside of which were a 20-megawatt generator made by General Electric in 1916 and two Corliss steam engines for which "the makers stopped selling parts [in] 1903."

Antiquated as it was, this was still finely engineered equipment, as were the Wood pumps and just about everything else installed during the S&WB's halcyon years. Now, a century on, it still got the job done 99 percent of the time, with its twenty-four major pumping stations capable of up to 50,268 cubic feet per second, or 30 billion gallons per day—the volume of the Ohio River. They drained 59,000 acres in New Orleans and another 2,250 acres in Jefferson Parish.[8] Superintendent Joseph Sullivan, inheritor of Wood's stoic dedication, was justified in saying, "Two hundred years from now it'll all still be here." Sullivan added, "The only thing that could mess us up is the city subsiding."[9]

KATRINA

On Tuesday, August 23, 2005, warm air upwelling over the Bahamas grew into a tropical depression and developed sufficiently over the next day for the National Hurricane Center to name it Tropical Storm Katrina. Late Thursday, Hurricane Katrina approached south Florida with winds of seventy-five miles per hour. Most New Orleanians hardly took note. Despite predictions of a busy season, activity in the Gulf had abated since early July, when a surprisingly strong tropical storm named Cindy (later upgraded to a Category 1 hurricane) roiled the city.

Hurricane Katrina's wind-driven rains killed nine people in north Miami Thursday night, after which the system picked up new fuel from 90-degree waters looping into the Gulf of Mexico from a Caribbean current. Computer models now had Katrina's track hugging Florida's western coast; later they shifted toward the Panhandle, then to Alabama. The farther west the track, the more Katrina drew from the deep warmth of that loop current, and the better organized it became. By Friday evening, forecast tracks pointed to Mississippi and Louisiana, and locals took note. "Katrina Puts End to Lull," read a secondary headline in Saturday's *Times-Picayune.* Later that day, computer models concurred on the strengthening Katrina making a New Orleans–area landfall.

Emergencies were declared in Mississippi and Louisiana, and people began to make plans. Authorities activated the interstate-reversing "contraflow" plan, which had been initiated for Hurricanes Georges in 1998 and Ivan in 2004, and was now key to evacuating a million people. Many left Saturday; more left Sunday, August 28, when the system became an enormous and spectacularly well-organized Category 5 hurricane. "KATRINA TAKES AIM," screamed the *Times-Picayune*'s banner headline.

That afternoon, the S&WB's executive director, Marcia St. Martin, called the EPA in Washington to clear her decision to dewater the sewage treatment plant in the Lower Ninth Ward, discharging wastewater into the Mississippi River to prevent its release in the event of a flood. Engineers also pumped down the drainage canals, to maximize storage capacity for storm-related runoff.[10] Over in city hall, Mayor C. Ray Nagin ordered a mandatory evacuation, though many could not abide even if they wanted to, one-quarter of households being without cars. Roughly 100,000 New Orleanians remained behind, of whom around 10,000 lined up Sunday evening outside the Superdome as a refuge of last resort. Feeder bands now swirled periodically though the weirdly yellow-gray skies, rendering an atmosphere that was foreboding, bordering on funereal. This was The Big One, and everyone knew it.

In the wee hours of Monday, August 29, 2005, plummeting barometric pressure swelled tides in the Mississippi Sound, and intensifying winds pushed the surge westward through the ungated GIWW/MR-GO funnel connecting Lake Borgne with the Industrial Canal. Surge also entered Lake Pontchartrain through the never-barriered Rigolets and Chef Menteur passes, raising water levels in the three ungated drainage outfall canals as well as the Industrial Canal.

Around 4:45 a.m., gauge data indicated a small breach had opened along the western floodwall of the Industrial Canal. Perhaps it was beneath I-10, or at a railroad gate farther south (which had also failed during Hurricane Betsy forty years earlier); wherever it was, it allowed external water to enter the sunken polder, and did not bode well for the hours ahead. In the predawn darkness, Gerald Elwood, in charge of the auxiliary generator on Florida Avenue in the Upper Ninth Ward, saw water quickly coming up the driveway. His job was to produce power for Pumping Station No. 19 to discharge runoff into the Industrial Canal. This was water coming *from* the Industrial Canal. "We never got a call from Station 19 to turn on the generators," Elwood recounted. He knew something was wrong.[11]

Katrina made landfall at 6:30 a.m. over Louisiana's Barataria Basin and proceeded straight north over Empire, Buras, and Hopedale, twenty-five miles east of downtown New Orleans. Although wind speeds had decreased to Category 2 levels, the surge retained Category 5 momentum, coming in at twenty-nine feet above normal sea level off Biloxi, fourteen feet in the "funnel," and over ten feet in Lake Pontchartrain.

By 5 a.m., surge inundated nearly all of eastern St. Bernard Parish's marshes as well as the Rigolets land bridge, and pressed against the city's perimeter levees. Even river levees saw rising waters, as the Mississippi backflowed and swelled from four to sixteen feet, cresting within five feet of the levee tops. By 7 a.m., surge began overtopping the circa-1960s hurricane-protection levees in the eastern New Orleans polder, as well as the MR-GO's guide levees, by now partially disintegrating. Six miles away, by some accounts, cracks developed in the floodwall of the Seventeenth Street Outfall Canal, and lake water trickled into similarly subsided Lakeview. It was barely daylight, and already Katrina was overwhelming the federal levee system.[12]

Along the Lower Ninth Ward side of the Industrial Canal, the fourteen-foot-high surge cascaded over the floodwall and scoured the inboard side of the earthen levee. At spots by North Claiborne Avenue and Florida Avenue, the erosive cascade softened the embankment enough for the concrete-encased sheet piling to lean inwardly, opening a wedge-like crevice into which more water seeped. The peaty soils below softened, and the whole embankment translated (sloughed) inwardly, taking the floodwall with it. Sometime between 7:30 and 7:45 a.m., two breaches opened in this manner, and soon expanded into catastrophic failures, nine hundred feet long at the North Claiborne end and two hundred feet by Florida Avenue—right where the 1920s drainage siphon had been, and where the 1895 discharge trajectory had been selected.

S&WB operator Richard Reese, running Pumping Station No. 5, had noticed water coming over the railroad tracks as early as 6:10 a.m. An hour later, it entered the station floor, and soon reached the top of the discharge gates to Bayou Bienvenue—the source of the incoming surge. Reese called for the power to be cut to the pumps: if water reached the 6600-volt charge, it would short out and explode. "Everything went black," Reese recounted, "and then the panic set in." The deluge swamped Reed's desk and paperwork. Radio communications from S&WB central control instructed Reese and his crew "to do what we gotta do to save ourselves." There was only so much more vertical space left in the station shed, lest they make it to the roof exposed to howling gusts.

Then, somehow, from somewhere, a rescue boat appeared, braving waves and winds, and brought the crew to the relative safety of the Claiborne bridge, itself an island of steel surrounded by mayhem. Where did the boat come from? "God sent it," Reese answered.[13]

Seawater with up to eighteen feet of head—that is, fourteen feet from the water height and minus-four from the sunken land—rushed into the Lower Ninth Ward, while more poured in from the "funnel." By 8:30 a.m., nearly the entire Lower Ninth Ward, Arabi, Chalmette, and Meraux lay under six to twelve feet of salt water, depending on their meager topography. Residents perished by the score under harrowing conditions, while hundreds more clung to wind-swept rooftops. To the north, the New Orleans East polder suffered numerous smaller levee failures and extensive overtopping. It too filled, and its thousands of houses took on water immediately, because most were built at grade level, as if to say flooding would never happen here. It had hurricane-protection levees.

So far most surge had entered from the east, through those three navigation canals dug between 1918 and 1968. Between 7 and 8 a.m., the drama shifted to the three lakeside outfall canals, initially dug in 1872 for drainage purposes. Some water had already seeped through an unfinished two-hundred-foot gap in the Orleans Avenue Outfall Canal levee, and while it caused some flooding, it may have actually spared a breach, by relieving pressure. No such luck for the London Avenue Outfall Canal: its two breaches had not resulted from overtopping, as had happened along the Lower Ninth Ward; rather, they failed from below. At the northern rupture, intense water pressure liquefied porous layers of ancient swamp humus, and the ground translated westward, cracking open the floodwall and releasing water into the Filmore section of Gentilly. The southern breach happened similarly, from below, also resulting in lateral translation, but here the subterranean softening exposed an underground sand layer—a five-thousand-year-old former barrier island known as the Pine Island Trend, which had played a crucial role in delta formation. Now the sand billowed up to the surface in the form of bizarre white dunes burying cars and houses. Floodwaters drowned the St. Anthony section of Gentilly and soon intermixed with those incoming from the northern breach. The greater Gentilly polder went under water, taking with it Pontchartrain Park, covering over the Gentilly Boulevard ridge, and creeping up the natural levee all the way to the rear of the Faubourg Marigny.[14]

Lakeview's Seventeenth Street Outfall Canal fared worse. Like their counterparts to the east, floodwalls here cracked open as foundational soils sloughed by fifty feet, a chunk of earth on the move. Damage may have started at dawn, at a location that aligned with the historic bed of a minor bayou which once flowed into the lake. Within the errant earthen mass was a thousand-year-old gumbo of silt, sand, plasticky clay, and thick layers of gaseous organic matter from ancient swamps and marshes, as well as artificial fill from the 1915 West End reclamation. As had happened on the London Avenue Canal, steel sheet piling within the concrete I-wall (that is, a vertical steel sheet positioned in the soil like an "I," without lateral support) had not been driven sufficiently deep in the 1990s to shield this peaty mass from intense water pressure in the canal. Now it shifted wholly, opening the third-worst breach that morning. Seawater jettisoned into Lakeview, a ten-foot-high source flowing into an eight-foot-deep bowl, just as the London Avenue Canal breaches were doing to Gentilly. The two abutting deluges rose and eventually merged, swamping City Park and, after many hours, advancing southward, over the Metairie-Gentilly Ridge, into the Mid-City and Broadmoor bowls, and up the natural levee of Uptown, cresting as far inland as St. Charles Avenue.[15]

DRAMA AT THE PUMPS

At great personal risk, more than three hundred Sewerage & Water Board personnel stayed on the job round-the-clock, most of them manning pumping stations. Early on Katrina morning, winds were strong and rains were heavy, but operators were able to remove the runoff—until the "runoff" started to rise faster than the rain was falling. How could that be? Bob Moenian, operator of the recently completed pumping station at a vulnerable I-10 underpass in the Navarre area, surmised that perhaps a vessel had ruptured a levee. "I did not know that the levees were *breached*," recalled Moenian of those moments. He tried with diminishing success to pump out the rising water *into* rising water. At one point Moenian got a radio call from an employee at Pumping Station No. 5 who asked, "*Mr. Mo, the water's coming up—we're about to drown—what do we do?*" Lifejackets on, the staffers scrambled up ladders at their respective stations to surveil the situation. Clearly something catastrophic had happened, but not until Tuesday did word get to stranded employees confirming that multiple levees and floodwalls had failed utterly, and any attempt to pump the water out was Sisyphean.

Reynaldo Robertson, in charge of the pumps at Melpomene No. 1 on Broad, was forced to stop relaying water to the lift pumps in Lakeview once the Seventeenth Street Canal breached, at which point his crew climbed up the catwalks of their circa-1900 shed. After a helicopter pilot was unable to perform a risky rooftop rescue, Robertson made radio contact with his colleague at the Algiers pumping station, Richard Alexander. In mid-storm, Alexander hitched a boat to his truck and drove over the Crescent City Connection to rescue Robertson and his crew. State police stopped him, forbidding any more boats to enter the floodwaters, even after Alexander informed them of his official status. Incredulous, Alexander radioed Deputy Chief Warren Riley of the New Orleans Police Department. Over the speaker phone, Riley shouted, "This is for you, or any other policemen, or anyone else listening: The Sewerage and Water Board has full access to the city—because without the Sewerage and Water Board, we don't have a city!"[16] Minutes later, Alexander got his boat in the water and made his way to Broad Street, passing people waist-deep in water, unable to help them. All were desperate. Some were angry. One was dead.

Alexander finally reached Robertson and colleagues, and together they set out to rescue others before making their way back to the truck and the West Bank. All had remained at their posts until they absolutely had to leave. "We stayed because we knew if we left," Robertson later said, "the City of New Orleans could potentially be lost."[17] His colleague Bob Moenian and coworkers found themselves in a similar situation at the I-10 pumping station, where they were rescued by canoe, after which they set out to save others. "We lost the battle," Moenian said. "But we didn't lose the fight."[18]

East Jefferson Parish saw flooding via two sources. One was from rainwater and lakefront-levee splash-over, which together accumulated in Metairie's low spots—undrained, because Jefferson Parish President Aaron Broussard made a highly controversial decision to evacuate his pump operators for fear of their lives. What resulted was not catastrophic flooding, but enough to make a costly mess—and infuriate residents who questioned Broussard's call. Old Metairie, meanwhile, flooded as deeply as Lakeview, as the deluge emanating from the Seventeenth Street Outfall Canal breach worked its way around the canal's southern terminus and found its way into the notorious Hoey's Basin.

It was just mid-morning, and Katrina's surge had already negated a hundred years of drainage and forty years of federal levee protection. Basins that once had standing swamp water now had standing seawater, only much deeper, and with tens of thousands of flooded buildings rather than cypress trees. Katrina triggered the surge; federal levees failed to block it; subsidence from drainage trapped it; and "a conspiracy of complicity" put residents in harm's way.[19]

DRAMA AT THE POWER PLANTS

Katrina made a second landfall near the mouth of the Pearl River, where it destroyed numerous Mississippi seaside communities before moving inland and weakening. Within New Orleans, the litany of levee failures had ceased around mid-morning, though strong winds persisted into late afternoon, and the flooding only grew deeper and broader.

Many journalists, focused on Katrina's diminishing intensity and eastward landfall, mistakenly reported Monday afternoon that New Orleans had "dodged the bullet." Some local reporters covering the story from affiliate newsrooms in places like Jackson, Mississippi, thought the flooding might have come from some sort of Sewerage & Water Board failure. Only those at or near the levee failures really knew what was going on, and that it would only get worse.

After the morning breaches, the afternoon winds, and that strangely placid Monday evening, flood levels steadily rose in the bowls. Even under clear blue skies on Tuesday and halfway into Wednesday, August 31, floodwaters still kept rising. Patients at Baptist Hospital in Uptown went to bed Monday night with a dry South Claiborne Avenue (which lies at sea level); they woke up Tuesday morning surrounded by water 3.5 feet deep. Water that covered Loyola Avenue by city hall (elevation 1.8 feet above sea level) on early Tuesday morning reached St. Charles Avenue (4 feet above sea level) Wednesday afternoon.[20]

It's at that point, two days after Katrina, that the floodwaters reached the S&WB's power-generating station in Carrollton, the electrical heart of all three subsystems. "We were standing in the driveway on Eagle Street looking towards Claiborne Avenue," remembered Gabe Signorelli, chief of Facility Maintenance, "when we saw this . . . heightening of water, increasing, and then rolling up the street toward the plant."[21] The turbines had already been silenced for the first time in memory, and now water was about to drown them. "We were rushing around trying to stop it," remembered operating engineer Damon Adams. "We had sandbags up and they washed away, so we actually got a backhoe fired up and took scoops of mud and put them in place to stop the leak into the plant."[22]

Keeping the water at bay, the engineers focused on their next challenge: Would they be able to cold-start the turbines? It had never been done before, and now was a perilous time to try. All the while, pressure weakened in the water mains, and life went out of the water-distribution and sewerage system. New Orleanians who remained in unflooded houses well remember that moment on Wednesday, August 31, when their tap water dribbled dry. Life in the city, what remained of it, was now untenable.[23]

As some S&WB personnel fought to save the system, others ventured to find their stranded colleagues. "I got in a boat to rescue the folks from Station A," said emergency manager Jason Higginbotham, in reference to the sewerage pumping station by the Municipal Auditorium in Tremé. "We went down Claiborne Avenue, to go towards Orleans Avenue, [and] I saw the bodies, just bodies . . . people who are just floating, dead." When they got to Station A, the pump operator refused to leave, for his dedication and also because he had arranged for his wife to meet him there as a last refuge if their Broad Street home flooded. It took much cajoling from Higginbotham and others to convince the operator to leave, for the good of everyone involved. Only later did the operator learn his wife had already perished in their home, after which he suffered a severe nervous breakdown.[24]

Elsewhere, everywhere, society seemed to be coming apart at the seams. "The unlawfulness," remembered Higginbotham, shaking his head. "People want to push that aside, but it happened. . . . It was chaos[;] at nighttime you can hear the gunshots."[25] S&WB workers manning the lakeside pumps in New Orleans East ran their trucks up and down Hayne Boulevard, rescuing people out of Little Woods, only to endure some individuals "taking shots at them while they were doing it [because] a lot of people blamed us for the flooding."[26]

In one tense moment at the S&WB water plant across the river, some residents of Algiers, which did not flood but lacked power and water service, converged on a FEMA truck bringing food and bottled water to the plant operators. Some brandished firearms, and violence nearly erupted over the desperately needed rations. Officials defused the situation by letting residents clean out the truck. "It hit home" just how bad the situation was, said Higginbotham, "when one of our plants was under attack."[27]

The bridge itself became a beacon of hope for desperate flood victims trying to flee the East Bank, even at night, drawn through a sea of darkness by the lights at the Algiers S&WB water plant. Most of the weary walkers were turned away by police, sometimes aggressively, but the lights still beckoned. "It looked like Oz; it had *power,*" said Robert Jackson, chief of communications at the S&WB. "Like the zombie stories, everyone goes to where the lights are. . . . During most of that period, the only lights you saw in the city were at the water plant in Carrollton and the water plant on the West Bank," where S&WB engineers were able to fire up the diesel-powered turbines.[28] In New Orleans's darkest moments, the S&WB kept alive the flickering light of hope.

"NO WATER, NO SEWER, NO DRAINAGE, NO CITY"

How could floodwaters continue to rise so long after a storm? How could they crest this high above normal sea level? The answer is threefold: drainage-caused subsidence, subsidence-necessitated levees, and inadequate levee engineering.

When a tropical storm dissipates inland, its surge typically retreats to normal sea level. But when a strong hurricane like Katrina pushes a huge surge into a labyrinthine delta, its retreat is slowed by ridges and marshes amongst wending bayous and shore-lined bays. Add to this the levees, floodwalls, railroad track beds, highways, houses, and other impedances of the built environment, and the backflow is further retained. Now add a series of giant topographic bowls, and surround them with a punctured brim, and the outgoing surge will continue to pour through the breached levees and into the lowest bowls by gravity alone, no storm energy needed.

Regarding the failures on the London Avenue and Seventeenth Street drainage outfall canals, "all three of these breaches rapidly scoured to depths well below mean sea level, so they continued to transmit water into the main Orleans East Bank [polder] for three days after the initial peak storm surge subsided."[29] Only when the swollen tide drained out sufficiently on Wednesday did external water levels finally drop below the height of the bottom of the breaches. For one bizarre moment earlier that day, floodwaters in the Lower Ninth Ward actually flowed back out the breaches—which is to say, their height surpassed that of global sea level. Once the Queen of the South, predicted to become the greatest city in the hemisphere, New Orleans now stewed its own watery plateau of filth.

At the unflooded S&WB headquarters on St. Joseph Street in the Central Business District, Executive Director Marcia St. Martin and General Superintendent Joseph Sullivan conferred over what to do. Their drainage system was now useless; even if the pumps could be activated, they would only eject water right back into the flooded city. This could only change if and when external water dropped below internal water levels. "Joseph Sullivan had calculated how long it was going to take for the water to equalize between the lake and city for the water to stop coming in," recounted St. Martin. "He projected it was going to equalize that Friday afternoon around 2 o'clock," four days out. "He was dead on the money," St. Martin marveled. "He did that on Monday afternoon probably around 3 o'clock. Brilliant engineer. Just brilliant." Not knowing whether the floodwaters might surpass St. Charles Avenue, St. Martin decided on August 31 to evacuate the S&WB headquarters and take up command at the Algiers water treatment plant, which had electrical power, running water, transportation access—and armed military protection.[30]

That was on the West Bank, which had its own problems. The fatal flooding, however, was entirely on the East Bank, and that's where electrical power was most desperately needed. It was at this point that the S&WB main power plant on Eagle Street in Carrollton became the single most important place to save New Orleans.

Its workers had had quite a week. They battled a bad electrical fire at Turbine No. 3 during the storm on Monday, fought to keep floodwaters at bay Tuesday and Wednesday, and later frantically loaded sandbags onto helicopters to help plug the breaches, while securing precious diesel fuel and cement in Algiers to build a provisional dam to save the Carrollton plant from flooding.

Now fueled up and (barely) dry, plant engineers prepared to do something never before attempted: cold-starting the turbines. "So many components have to come on and stay on, and dovetail into another operation, and have to fall in place just right," explained operating engineer Damon Adams. "It took four tries and three days," each attempt requiring the refilling of the boilers, reheating of water, releasing of steam, and rolling of the turbines. Finally, "on the fourth try it stuck, and we got it to hold, [and] we got it all right." Steam billowed out of the powerhouse stacks, the turbines roared to life, and the three critical century-old systems regained their electrical current. It was a historic moment, but no one present had time to reflect, and few elsewhere would ever recognize it.[31]

At the auxiliary power station in the completely flooded Upper Ninth Ward, Gerald R. Elwood never got the call to fire up his turbines from nearby Pumping Station No. 19. Looking out for his colleagues from the third floor of his Florida Avenue plant, all he could see was water, water, water. "The only evidence of life," he recounted, "was myself and my coworkers." The next day, some bedraggled men in life preservers "came up the stairs [and] knocked on the door." They were the pump operators, and they had just swum through deep filth strewn with debris. "We were *happy* to see them," Elwood said. "[Now] let's try to get the generators going." As in Carrollton, dozens of preliminary steps first had to be taken, including solving a problem with the air compressors. "At 2 or 3 o'clock in the morning, we were lifting 55-gallon drums putting oil into the generators. . . . We *had* to get the generators up. We *got* to get the water out."[32] They succeeded, on both accounts, at both generating stations, downtown and uptown. "That was the biggest challenge," said Adams of those dramatic moments. From those plants on Eagle Street and Florida Avenue, the heartbeat of hope pumped back to life. "Without the Sewerage and Water Board, *there is no city*," said Adams, "because without water, there is no society[;] there is anarchy, pestilence, disease. . . . It *had* to come back."[33]

Soon, some three hundred essential S&WB personnel who remained on the job went from deadly crisis management to round-the-clock operations, while also coordinating their own family predicaments. Eighty percent of the agency's 1,200 employees lost their homes to the flood, including Executive Director Marcia St. Martin, and because most were native-born New Orleanians, most had extended families who were also flood victims.[34] Everyone thinks of police, fire, and military personnel as "first responders," noted Robert Jackson, communications director of the S&WB, but "in fact, we're 'zero responders.' We never left. . . . We never stopped working."

Few would ever credit S&WB workers for being the first people to save New Orleans, but that's exactly what they were. They were the latest, truest drainage kings—courageous heroes in the centuries-old saga of the dewatering of New Orleans, warts and all. "It's an untold story because we're practically an invisible force," said Jackson. "We're all underground; everything we have is underground. [But] no matter how many billions you put into a city—you can put gold roofs on every house—if you don't have sewer, water, and drainage, you have to leave. . . . No water, no sewer, no drainage, no city."[35]

A RECKONING

As forensic investigators pieced together exactly what happened, New Orleanians came to comprehend a terrible new truth. The Katrina deluge turned out to be not the inevitable outcome of a noble system overwhelmed by a superlative storm, but rather a scandalous failure of a piecemeal amalgam of inadequate, under-engineered, obsolete, poorly inspected, and underfunded levees and floodwalls. They were all federal charges, as stipulated by Congress first in 1955 and more so in 1965, and once the facts became indisputable, the US Army Corps of Engineers assumed full responsibility.

The subsequent outrage and derision of the Army Corps soon became practically part of New Orleans culture. Activists committed to keeping the pressure on corps officials have since adopted a reductionist narrative of the 2005 catastrophe, insisting that it be called "the Federal Flood," dismissing Katrina itself as a no-big-deal storm, and bristling at any mention of local complicity or culpability—despite the resistance of local agencies to gates on the outfall canals, environmentalists' opposition to the Barrier Plan, and the Port of New Orleans's advocacy for the Industrial Canal, GIWW, and MR-GO. Understandably, the passionate public discourse following Katrina focused on federal levees and floodwalls, as their failure was indisputably the proximate cause of the deluge. Better levees became a civic mantra, and "MAKE LEVEES NOT WAR" and "CATEGORY-5 LEVEES NOW" became popular bumper stickers.

What was harder to sloganize was the ultimate cause of the 2005 catastrophe, which was its hundred-year backstory—that is, how flood protection became ever more challenging vis-à-vis an expanding human domain upon a deteriorating delta.

How the manmade levees along the river had deprived the delta of fresh water and sediment.

How canals dug for navigation and petroleum extraction had led to the intrusion of seawater.

How fossil-fuel combustion had affected global temperatures, sea levels, and climate patterns.

And how municipal drainage triggered soil subsidence, lured development into low places, exposed people to mounting hazard, and helped turn a twelve-hour storm into a month-long flood.

According to the most thorough scientific investigation of the Katrina fiasco, "approximately 80% to 90% of [the flooding] came from the three catastrophic failures *along the drainage canals* [and] resulted in approximately half of the 1,293 deaths attributed (to date)."[36] Yes, failed levees and floodwalls actually let the water in, and had they held, the drainage system would have won the day—or days. "The aggregate pump capacity," wrote the investigators, "could have cleared the city of flood waters in less than three days if the levees had simply been overtopped without failing."[37] But because the levees and floodwalls did fail, the subsidence that drainage had caused made the subsequent deluge much worse, by trapping the waters in human-made, human-occupied bowls.

Those failures made the world's toughest urban drainage problem immeasurably tougher, and nearly vanquished the world's second-largest drainage system—until a heroic team of S&WB engineers turned the tide at the turbines. Diesel burned, water boiled, steam formed, generators rolled, electricity flowed—all 25 cycles of it—and the great Wood screw pumps and the Sullivan pumps churned to life, sucking up floodwaters and discharging them through patched outfall canals. Incredibly, the engineers and operators managed to remove most of the water from the city's main polders in just eleven days, far faster than the ninety days many predicted.[38] Helping drain other areas was the Army

Corps of Engineers' Task Force Unwatering, which brought in truck-mounted pumps to remove a quarter of a trillion gallons of floodwater over fifty-three days, into late October 2005. In the interim, the powerful Hurricane Rita swept into Louisiana and mercilessly reopened the Ninth Ward breaches, as if to hammer home the searing lesson that, as one critic put it, "New Orleans was a beautiful machine that was left to rust."[39]

Water destroyed nearly everything it touched in thousands of houses and offices, but it only soiled or slightly damaged the steel pumps inside the sturdy S&WB sheds. Electrical equipment was a different story, but even then, skilled electricians swiftly rewound motors and repaired consoles. "The beauty of our 25-cycle system is that all this stuff can be done in place," said facility maintenance chief Gabe Signorelli. "You can pull the coils off the motor, send it out, get it rewound, bolt it back on, and then tie it into the next one. . . . All of our old equipment[,] our 1900-vintage pumps? All we had to do was drain the oil out of them and clean the reservoirs out, put the oil in, get the motors finished up . . . and put them back in service."[40]

The obsolescence of 25-cycle electricity, the antiquity of the machinery, and the sheer idiosyncrasy of New Orleans's uphill hydrology meant that expertise to re-dewater the flooded city had to come from in-house. New Orleans's infrastructure was so thoroughly indigenous in its ingenuity that no one else on Earth could fathom it. "You can't just hire a technician to fool with 25 cycle; they have to be trained on it," said Richard Reese, who had manned the pumps in the Lower Ninth Ward until he had to be rescued by boat. "It's unique to the Water Board"; [we've] had this since the turn of the century." With the nation's toughest recovery facing the world's toughest drainage problem, one can understand how some S&WB operators felt overwhelmed. "Where do I start? Who do I contact?" Reese pondered. "How do I get these machines repaired?"[41]

Assisted by the National Guard, FEMA, the Army Corps, and other recovery workers, S&WB leaders devised a plan to assess the damage, repair what they could, restore services to the least-damaged areas first, and work outwardly. They succeeded in getting tap water and sewerage services back to unflooded neighborhoods within a few weeks. "When people returned to their homes in October," said former Executive Director Marcia St. Martin, "they had drinking water. They had wastewater."[42] S&WB workers next tackled repairs in neighborhoods that had been lightly flooded, and then those deeply flooded, testing water quality before giving the greenlight. In steady increments, taps flowed, toilets flushed, and runoff got removed, tenuously, imperfectly, sometimes under flickering lights. The people of the S&WB brought post-Katrina New Orleans back to life, despite enduring $300 million in infrastructure damages and incalculable personal loss.[43]

STATUS QUO ANTE WINS THE DAY

In impassioned public meetings of late 2005 and early 2006, citizen-activists and professional planners alike invoked a certain metaphor to reconcile the tragedies of the past with the promises of the future. Katrina, they said, had *wiped the slate clean.* Yes, it had destroyed so much that was good, and we lament and mourn that loss. But, the argument went, Katrina also subverted much that was bad, and thus created a rare opportunity to solve old problems, rebuild sustainably, diversify the economy, and *make it right.* So uplifting was this lemons-to-lemonade spirit that "Make It Right" became the name of actor Brad Pitt's nonprofit missioned to rebuild the Lower Ninth Ward sustainably and justly, wiping the slate clean of history's mess.

But Katrina had not wiped the slate clean, and even the Make It Right Foundation fell short of its eponymous mission. Probably the only disaster in New Orleans history that *did* "clear the slate," figuratively speaking, was its first hurricane. That 1722 storm had obliterated four years of slapdash construction and enabled our first drainage king, Adrian de Pauger, to proceed with orderly urbanization in the form of today's well-designed French Quarter. Prior place-making, from 1718 to 1722, was simply too brief and desultory for residents to inscribe any real value into the soil, fiscal or structural or emotional. A few dozen wrecked hovels were hardly worth rebuilding identically. The promise of a better future trumped reversion to a mediocre past.

Contrast the 1722 hurricane with Katrina in 2005, by which time nearly three centuries of prior investment had created a deep wellspring of place-based value. Some of it was socioeconomic, in the forms of real estate equity, businesses, jobs, institutions, infrastructure, and architecture. The rest was cultural, in the form of social relations, historical memory, and the psychological attachments that geographer Yi-Fu Tuan described as "topophilia"—love of place. In the rustic outpost of 1722, a single morning of gusting rain cleared the way for a radical urban overhaul, a true opportunity to make things right. But in 2005, ten feet of seawater could not wash away this 287-year-old repository of civic wealth. Once those fatal waters were pumped out, that largesse not only regained its value; it became *priceless*. What ensued was inspirational to behold, as hundreds of thousands of New Orleanians fought for their city, rebuilt their neighborhoods, and reclaimed their lives.

That proclivity to return to normalcy, however, tapped into a reactionary impulse, the type that fears change, suspects "outside" experts, casts opportunity as opportunism, and questions all motives except its own. Topophilia can also mean inertia. Fighting to rebuild every wrecked and sunken neighborhood, however beloved, can also mean slouching down a path of least resistance. Defiance in the face of adversity can also betray a deficit of wisdom, and end up in a dubious victory of *pathos* over *ethos* and *logos*.

As a result of these as well as pragmatic factors—like the lack of legal authority and hard cash needed to carry out thousands of expropriations—all the loftiest rebuilding visions fizzled, the grandest recovery plans flopped, and the *status quo ante* won the day. The Louisiana Road Home Program, funded by the federal government and administrated by the state, officially sanctioned those who wanted to rebuild in place, and thus essentially delegated the recovery to flooded homeowners themselves. With the exception of those in the Lower Ninth Ward, the vast majority decided to rebuild in place. No neighborhoods, no matter how damaged, sunken, or depopulated, were closed after Katrina.

It's hard to imagine Pauger or Bienville allowing colonists to decide how to rebuild *La Nouvelle-Orléans* in 1722. Certainly the French Quarter would look different today if they had done so, had it survived at all. But then again, Bienville and Pauger were imperialists making dictatorial decisions on behalf of an absolutist monarchy, and one of their greatest powers was to decide when to "wipe the slate clean" and how to "make it right." Democracies do things differently, and American democracy, for better or worse, tends to do things from the bottom-up. For years to come, countless commentators offered widely varying judgments of the Katrina recovery, some seeing it as a triumph of people and culture, others an abrogation of leadership and expertise, still others an injustice redolent of systemic racism and a cry for centralized planning. If contested narratives and multiple truths ever had a name, it is Katrina—or rather, *the Federal Flood.*

And so neighborhoods returned largely as they were before Katrina, only with fewer people, more empty lots, higher municipal costs per resident, and fewer overall opportunities, while the higher-elevation historical areas saw intensified gentrification and inflated real estate prices. The main "learning" from

Katrina took the form of stricter building codes calling for houses to be raised three feet above the grade or base flood elevation, whichever is higher. Private home-builders also saw a rise in demand for traditional pitched roofs, particularly hipped in shape and with multiple rooflines, in part for their historical look but also to shed rainfall and buffet winds.[44] That made things slightly safer for those who resided in those houses, but did little for the rest of the community.

The burden of reducing that risk—to the magic number of 1 percent, the minimal yearly chance of flooding needed for homeowners to qualify for the National Flood Insurance Program, and thus gain access to the mortgage market—got placed in the hands of the US Army Corps of Engineers, the very federal department responsible for the breaches. No one else could do it. Critics spent the next few years simultaneously bashing the Army Corps and entrusting their lives to it.

THE HURRICANE AND STORM DAMAGE RISK
REDUCTION SYSTEM

The Army Corps delivered, at least to the extent of its allocated resources. During 2006 to the early 2010s, the department pulled off "one of the largest civil works projects ever undertaken," accomplishing everything it had failed to do—or had been stymied from doing—the previous fifty years.[45]

After the dewatering ended in October 2005, the corps's Task Force Guardian set about repairing the ruptures and other weak spots (there were twelve just on the three outfall canals) to prepare the city for the 2006 hurricane season.[46] In the meanwhile, the state legislature disbanded or disempowered local levee boards—including the Orleans Levee Board, which was relieved of the levee-specific duties it had had since 1890—and replaced them with two new Southeastern Louisiana Flood Protection Authorities, one for each bank of the river. Those entities partnered with corps engineers and contractors to install, using the speedy "design/build" contracting approach, a true *system* of flood defenses, one that was holistically envisioned, held to the same standards, and maintained and operated as an integral entity.

Engineered for a storm magnitude having a 1 percent chance of occurring in any given year (Katrina was deemed a 0.25 percent storm, specifically once in 396 years), the new system had three components: the Lake Pontchartrain and Vicinity Project on the East Bank, the West Bank and Vicinity Project across the river, and the New Orleans to Venice Project for lower-river communities. Realizing it now ought to avoid the word "protection" unless it could legally stand behind it, the corps officially named the project the Hurricane and Storm Damage *Risk Reduction System,* or HSDRRS.[47] Importantly, "the entire $14.6 billion" of the HSDRRS budget was made "available at the beginning of the project, mark[ing] a dramatic change from the previous 40 years, which saw individual levees and floodwall projects built in fits and starts."[48] No wonder the prior system was piecemeal and irregular. It was funded that way.

As work got underway, contractors trucked into five Louisiana parishes ninety-three million cubic yards of clay from as far away as coastal Mississippi, and engineers designed and ordered specialized components from manufacturers nationwide. Into the early 2010s, workers constructed 350 miles of heightened levees and stronger floodwalls, fifty-six flood gates, a massive storm-surge barrier, nine new drainage pumps, and seventy-three nonfederal pumping stations, including those repaired and storm-proofed.[49]

Spatially, the vast majority of the HSDRRS effort comprised levee improvements, most of them also featuring concrete floodwalls built with deeper and sturdier steel T-beams, rather than shallow and flimsier I-beams. Financially, the effort focused on three key components: the Outfall Canal Closure Projects, the West Closure Complex, and the Inner Harbor Navigation Canal–Lake Borgne Surge Barrier. Each aimed to mitigate vulnerabilities initiated or completed in 1871, 1895, 1915, 1943, and 1968.

First priority was that triple hobgoblin of the ungated outfall canals. For 135 years, the Seventeenth Street, Orleans Avenue, and London Avenue drainage canals had allowed high water in Lake Pontchartrain to intrude inland, creating a threat as the surrounding lands sunk in elevation and rose in population. Infighting between the Orleans Parish Levee Board and the Sewerage & Water Board led to a stalemate over the question of gating their mouths, which forced the Army Corps of Engineers to resort to the alternative of raising the levees and floodwalls. After they breached during Katrina, *everyone* became pro-gate, and those local powers which had once been anti-gate kept nervously silent as public outrage flowed almost exclusively to the Army Corps over the failed floodwalls.

Working speedily, the corps and its contractors spent $400 million installing three Interim Closure Structures (ICS) in time for the 2006 hurricane season. Each temporary device comprised closable gates fronting two banks of pumps, with bypass pipes snaking around the gates connecting canal water to lake water. The gates served to prevent lake surge from entering the city, and the pumps and pipes served to eject cityside runoff otherwise blocked by the closed gates. Capacity varied with catchment size: the Seventeenth Street Canal ICS, which drained as far away as Uptown and Old Metairie, could remove up to 9,200 cubic feet per second via eighteen hydraulic pumps, eleven direct-drive pumps, and fourteen bridge pumps. The Orleans Avenue ICS's ten hydraulic pumps could remove 2,200 cfs from Lakeview and Mid-City, and the London Avenue Canal ICS's twelve hydraulic pumps and eight direct-drive pumps handled 5,200 cfs from throughout greater Gentilly and the lower wards.[50]

In 2013, the department let contracts worth $615 million to replace the Interim Closure Structures with Permanent Canal Closures & Pumps (PCCPs), which were "permanent gated storm surge barriers and brick façade pump stations . . . equipped with a stand-alone emergency power supply capacity so that it can operate independently of any publically provided utility." Total capacity of the Seventeenth Street Canal PCCP would be 12,600 cfs; Orleans Avenue, 2,700 cfs; and London Avenue, 9,000 cfs. "Constructed to withstand 200 mph winds at three-second gusts and 155 m.p.h. sustained winds, [each PCCP] includes a control building with safe housing for support staff and onsite fuel storage capacity . . . for five days of continuous operation."[51] Described as "the most complicated pump stations in Louisiana," the three PCCPs were completed in late 2017. A year later, the three local sponsors of the HSDRRS—that is, the Coastal Protection and Restoration Authority, the Southeastern Louisiana Flood Protection Authority–East (SLFPA-E), and the New Orleans S&WB—agreed the SLFPA-E would be responsible for operation and maintenance of the three PCCPs, and that the S&WB would pay half the estimated annual costs of $4 million. After nearly 150 years, the three outfall canals were finally closable.[52]

What the PCCPs did to mitigate the damages of manmade drainage, other HSDRRD components did for the harm wrought by artificial navigation canals. At the far eastern flank of the metropolis, the Inner Harbor Navigation Canal–Lake Borgne Surge Barrier aimed to barricade the "funnel" formed by the GIWW and MR-GO shipping channels, completed in 1943 and 1968 respectively. Unique in the nation and evocative of the Netherlands, this $1.1 billion mega-barrier is made of 1,271 high-density hollow high-density concrete pilings, each 5.5 feet wide and 144 feet in length. Floated in on barges in 2009, each piling was lifted vertically and hammered by a steam-operated pile-driver 130 feet into the subal-luvial surface, then augured out and filled with concrete to 80 feet down. Spanning 1.8 miles across the Bayou Bienvenue marsh, the 1,271 "soldier pilings" were then topped with a concrete pediment bringing their height to 26 feet above mean sea level, and reinforced with angled "batter pilings" for extra strength against incoming surges. Closeable sector gates were installed at the GIWW for the passage of shallow-draft barges, and a smaller vertical-lift gate was put at Bayou Bienvenue for fishing and recreational vessels. "The largest design-build civil works project in the history of the Corps," the IHNC–Lake Borgne Surge Barrier was completed in 2011, served well against Hurricane Isaac's surge in 2012, and has since been turned over to the SLFPA-E for operation and maintenance.[53] As for the MR-GO, it would get no gate. The ill-conceived seaway, opened with fanfare in 1968 and ignominiously closed in 2008, was physically blocked by the surge barrier, as well as by a rock barricade at Bayou La Loutre further east. Welcome as they are, these two slender sutures are dwarfed by the gigantic scar on the surface of the planet that the MR-GO will forever remain— greater New Orleans's greatest mistake.

Seventeen miles southwest of the surge barrier, across the Mississippi, is the equally remarkable West Closure Complex, located where the Harvey and Algiers canals fork off from Bayou Barataria. All three of these waterways also host the GIWW, and their juxtaposition with respect to the Barataria Basin makes them roughly the West Bank's equivalent of the East Bank's "funnel." The West Closure Complex entails a 530-foot-long steel shed housing eleven 5,000-horsepower pumps capable of ejecting 19,000 cubic feet of stormwater runoff per second—the world's largest pumping station, costing $1.1 billion, same price as the surge barrier. These "flowerpot pumps," named for their tapered shapes, are powered by four diesel generators and fronted by debris-removing raking devices, and are positioned astride a 225-foot-wide channel with two sector gates—themselves the largest in North America. Like those at the surge barrier, these gates remain open for the routine passage of barges and other vessels, and swing closed when a storm nears.[54] They are located only a mile and a half from the site where, on that festive day in February 1915, George Hero first fired up his pumps to drain the West Bank swamps. In doing so, the Drainage King had also unwittingly triggered a sequence of decisions and developments that, directly and indirectly, led to the need for the world's largest pumping station. Contractors began work on the West Closure Complex in 2006 and finished in 2011, at which point the Army Corps turned over maintenance and operation to the Southeastern Louisiana Flood Protection Authority–West.[55]

THE POSTDILUVIAN DRAINAGE SYSTEM

The political hurricanes that followed Katrina resulted in an administrative up-heaval at the S&WB, not to mention the Army Corps of Engineers and Orleans Parish Levee Board. Since its inception, the S&WB has had an unusual place in Louisiana government. It was created by state law but operates as a city agency, yet it is not a city department, nor does it fall under a department. It is "one of 10 'unattached' boards and commissions placed under the executive branch by New Orleans's Home Rule Charter, [and] both the city and the state have some amount of control over the agency's powers."[56]

After Katrina, both city and state had much to say about that control, and changes were made—lots of them. "The Board [of Directors] has changed, the governance structure [has changed], and the executive director has changed," said S&WB's executive director, Cedric Grant, on the tenth anniversary of Ka-trina. "That's only happened maybe twice in the 116-year history of the Board."[57] Today, governance of the S&WB comes from a board of directors comprising the mayor of New Orleans, two members of the city's board of liquidation, one city council member, and seven members of the public selected from the five coun-cil districts, along with two customer advocates.[58] A subsequent charter change made the position of executive director accountable for day-to-day operations as well as higher-level leadership—one person, connecting all the dots, coordinat-ing all the projects, singularly answerable to the board of directors, the mayor, and the city council, which has regulatory powers. Rate increases followed the ministerial changes, and along with federal funding from FEMA and the corps's SELA project, the S&WB embarked on major drainage improvements.

Whereas before Katrina the priority was operations, now the S&WB focused additionally on emergency planning and "future proofing"—that is, building re-silient and redundant systems for uncertain times ahead. "We're doing more work in New Orleans now than we've done in a generation," said Grant of some fifty different infrastructure projects. Among them were repairs to every pump-ing station on the East Bank; raising critical infrastructure above base flood el-evation; $150 million in upgrades to internally generated electrical power and the unique-in-the-nation underground power grid that gets electricity to the pumps; and new backup power sources from Entergy to each station, should the in-house system fail. A subsequent S&WB general superintendent, civil engineer and longtime levee authority Bob Turner, continued the shift toward Entergy-generated electricity, and discussions over national infrastructure spending, which included a visit by President Joe Biden to the Carrollton plant in May 2021, spoke favorably of a full switch to municipal power.[59] But when Hurricane Ida struck on August 29, 2021, and wrecked Entergy's transmission system, it was the S&WB's in-house generators that saved the day for New Orleans, keeping water flowing, toilets flushing, and just enough drainage capacity to prevent most stormwater flooding—a reminder of the value of internal redundancy.

Recognizing that it needed its own emergency operations center, rather than piggybacking off the city's, the S&WB also built a command center at the Carrollton plant to centralize all communications, decision-making, and dispatching, while augmenting the capacity and resources at the Central Yard on Peoples Avenue. There were also corps-funded upgrades directly to the drainage system, including $23.8 million of improvements for Pumping Station No. 5 in the Lower Ninth Ward, $15.9 million for the storm-proofing of pump stations nos. 1, 2, 3, 4 and at I-10, and the continuation of the circa-1996 SELA project to augment canal capacity on four Uptown avenues plus Florida Avenue, costing well over a quarter of a billion dollars. "It's been a quantum leap from 2010 to now," said Gerald R. Elwood of the progress since Katrina.[60]

One would be hard-pressed to find more than a handful of generous New Orleanians who agreed with that assessment. Grant, Elwood, and others would be the first to acknowledge the S&WB has something of a brand problem; indeed, it probably has the most fraught reputation of any entity in local government. The rebukes seem to transcend all divisions in local society, emanating from downtown merchants and uptown doyennes, from Gentilly pastors and Bywater hipsters. Their grievances invoke problems ranging from ruptured water mains to flash flooding, from erroneous bills to boil-water advisories, from deep potholes to bloated pensions.

Thought seemingly antithetical, the criticism and the achievements of the S&WB are not hard to reconcile. On the critical side are 400,000 users with a rightful expectation of having their utility needs delivered reliably. After all, these are services for which they pay steep monthly user fees (for water and sewerage), and a tax millage for drainage (borne by only 57 percent of property holders, nonprofits exempted; all other S&WB funding coming from bonds). On the laudatory side is nothing short of the world's toughest urban drainage problem, resolved when civic perseverance and local ingenuity pulled off an engineering miracle.

But then New Orleans fell on hard times. The economy stagnated; populations fled; revenue shriveled; and civic spirit flattened. Infrastructure rusted; equipment aged in place; and standards and technologies changed. The world moved on to satellite-based measurements to calibrate elevation data, and the S&WB still uses the 1879 Cairo Datum. The nation moved to 60-cycle electrical frequency, and the S&WB remains on century-old 25-cycle. The world moved to digital controls, and the S&WB still uses analog knobs and gauges.

Other critiques were rooted in nineteenth-century understandings of the environment. When the SW&B was established in 1899, few perceived the urban landscape—that is, the city's lithosphere, hydrosphere, and built environment—as integrated domains of coupled natural and human systems. The S&WB was thus never granted control over watershed management, though it is very much responsible for managing waters shed therein. It has no ownership of groundwater, though its task gets harder when groundwater is lowered. It had no legacy of urban planning, zoning, and land use—all that went to the City Planning Commission—yet it must get water in and out of every used parcel. And so the S&WB focused instead on daily operations of three utilities handled separately—pluvial drainage, potable water treatment and distribution, and sewage removal—despite their interrelations. All three use the same generated electricity to push and raise water, with intakes and discharges into local water bodies—the physics of which will all fluctuate as climate changes, affecting rainfall patterns, soil salinity, and sea levels. All of that needs to be planned for, by an agency with no heritage or authority for master planning.

Still other critiques are well warranted. The same pension-based civil service system that retains dedicated employees and breeds local expertise can also nurture lethargy and indifference (and repel brilliant young minds who, for some unfathomable reason, might *not* want to spend their entire careers in New Orleans). The same separation of drainage assets between the S&WB and the Department of Public Works that remedied the 1991 defeat of the millage renewal also led to confusion over things such as clogged catch basins and broken conveyance systems. The S&WB's inability to issue consistently accurate bills seems to be a problem not shared by counterparts in other cities, and attempts to secure outside help are too often met with resistance from career employees fearful of change.

It was the "date storms" that turned everyday criticism into actionable crisis. These intense downpours—"rain bombs," some people call them—seemed to be occurring more frequently, and the micro-deluges they caused seemed to get worse despite the post-Katrina upgrades. It wasn't just an impression; research on precipitation patterns in New Orleans during 1960–2017 showed that rainfall, while holding steady in terms of annual magnitude, occurred in shorter and more intense bursts, and that extreme storms now dumped more rain—with "'huge' implications [for] urban drainage systems, like New Orleans's pump stations[,] built around a 24-hour estimation of rainfall."[61] Worst-hit spots saw rainfall far exceeding the vaunted "1 inch in the first hour, half-inch per hour thereafter" understanding of what New Orleans's drainage system could remove. But could it? Even areas getting within that range were seeing street accumulations. "Everyone from mayors to meteorologists has parroted that capacity assessment," reported investigative journalist Katie Moore, "but . . . that old adage may be nothing more than a myth"—else a pithy guesstimate that gained traction through sheer repetition.[62] True capacities varied widely, as some polders retained or absorbed more rainfall than others, and all bets were off if a generator failed or a pump went offline.

THE AUGUST 5 FIASCO

The worst of the date storms, May 8 in 1995, served as the benchmark for what happened on August 5, 2017. The steamy Saturday began with the usual chance of afternoon showers, as folks made their way to the popular Louis Armstrong Festival in the French Quarter. It ended with a "no-notice rain event" later deemed to be "once in a century." Starting around 2 p.m., a thunderstorm moved over the metropolis and dropped normal precipitation. But as it stalled over the heart of the city, the system began regenerating in place, its rainfall intensifying without moving. At 3:17, dispatchers started to field 911 calls from motorists caught in floodwaters. At 3:30, calls came in from people trapped in buildings. At 3:45, the National Weather Service issued a Flood Advisory, and upped it to a Flood Warning at 4:02, by which time the city's Emergency Operations Center went into action.

Cars flooded by the thousands, and homes and businesses by the hundreds. Motorists got stranded for hours, for the rapidity of the rise and lack of warning about impassable streets. Police and fire services responded to over two hundred calls for emergencies. Two, three, even four feet of water accumulated in parts of Mid-City, Lakeview, Gentilly, and the Seventh Ward—so deep that social-media users confused real-time photos with those dating from Hurricane Katrina. Even some streets in the French Quarter, among the highest in town, saw a foot of water or more. A total of 6 to 7 inches of rain had fallen over the heart of the East Bank within a few hours, and over 9.7 inches fell in Mid-City within three hours—12 percent of annual precipitation in 0.034 percent of the year.[63]

Just two weeks earlier, on July 22, another flash flood had caused its share of mayhem, raising questions about S&WB operations. Were all the pumps activated? Were all generators online? The August 5 storm, worst pluvial deluge since 1995, further stoked such suspicions, as some residents reported seeing water rising well after the rain had stopped, and others heard only silence at certain stations. Twice in a month, moviegoers departing the Broad Street Theater, located only five hundred feet from Pumping Station No. 2, had to wade through a swamped lobby—in a building that had last flooded during Katrina. Ditto for the nearby headquarters of the Zulu Social Aid and Pleasure Club.[64]

Something certainly seemed wrong. But during a press conference outside city hall on the evening of August 5, S&WB Executive Director Cedric Grant assured reporters that "all the [pumping] stations are being manned and operational." Colleagues concurred, as did other agency leaders in subsequent media exchanges. Those repeated assurances, and the fourteen hours it took to finally discharge the floodwaters, suggested that maybe this was just a case of nature winning a round—a fully operating system overwhelmed by a freak event.[65]

The complicated truth began to surface at an excruciating three-hour special city council meeting held on Tuesday, August 8. In his testimony, Cedric Grant acknowledged that he had since learned that "some pumps were not operational during the weather event, and that there were some power generation issues that impacted our ability to fight the flood at the highest capacity. [This] conflicts with information I was given to provide to the public[;] our staff was not forthright, which is unacceptable." To stunned council members and a jeering audience, Grant announced his retirement on the spot, after six years on the job and forty years in utility management.

Next to testify was General Superintendent Joseph Becker, an accomplished engineer with thirty years at S&WB. Becker said he now understood that 8 of the 121 pumps were malfunctioning, yet nevertheless asserted to the council that "all of the pump stations were working at their maximum capacity." Parsing his words to puzzled council members, he iterated, "all of the pump stations were working at the capacity they had available to them." When Councilman James

Gray pressed for the hidden meaning, the engineer admitted that the Lakeview stations were down 37 to 43 percent of their capacity. "How did you decide to tell the public that the system was fully operational," asked Gray, "when you knew that you were 43 percent short at that location?" "What I was trying to say," Becker replied, "was that we used the *available* pumping capacity to its fullest extent." Gray completed the sentence: "The available pumping capacity *meaning only 57 percent intended for that location.*" Further questioning brought out that turbine problems at the generating plant had actually reduced that figure of 57 percent down to 52 percent.

The only thing now going down the drain was the S&WB's credibility. "I started this meeting," sighed Gray, "actually feeling *bad* for Mr. Grant, because I thought people were taking little bits of information out of context and reaching the wrong conclusion[.] I thought we were on a witch hunt. *But then we found witches.*"

The city's worst flood since Katrina was about to incur the biggest administrative shakeup since 2005, only this time, S&WB staffers were the goats rather than the unsung heroes. Grant, Becker, and the agency's chief spokeswoman, dedicated civil servants with years of experience, were all forced into resignation.[66]

Later investigations confirmed that key pumps had been off-line at two stations in the worst-hit polders on August 5, and that sixteen pumps had not been operating elsewhere that day. Additionally, three of the five turbines had not been available during the storms of July 22, August 5, and another downpour on August 8—same day as the city council special meeting. The malfunctioning assets together "relegated pumping capacity to 45–70% of design capabilities during these rain events," reducing removal rates to a fraction of the putative pace. Exacerbating the deluge was the fact that "numerous catch basins and drain lines within impacted drainage basins were clogged, undersized, or otherwise compromised"—a responsibility that once pertained to the S&WB, until that 1991 loss of funding shunted it to the Department of Public Works.

It got worse. Engineer Matt McBride, a citizen watchdog and highly qualified S&WB critic, documented that some pumps had actually spun in reverse, drawing outfall water and injecting back into flooded streets. Four days later, on August 9, an electrical fault ignited a fire at one of the five "ancient" turbine generators at the Carrollton plant (three others being already off-line), "resulting in no capacity to self-generate electricity for the city's drainage system for a short period of time."

As meteorologists predicted more storms on Thursday and Friday, Mayor Mitch Landrieu declared a state of emergency. Schools closed; businesses sandbagged doors; and motorists parked on neutral grounds in preparation. If rains like those of July 22, August 5, or August 8 fell on August 9, the city would have met them with circa-1880s pumping capacity—and all of this in the middle of hurricane season.[67]

Pure luck kept the storms at bay, buying precious time for the S&WB to bring the system "back to abnormal," as some wags put it. "The August 5 Flood" became part of the city's vernacular, but unlike other date storms, this shorthand meant more than ruined carpets and insurance claims (681 filed, counting just those with the National Flood Insurance Program). The drainage debacles of the summer of 2017 signaled that an intertwined mass of decades-old factors had flummoxed the S&WB in their task of draining New Orleans. According to an external analysis, those "root causes" ranged from inadequate planning, inconsistent leadership, and insufficient funding, to aging turbines, improper maintenance, and "bureaucratic inefficiencies and limitations [which] hamstrung the ability to more proactively and expeditiously clean clogged drain lines."[68]

When comparably violent storms swept through the city on May 18, 2018, trashing the Bayou Boogaloo Festival and crashing Tulane's Wave Goodbye graduation party, frustration took on nearly polemic tones. In an interview that evening, Mayor LaToya Cantrell duly updated the press on pump operations and generator capacity. Twice during her talk, she departed from her talking points to make a higher-level statement. "We are a city that floods," said the mayor of the City of New Orleans, emphasizing every word. "We are a city that floods."

The mayor's assertion—or rather, acknowledgment—channeled the Native understanding of water as a condition of this fluvial delta—a given, an expected and even necessary element, an agent of the land's transience. European colonists viewed water differently, as a technical problem to be solved rather than tolerated or adapted. They were subjects of a real king, and every subsequent drainage king commandeered water relentlessly, with shovel, paddle, and pump; with ditch, canal, and levee; with gate, wall, and barrier. Yet precisely three hundred years after Bienville founded New Orleans and started that problem-solving, here was the mayor of New Orleans declaring that, *still*, "this is a city that floods." And she of course was right. So were the Natives.[69]

12

Rewatering New Orleans,

2010S–2020S

If water in New Orleans is truly a condition and not an abnormality, then assumptions underlying the city's centuries-long dewatering need to be reexamined—starting with the very notion of *dewatering*. That man-against-nature mindset harked back to the days of miasmas, yellow jack, and black vomit, when the best science emphatically maligned moisture. It did so in the clearest and most urgent terms, as expressed by Dr. Edward Barton in 1854: "*The drainage . . . should be so effectual that no water should exist within two to three feet of the surface. . . . The swamps . . . must be effectively drained . . . at first, thorough and complete, [and covered with] a perfect pavement . . . of materials that would neither admit of absorption nor evaporation. . . . An extensive, dense forest growth not only invites moisture[,] but retains it. . . . Clearing the low country then, and thoroughly draining it . . . greatly tends to improve its sanitary condition, [and] is urgently demanded here.*"[1]

Civil engineering and city planning practices had inherited these notions, even as medicine moved beyond miasmas, and "the world's toughest drainage problem" remained so committed to a hardened approach that only a visionary outsider could give voice to alternative thinking.

Enter the unlikeliest of drainage kings: a man from Plain Dealing, in the piney hills of Bossier Parish—about as far as you can get from New Orleans and still be in Louisiana—who came to realize the dewatering mindset needed to be flipped.

DAVID WAGGONNER

A Yale-educated architect, David Waggonner brought a diverse array of complementary perspectives to New Orleans's troubled ambit. He had an ecologist's sense of interconnectivity, a geographer's understanding of landscape, a designer's eye for form and function, and a philosopher's nous of transience. That cognizance steered him to think beyond the static elements of the built environment and focus on its circulatory systems, namely water: where it was, where it wanted to go, how it could threaten, but also how it could nourish and sustain. That holistic perspective came from his background as an architect, but also from philosophies informed by world cultures and religions. Daoism taught him to "see the void"—to pay attention to the seemingly empty space *around* the hardware garnering all the attention. "Instead of looking at the networks [and] machines first," he said, "take the Native perspective, the Indigenous perspective, and think how it works without us. . . . That Daoist perspective has driven all of this, that 'the container,' 'the void,' is the useful piece. It sounds poetic, and it sounds religious almost, but it's practical. Which is the best religion."[2]

Waggonner proposed thinking about pumps as the back end of the drainage strategy, a radical departure from the century-old stance of viewing them as the system's hearts. What would go at the front end, he posited, was the void—that is, the urban landscape, redesigned for a slow-and-store approach to stormwater runoff, leaving the piping and pumping only for whatever remained.

Ergo, let gravity and nature do the routine work:

- Slow the movement of runoff across the cityscape to avoid overwhelming pipes and pumps.
- Store rainfall in retention ponds to keep it off the streets and reduce infrastructure dependency.
- Design buildings to capture runoff in cisterns, sumps, and bioswales.
- Widen underground pipes to hold more water, and rework their connections to minimize groundwater removal.
- Maximize permeable surfaces and use pervious pavements, so that precipitation can recharge groundwater and reduce future subsidence.
- Widen outfall canals to store more runoff; remove their floodwalls in favor of lower earthen berms; and landscape their flanks into ecological amenities.
- Circulate impounded rainwater and groundwater throughout natural water bodies.
- Plant trees to intercept and store water droplets, reduce urban heat, increase wildlife habitat, and create beauty.

These principles reflect the understanding that, while natural environments absorb 50 percent of rainfall via soil and vegetation as another 40 percent disappears through evapotranspiration, leaving only 10 percent as surface runoff, urbanized environments handle only 15 percent in soil and vegetation, and 30 percent in evapotranspiration, leaving fully 55 percent as runoff. In other words, urbanization more than quintuples runoff, even more in an urbanized fluvial delta in the subtropics, and during intense rains, it yields a water volume far greater than pump capacity—thus the infamous "date storms." Any strategy that increases the first two variables (soil/vegetation absorption and evapotranspiration) will reduce the third (runoff).[3]

Early drainage kings understood this relationship, instinctually if not explicitly, and had no choice but to work with nature in upping those first two variables. Hector Carondelet, for example, capitalized on gravitational flushing when he had his canal dug in 1794. Barthélemy Lafon knew about the benefits of water retention and urban forestry when he designed Coliseum Square in 1806. The Auxiliary Sanitation Association understood how storing "surplus water in the vast 'reservoir canals' [would] relieve the draining wheels" in 1881.[4] Even those who eschewed every drop of water recognized that stagnation was the problem, and circulation the solution. But that cognizance began to evaporate once newfangled engineering promised total victory over the overlearned misunderstandings of miasmatic theory—such that, finally, every drop really *could* be pumped out.

And now here was a man hoping to restore that cognizance. Waggonner brought to that task a patrician's sense of civic duty and an appreciation that progress often comes from personal relationships, something he gleaned as the scion of a political family. He also had a flourishing architectural practice with his business partner, Macnaughton "Mac" Ball, which had focused on structural design but shifted to planning after the Katrina flood. Waggonner had become dismayed at early recovery discussions, in which some experts needlessly agitated the public with sloppy maps and flip proposals based on the faulty presumption of the "blank slate." Others, particularly architects, turned the moment into red-herring referendums on New Urbanism and historicity-versus-Modernism debates.

Waggonner & Ball had a chance to do things differently when the firm won a contract to develop a recovery plan for St. Bernard Parish. That effort focused mostly on rebuilding strategies, but it gave the staff a chance to think how drainage and urban water management could be reworked to make the devastated parish, flooded to a degree of 99 percent, something worth returning to.

At around that time, Louisiana Sen. Mary Landrieu got a call from Dutch Ambassador Boudewijn van Eenennaam and senior economist Dale Morris, an American at the Dutch Embassy eager to engage his host country and his home country on a Katrina-inspired international conversation on coastal sustainability. The Netherlands, they told the senator, had its own catastrophe, in 1953, when the North Sea inundated much of the low country; perhaps Louisianians would benefit by seeing the Herculean system of barriers and dykes the Dutch had built in response.[5]

Senator Landrieu took the diplomats up on their offer, and assembled a delegation of local and state officials, experts, and civic leaders to take a sort of pilgrimage to the hydrological Holy Land. Among the participants was David Waggonner, a personal friend of the senator, who, like many people on junkets, at first did not quite know why he was there. "It was a very strange trip," Waggonner reflected fifteen years later—deeply influential, as it turned out, but not in ways initially expected. The main goal was to see and learn how the Dutch barricaded themselves from North Sea storms, the presumed paradigm for post-Katrina New Orleans's "Category-5 Levees Now!" mantra. After arriving and meeting with officials, off went the delegation into the Dutch countryside. "We were taking long trips out to see those flood defense structures," Waggonner recalled, "but those long bus rides gave an important opportunity to talk to people." That was one lasting outcome: personal relationships—with Dutch diplomats, with Louisiana leaders, with city councilmembers and civic luminaries, all of whom would play key roles later. Sociologists use the term "social capital" to describe these valuable networks of personal relationships; Waggonner uses the word "friends."

Another outcome of the trip appeared right out the bus window, en route to the dykes and barriers. "I'm looking at the landscape," Waggonner recounted, and "I'm seeing all this green and water": landscaped retention ponds and canals that were beautiful; farmed polders that reserved space for water to do what it wanted; room made for rivers to overflow if they needed to. As much as the Netherlands had to battle North Sea storms, it nevertheless learned to live with water—palustrine, fluvial, and pluvial water—and, most remarkably, marshalled it to the benefit of keeping out the sea.

New Orleans too battled the sea, but it did so by making war on the very fresh water needed to keep the delta robust and buffered. What Waggonner saw out the bus windows "made a huge impression."

Other delegates focused on the intended purpose of the trip. "The Louisianians went looking for the flood protection," recounted senior project designer Ramiro Diaz. "But what David realized was that flood protection was more than just walls and dykes and surge barriers. [The Dutch] really think about water at all levels and all scales, whether it's groundwater, or rain water, or riverine management. . . . There's no line between [political jurisdictions]; in fact, democracy within the Netherlands was borne out of their water boards."[6]

Waggonner was also impressed by how Dutch culture valorized its water experts. Their offices were "hallowed places," with "incredible maps on the wall." The dignity of the governmental endeavor spoke to the gravitas of its charge, and the national commitment behind it. At the end of the trip, he had a conversation with famed geographer Woody Gagliano, known in Louisiana as the father of coastal restoration. A bit at sixes and sevens over what was now expected of the delegates, Gagliano pulled over Waggonner and asked him, "David, what do you want out of this?" Waggonner replied, "I just want to draw a plan."

He got his chance by participating in the Rockefeller-funded Unified New Orleans Plan, a 2006 citywide mega-charette that put a lot of emphasis on community yet little on managing water for community safety. Needing to learn more from the Dutch, Waggonner returned to the Netherlands in November and met again with Dale Morris at the embassy.

Morris was just the right person to connect Waggonner with the water-management brain trust in Amsterdam and Rotterdam, the two Dutch cities most akin to New Orleans. Their conversations veered not toward barricading external water, but living with internal water, and the remarkable way that the latter ameliorated the former. More trips followed in 2007, on which Waggonner walked canal-lined streets, rode trains across the countryside, and observed how "the void" in Dutch cities was anything but. Green, blue, and gray infrastructure (that is, parklands, waterways, and structural devices) were designed integrally, and gravity did as much or more work as the pumps. He brought with him colleagues and friends, such as Shreveport-born Paul Farmer of the American Planning Association, and met with key figures like Piete Dircke of the design firm Arcadis, Han Meyer of Delft University of Technology, and others from the Netherlands Institute of Spatial Planning as well as Deltares, the Dutch headquarters for water research. Waggonner was amazed by how these strangers had been affected by the calamity in New Orleans, and how much "they cared about us." The Netherlands had suffered its own Katrina-like trauma, and since 1953 its people vowed to never let it happen again. Now these strangers-turned-friends were offering their expertise to a sister city in need.

That offer needed funding. It was too soon to seek a government grant; a corporate gift didn't quite work out; and Waggonner could no longer pay for the trips out of his own pocket, as he had been. So he picked up the phone and started calling friends and colleagues—or as a sociologist might put it, leveraging social capital to access fiscal capital. One donor paid for breakfast. Another took care of the coffee breaks. The Port of New Orleans offered an auditorium. Louisiana Economic Development paid for dinner and buses. Co-organizer Dale Morris arranged for the Dutch government to pay airfare, while experts' time from both countries was offered in-kind. Somehow, it worked.

What were they going to call this event, these dialogues between American and Dutch experts about New Orleans? pondered Dale Morris as they worked on the agenda. Waggonner seized upon Morris's cue and reworked it into a crisp alliteration, and *Dutch Dialogues* was born.

DUTCH DIALOGUES

The first Dutch Dialogues event took place at the Port of New Orleans headquarters on March 8, 2008. Forty attendees learned of Dutch experiences and methods, and, upon hearing from locals and taking tours, eagerly listened to the visitors' initial impressions of New Orleans. One reaction came in the form of a bemused rhetorical question: *"Where's the water?"* The visitors came expecting to see an urban delta interlaced with water bodies, but instead found high levees, concrete walls, and dried-out land. Lesson one: deltas are supposed to be wet, and delta cities need to be as wet as possible to be as safe as possible. In attendance was Paul Farmer, director of the forty-thousand-strong American Planning Association, who, intrigued by the challenge, decided to adopt Delta Urbanism as the central theme of the APA's 2010 national conference.

Dutch Dialogues needed a follow-up workshop where pens would be put to paper. That second session was scheduled for October 2008, and as the date approached, so did a powerful storm. Category 4 Hurricane Gustav triggered a mandatory evacuation and struck on September 1, weakening along a northwesterly track. While the patched levees managed to hold around New

Orleans—this was well before the HSDRRS had been completed—the city had been disheveled, and residents' lives once again disrupted. Organizers pondered whether they should cancel the workshop, but the consensus was clear: this is the *ideal* time to think about living with water!

Dutch Dialogues II, sponsored by the Netherlands government and the APA, was held October 10–14 at the Tulane University School of Architecture. Waggonner and his team devised a methodology they would later refine and replicate nationwide: participants were broken up into specialty subgroups, assigned problems at different scales (regional, by bank, by polder; by outfall canal, ridge, or bayou), and equipped with maps, colored markers, and plenty of paper. Their charge: think through how stormwater runoff could be stored and circulated to unburden pumps and minimize pluvial flooding, while creating beautiful amenities with ecological services. Each group would then report back in a plenary session for critique. Teams were interdisciplinary, international, and intergenerational, and the results were innovative, beautiful, heavy on the blue and green, and light on the gray.

Indeed, the word "drainage" did not come up often in Dutch Dialogues, nor did "pump" or "dewater," because the Dutch approach was to lessen the need for all of the above. In a city proud of tackling the world's toughest drainage problem, all this was quite revolutionary. It resonated with the local environmental, planning, design, and nonprofit communities, who in this era were starting to embrace words like resilience and sustainability in their missions.

But many members of the public did not quite understand why these curious folk from Holland were busy sketching ponds and parks, when everyone here had been calling for Category 5 levees. As for the city's drainage establishment, namely the Sewerage & Water Board, reactions to Dutch Dialogues ranged from ambivalence to skepticism. "The Water Board never opposed what we were doing," Waggonner made clear in an interview. "Marcia [St. Martin] never opposed it. They did tend to give it lip service. They would point out that in New Orleans East, we have a Dutch system," with ornamental lagoons that stored water, canals that flowed below-grade, and pumps along the periphery. But "they were only looking at the piece of the lesson they could feel good about." Worse yet, "they didn't look at the water level in the canals as a problem." That is, if viewed from above, New Orleans East does indeed have an impressive network of storage lagoons. But a cross-sectional view shows that water levels in the drainage canals are typically drawn down very low, to fifteen to sixteen feet below sea level, to create extra storage space for intense rainfall. That declivity had the deleterious effect of drawing down groundwater so severely that New Orleans East now has the lowest elevations of the metropolis, to ten to fourteen feet below sea level. The one urban polder that ought to have benefited from good surface water storage instead got the most subsidence. Had water levels been kept higher from the beginning, and ponds and lagoons designed to store excess runoff, then soils would have been kept wetter and less prone to sinkage. Katrina flooding would thence have been that much shallower.

Waggonner particularly regretted that Joe Sullivan, the revered engineer and long-time general superintendent, "was pretty upset" with Dutch Dialogues. "For Mr. Sullivan, it was *personal.* This was a great man, *great* man, had been a prisoner in the Japanese war; I had all the respect for him. But he saw these canals as *utility corridors*[;] they were really not public space. . . . The idea that you were not going to rely solely on drainage" seemed unfathomable to him. Sullivan retired in 2008 and died in 2011. Pumping Station No. 6 in Lakeview now bears his name, an honor shared by Albert Baldwin Wood, for whom the Melpomene Pumping Station No. 1 is named.

Another common S&WB response ran along the lines of, in Waggonner's paraphrasing, "This would have been fine if we would have taken that direction from the beginning, but we took the direction to drain the city." Now that that direction had proven insufficient, Waggonner countered, "they want every penny to reinforce a system that can't do [the job]." That was the essence of Waggonner's argument: that the S&WB could not, by its own acknowledgment, meet the drainage needs of the city, because too much rain fell too intensely for the system to keep up—and that's when everything was running properly. So why not consider spreading out that excess water, temporally and spatially? And why not make those interventions beautiful and useful?

The S&WB would eventually come around and embrace Dutch Dialogues' Living with Water approach, particularly when Cedric Grant became executive director. Why? "Cedric had been exposed to the Dutch," said Waggonner. "He had gone to the Netherlands a couple of times. He felt it was part of his identity." In other words, it had become personal.[7]

It was at the 2010 conference of the American Planning Association, held at New Orleans in April during French Quarter Fest weekend, where Dutch Dialogues truly gained momentum. APA director Paul Farmer had invited Waggonner to arrange for the third session to be held in unison with the conference at the Convention Center (itself on reclaimed land—the old St. Mary Batture). With nearly double the number of participants from 2008, including twenty-six visiting experts, Dutch Dialogues III focused thematically on water circulation among the ponds and canals, and spatially on Gentilly, the archetypal New Orleans polder, with a demographic composition representative of the city.

In addition to plenty of specific design ideas, Dutch Dialogues III had two major outcomes. It put Gentilly on a path toward becoming an actual living-with-water pilot project, and it gave the larger effort a perfect professional audience—thousands of planners learning the lexicon of resilience, sustainability, and climate-change adaptation, with New Orleans as their case study. "The band was growing," recalled Waggonner after the conference. "People wanted in—people from elsewhere in the country who had heard about this."

Thus was born one of the most influential nonpolitical movements initiated by a single citizen in recent New Orleans history. While Waggonner is quick to deflect credit to numerous friends and colleagues, those folks point out that none of it would have happened without Waggonner. The name "Dutch Dialogues" played no small part in spreading the living-with-water message, by giving the abstract concept a euphonic moniker. "I always had an ear for a hook in a song," he later said with a smile. Waggonner & Ball trademarked the name, not to capitalize on it, but to prevent Dutch corporations from doing so, as they savored a potential new market for their specialized know-how.

THE GREATER NEW ORLEANS URBAN WATER PLAN

The success of Dutch Dialogues III led Waggonner to seek funds to keep the momentum going, instead of bootstrapping gifts through what he jokingly called "Southerners depending on the kindness of strangers." He had a connection to the Louisiana Office of Community Development, which had secured funds from the US Department of Housing and Urban Development. Planners in its Disaster Recovery Unit had considered the ideas of Dutch Dialogues to be worthy of investment, but because the temporary agency was about to sunset, it had to transfer the funds to a nonprofit to oversee an open competition. That was a perfect role for Greater New Orleans, Inc., a regional economic-development organization with parallel interests in social and environmental improvements. Under the leadership of Michael Hecht, who had been intrigued by reports of Dutch water experts in town, GNO Inc. agreed to coordinate a competition for over $2 million in funding to formalize the living-with-water approach. Waggonner & Ball had the advantage of three prior years of experience through Dutch Dialogues, and readily formed a team with its participants. But by no means did that give it a leg up in the competition, which GNO Inc. ran fair and square, with "real tough" requirements and interviews. It was a bit of an odd situation, Waggonner confessed: "So you create this child, and now there's a little bit of money to clothe it and to feed it, and now there's this big thing," with outside competition—"100,000 people at AECOM were going against our 25-person firm."[8]

An open competition, of course, was the proper thing to do, and Waggonner & Ball won the grant. Their deliverable: a full-blown Urban Water Management Plan for the east banks of New Orleans and Jefferson Parish plus urbanized St. Bernard Parish. GNO Inc.'s interest went beyond flash-flood reduction and urban beautification; they saw jobs in water, just as the Dutch had turned their expertise into an exportable commodity, and Baldwin Wood had done with his pump patents a century ago. "We view water management as critical to protecting our economic base," said Robin Barnes, executive vice president of GNO Inc. "We have a mayor [Mitch Landrieu of New Orleans] and two parish presidents [John Young of Jefferson and Craig Taffaro of St. Bernard] who are excited about going forward with this."[9]

With sixty people hailing from two dozen firms, universities, associations, and agencies, Waggonner & Ball began work in the spring of 2011. Six principles guided them, two regarding water (*when it rains, slow and store; when it's dry, circulate and recharge*), two about ecology (*live with water; work with nature*); and two regarding people (*work together, and design for adaptation*). Local officials and drainage operators were not among the team members, as they were viewed as the target audience.[10] That is to say, this was not the forum for drainage kings; rather, it was to issue a new charter for the kingdom. "Noticeably absent," pointed out one journalist, "was the Army Corps of Engineers, which is carrying out major canals improvement projects in Uptown New Orleans"— that is, the ongoing SELA project, the very embodiment of the traditional gray approach of pumping out every drop.[11]

Over the next two years, team members carried out an extraordinary amount of research and design work. Their domain ran from groundwater to rainwater, river to lake, neighborhood to region, and local to global. Throughout the autumn of 2013, Waggonner & Ball held a series of public meetings to unveil its Greater New Orleans Urban Water Plan. Richly illustrated, loaded with pull-out quotes and boxed asides, and written in tones ranging from technical jargon to soaring rhetoric, the twenty-nine separate reports ran the risk of overwhelming readers. So team members designed their reports to communicate effectively with varied sub-audiences: planning professionals, policy-makers, implementers, and most importantly, ordinary citizens. It was the latest of a long list of plans in the drainage history of New Orleans, dating back to Dunbar's report in the 1830s and Pauger's plan from the 1720s. But this one was fundamentally different in how it beheld its be-all and end-all, water.[12]

The Greater New Orleans Urban Water Plan regarded the East Bank as three super-polders, dubbed the Jefferson-Orleans Basin (Kenner to the Industrial Canal), the Orleans East Basin (New Orleans East to Highway 11), and the St. Bernard Basin (Lower Ninth Ward down through Arabi, Chalmette, Meraux, and Poydras down to the HSDRRS floodwall).

For the Jefferson-Orleans Basin, planners recommended steering runoff from the French Quarter and adjacent wards eastward to the Industrial Canal, rather than to the lake, a corollary of the 1895 plan to discharge eastward instead of northward. They also ascribed importance to the Gentilly Ridge as a natural watershed, recommending that areas to its north would discharge into the lake, and areas to its south into the Industrial Canal or GIWW. Both these recommendations made it feasible, at least on paper, to reconceive the former Agricultural Street Landfill and adjacent Florida-Desire areas as "Desire Parklands."

Another bold concept for the Jefferson-Orleans Basin involved the three circa-1872 outfall canals in Lakeview and Gentilly. The planners viewed the new post-Katrina gates and bypass pumps at their mouths as negating the need for their floodwalls. Removing them in favor of grassy berms distanced farther apart would transform these narrow, ensconced assets into capacious, multipurpose water parks, with a far greater water capacity at the level of the lake. The berms could curve and widen where adjacent public spaces allowed, or where vacant properties could be bought up, creating bayou-like water bodies wide enough for boating and other recreational activities. This water, higher than adjacent neighborhoods, could then be circulated through retrofitted neutral grounds, lagoons in City Park, and in Bayou St. John. Come bad weather, the gates would be closed, the bypass pumps activated, and the water parks converted to stormwater storage. Pumps would play a secondary role all along.[13]

Yet, as if in acknowledgment that pumps were still important and that nature can't always be accommodated, the plan also encouraged a pump-to-the-river approach for certain parts of Jefferson Parish, such as River Ridge and Old Metairie in Hoey's Basin, whose lakeward discharge had long been hampered by the Metairie Ridge. More radically, the planners suggested redirecting East Jefferson stormwater runoff westward into the LaBranche Wetlands, where saltwater intrusion and erosion had long been problems, rather than northward into the lake, as had been done for a century.

Other, more modest recommendations included redesigning vacant lots, highway shoulders, and other idle land into retention ponds; retrofitting streets with permeable pavement, rain gardens, bioswales, underground storage chambers, and catch basins; beautifying utilitarian drainage ditches; and expanding scenic ponds and lagoons for water storage on green spaces such as Lafreniere Park and the Lafitte "Blueway" (a take on the Lafitte Greenway under construction at that time, formerly the Carondelet Canal from 1794).[14]

Similar retrofits were also recommended for the New Orleans East and St. Bernard basins, where suburban environs and adjacent marshes offered additional opportunities. Planners envisioned reconnecting stagnant Bayou Sauvage, the last segment of the abandoned Bayou Metairie–Gentilly distributary, with the artificial Bayou Michoud to the west and Chef Menteur Pass to the east. Akin to their suggestion for Jefferson Parish, planners contemplated rerouting New Orleans East's runoff eastward into the freshwater-starved marshes of Bayou Sauvage National Wildlife Refuge, rather than northward to Lake Pontchartrain. The Hayne Boulevard corridor, meanwhile, offered spaces for waterfront access and lacustrine wetlands, for recreation, wildlife, and stormwater storage. In the St. Bernard Basin, the Bayou Bienvenue wetlands begged for a new freshwater source, which could be attained through a "Chalmette Blueway" diversion of Mississippi River water. That tactic could also reactivate Bayou Terre Aux Boeufs as a distributary and push back saltwater intrusion coming from the east.[15]

Such massive reconfigurations lead some to dismiss the Urban Water Plan as an impractical vision blind to fiscal realities. Others contend it won't work, or at least not well enough, else that stormwater flooding does not existentially threaten the city, and that the needed $6.2 billion would be better spent on stronger perimeter levees and coastal restoration against surge flooding.

Advocates counter that the Urban Water Plan never pretended to be a shovel-ready blueprint, with budgets and bulldozers at the ready. Rather, it was a conceptual plan, based on the notion that you have to first imagine a better world before you can build it. As for viability, the whole plan did not have to be installed all at once, nor did it have to "succeed" at once, like a moonshot, be-cause every drop handled by nature was one less drop needing pipes and pumps. Gains could be incremental, across scales and jurisdictions, and not necessarily overseen by a centralized node.

Collaborators from the Netherlands, hailing from a small affluent country with a national commitment to water management, did not necessarily under-stand the bottom-up American way. "The Dutch were thinking, ok now, when do we get started?" recalled Waggonner. "But of course we don't have anything like that organizational system."[16] In fact, scores of micro-projects are already underway, ranging from tree planting to rain gardens to cisterns, and many are driven by nonprofits working with neighbors on private land, in that bottom-up American way.

Plenty of challenges lie ahead for the green and blue water-management movement. Utilities and infrastructure agencies are inherently conservative, given their legacy equipment and operational charge, and cannot be expected to undo a century in a year. Some activists, meanwhile, are prone to "passion," which can translate to overpromising and burning out. Some players hop on the water wagon solely for the funding opportunities, which are limited to say the least. As for rewatering per se, monitoring efficacy is extremely difficult, given the entanglement of variables, and if advocates cannot prove that benefits ex-ceed costs, the whole movement may be reduced to symbolic gestures. Regard-ing civic engagement, some citizens have the mistaken understanding that the Urban Water Plan will abate Katrina-style deluges (it won't, hardly), or that their new bioswale will forever keep their parlor dry (the risk of overpromising). Fi-nally, community resistance may intensify once folks are asked to make real sacrifices—to tolerate months of torn-up streets, or to expropriate homes for water parks. One person's "void," after all, might be another's neighborhood, and not quite the blank slate perceived by outside experts.

But the *status quo* has its costs too, as well as its share of overpromises, misun-derstandings, mismeasurements, and misgivings. "The plan estimates it would secure $22 billion in avoided costs and economic benefits," wrote Lorena O'Neil in an *Atlantic* article titled, "Why Doesn't New Orleans Look More Like Amsterdam?" That figure included the avoidance of $8 billion in flood damages, $2.2 billion in avoided subsidence damages, and $609 million in saved insurance premiums. "The Water Plan proved about a 4-to-1 benefit-to-cost ratio," said Waggonner.[17]

The Greater New Orleans Urban Water Plan garnered widespread media attention, and got the public thinking about water's place in a delta city. The

Living with Water approach made the circa-2005 call for Category 5 levees now sound like only part of the solution, just as it did for the circa-1900 call for massive pumps. The plan even impressed the toughest and most critical audience, the Sewerage & Water Board. "One of the problems of Dutch Dialogues," said Ramiro Diaz of Waggonner & Ball, "was the Sewerage and Water Board and the Army Corps of Engineers would discount it as 'That's just something those crazy architects put together.'" Now, said Diaz, "the Sewerage and Water Board are advocating for parts of the plan."[18]

David Waggonner is happy to meet them halfway, because he is the first to recognize that pumping will always be key to the New Orleans drainage system, even if we learned to live with water as he wishes. We just need to acknowledge the past damages and future limitations of the old gray approach, and add blue and green tactics as supplements, not replacements. As geographer Adam Mandelman pointed out in regard to other deleterious delta escapades, the problem was not the invention of brilliant engineering technologies, nor of their steadfast application; rather, it was our failure to take responsibility for their effects. "Frankenstein's crime was not that he had invented a creature," he wrote, citing French sociologist Bruno Latour, "but that he had insufficiently cared for it."[19]

What Waggonner wants is to take that responsibility and exercise that care. "In 1950," he said proudly, channeling his inner New Orleanian, "the *best engineering job in the United States was General Superintendent of the Sewerage and Water Board.* It was the most interesting engineering contraption in the world. And I think it still is. We just have to figure out where the pipes are."[20]

"THAT ALL MAY BE ONE"

The most successful dialogues are those that lead to action. Everyone knew the Urban Water Plan was not authorized for deployment, but those involved hoped it would at least launch one major pilot project. Where to put it? Water needed space, and on account of the individualized, homeowner-based nature of post-Katrina rebuilding, those vacant lots now owned by the New Orleans Redevelopment Authority (NORA) were scattered hither and yon, each no more than a few thousand square feet. What was needed was aggregated space, and the legal right to redesign it for water.

Prayers were answered by the owners of a tract in Gentilly, the very polder that Dutch Dialogues III had previously focused on. It pertained to the Sisters of St. Joseph, a congregation founded in France in 1650 which had established a novitiate in New Orleans in 1863. Ministering mostly in the Gentilly area, the sisters had acquired twenty-five acres on Mirabeau Avenue just east of Bayou St. John, and in 1952, dedicated a provincialate headquarters on the property. Equipped with a convent, offices, chapels, and classrooms, the substantial Mirabeau campus reflected the mid-century growth of the Catholic Church in the United States.[21]

A half-century later, that trend had reversed, and the congregation's aging ranks struggled with upkeep of the sprawling three-story complex. Then came a nearly biblical sequence of traumas. The Katrina flood destroyed the first floor, and took a million dollars of the sisters' retirement fund to restore. A year later, lighting struck and burned the third floor, and because of the weakened water pressure at the S&WB plant, the blaze could only by extinguished by helicopter bucket-drops of Bayou St. John water—which destroyed the second floor. Charred and soggy, the wreckage had to be completely razed. "It was like watching someone you love die," said Sister Barbara Hughes.[22]

Resigned to a diminishing domain but buoyed by their motto, *That all may be one,* the sisters endeavored to do something beneficial with their land. That's when they met David Waggonner, and what ensued can only be called a match made in heaven. The congregation found his living-with-water message attuned to their Christian mission, and the architect found the nuns delightful—and their parcel perfectly positioned hydrologically. After a series of discussions, the Congregation of the Sisters of St. Joseph turned their land into their ministry, and agreed to lease all twenty-five acres of it, worth $11 million, to the City of New Orleans for one dollar a year.[23] The Mirabeau Water Garden became a marquee project in the Urban Water Plan.

It was good timing. Mayor Mitch Landrieu had taken office in an era when environmental issues and climate change were rising as public concerns, and his administration was the first to turn resiliency and sustainability into job titles in city hall. Behind him were key people such as Andy Kopplin as chief administrative officer, MIT-trained planner Jeff Hebert as director of NORA and later the city's first chief resilience officer, and Prisca Weems as the city's first stormwater manager, among others.

Water featured prominently in every environmental discussion. The city released its own Resilient New Orleans Plan in 2015, which largely incorporated the recommendations of the 2013 Urban Water Plan. Both efforts enabled the city to win major funding for a "Gentilly Resilience District" from the US Department of Housing and Urban Development's National Disaster Resilience

Competition. This $141 million grant, which built upon prior water-related investments from a FEMA Hazard Mitigation Grant, focused on implementing the Mirabeau Water Garden as a recreational park that stored and circulated up to ten million gallons of runoff. It also entailed ten other smaller water-based landscape projects, including "blue & green corridors" on neutral grounds, a swamp preserve at a remnant patch of forest near Dillard University, and new green infrastructure in the Gentilly-area neighborhoods of St. Bernard, Pontchartrain Park, and Gentilly Woods.[24]

Looking back on the journey, Waggonner said his 2013 plan "was not mandated or sanctioned by anybody, but it was adopted in pieces by everybody"—by the city, by the federally funded Gentilly Resilience District, by the state in its program named Louisiana's Strategic Adaptations for Future Environments (SAFE), and by various private entities.[25] Perhaps most important was public education on "water literacy," starting with the idea that urban water is not some static refuse to be disposed of, as the drainage kings saw it, but a circulatory life force, like blood in veins. "Think how much the citizens have learned since we started," Waggonner marveled in 2020, by which time many public officials had been enlightened on living with water. Among them were leaders in city government, such as in NORA, which has since established numerous rain gardens and launched its Growing Green program to let people lease lots as gardens and pocket parks. The DPW began installing rain gardens and pervious parking lanes on streets due for repairs, and put traffic lanes on "diets" in favor of bike lanes and wider sidewalks with bioswales. The S&WB invested a half a million dollars per year into green-infrastructure projects, and hired a new executive director, Ghassan Korban, for his civil engineering proficiency as well as his blue and green commitment. The Jefferson Parish Drainage Department installed a levee-ringed retention pond at Pontiff Playground to relieve flooding in Hoey's Basin, and a pump-to-the-river pipe to send Harahan's excess water to the nearby Mississippi rather than the distant lake, two concepts also recommended by Dutch Dialogues. The City of New Orleans, meanwhile, passed a zoning requirement that all development must "retain, detain and filter the first 1.25 inches of stormwater runoff during each rain event," and other ordinances calling for permeable materials on parking lots and other surfaces.[26]

In academia, the Tulane School of Architecture devoted numerous teaching studios to Urban Water Plan projects and incorporated stormwater management into its curriculum, while its students helped design drainage improvements in Gretna and elsewhere. The school also hired a new dean, Iñaki Alday, in part for his international expertise in water landscapes.

The nonprofit sector has thoroughly embraced the urban water movement, through organizations such as the Water Collaborative, the Urban Conservancy, the Urban Water initiative of the Greater New Orleans Foundation, Propeller, and the Ripple Effect, which was cofounded by former Waggonner & Ball designer Aron Chang to teach water literacy to the next generation of New Orleanians. Waggonner & Ball itself did well, replicating its New Orleans–honed Dutch Dialogues methodology in other coastal cities, including Norfolk, Bridgeport, and Charleston, in a way reminiscent of how Baldwin Wood exported his screw-pump technology to cities worldwide.

At the state level, Sen. Mary Landrieu's trips to the Netherlands led her to realize that Louisiana needed something like the independent Dutch water research institute Deltares. Through a state collaboration with the Baton Rouge Area Foundation, Senator Landrieu established The Water Institute, with a mission to connect "academic, public, and private research providers" and conduct "applied research to serve communities and industry." The Water Institute, which became a Center of Excellence in 2014, now operates in an impressive new riverside campus in Baton Rouge, and its director of strategic partnerships is Dale Morris, cofounder of the Dutch Dialogues partnership. In 2019, The Water Institute opened a campus at the University of New Orleans Research and Technology Park, on the same soils that Col. Marcel Garsaud had dredged from Lake Pontchartrain ninety years earlier.

As the latest chapter in the history of the dewatering of New Orleans, the rewatering movement can be traced back largely to Dutch Dialogues, and to that "strange trip" to the Netherlands in the bleak postdiluvian winter of 2006. The ensuing international collaboration, launched on sheer goodwill, will be successful even if, figuratively speaking, it amounts to no more than a few drops in the bucket—because, literally speaking, that's more drops keeping the delta wet, and fewer drops needing to be pumped out.

None of it would have happened were it not for the man from Plain Dealing. "One of the things architects don't do very well," David Waggonner mused, "is to think about *transience.* We're taught in statics, but we don't see movement. We photograph buildings, but with no people. We calculate water in cubic feet, but it moves per second. These are not fixed conditions. Life is not stable; it's moving. Water is the source of it all."[27]

ACKNOWLEDGMENTS

This book took two years to write but over two decades to research, because its complex subject has undergirded the content of nearly all my previous books and articles about the geography of New Orleans. My gratitude goes first and foremost to the city and its people, who have both achieved and endured this centuries-long negotiation with water, particularly to those "drainage kings" who tried their best given the circumstances of their times. My thanks go to my sources, be they the nameless scribes of primary documents, the engineers behind those maps and plans, the authors of academic studies, or the informants who edified me personally, among them David Waggonner, George Hero III, the late Joe Sullivan, Marcia St. Martin, Ghassan Korban, Tonja Koob, Dennis Lambert, Jaime Ramiro Diaz, Aron Chang, Josh Lewis, Jeff Adelson, and David Hammer. I express gratitude to the many institutions and agencies that provided access to data, documents, and maps, including the City of New Orleans, New Orleans Sewerage & Water Board, US Army Corps of Engineers, *The Times-Picayune/New Orleans Advocate* and its predecessors and competitors, The Historic New Orleans Collection, Tulane University Southeastern Architectural Archive, US Census Bureau, US Library of Congress, Howard-Tilton Library at Tulane University, Louisiana Collection and Special Collections of the Earl K. Long Library at the University of New Orleans, Louisiana Division of the New Orleans Public Library, Louisiana Research Collection at Tulane University, Louisiana State Land Office, Louisiana State Museum, New Orleans Notarial Archives, and Waggonner & Ball Architects. A fund from Tulane University's ByWater Institute, where I am an affiliate, allowed me to secure the services of my talented colleague Andrew Liles to draw the four bird's-eye illustrations in this book. Marco Rasi, a drone photographer with whom I have collaborated on prior projects, generously contributed aerial photographs from our field trips in

October 2021. Special thanks go to Dean Iñaki Alday and Associate Dean Scott Bernhard of the Tulane University School of Architecture, where I hold a faculty appointment and serve as associate dean for research, and to the staff of Louisiana State University Press, including the editor-in-chief, Rand Dotson; the managing editor, Catherine L. Kadair; the senior editor, Neal Novak; the book designer, Michelle Neustrom; and the marketing director, James Wilson, as well as copyeditor Stan Ivester, for their ongoing support of my work.

Deepest appreciation goes to my wife, Marina Campanella, and to our young son Jason Campanella, who can dewater (and rewater) a batture lagoon in minutes flat.

TIMELINE

Prehistoric	Indigenous groups inhabit *Balbancha* in conditional manner, adapting to or retreating from watery landscapes.
1718	Bienville establishes New Orleans for French; colonial engineers view water as problem to be solved rather than condition to be tolerated.
1719	New Orleans's first flood prompts construction of first levees.
1721	Adrien de Pauger designs street grid of today's French Quarter.
1722	New Orleans's first hurricane destroys initial development; allows Pauger to lay out new streets and gutters, commencing "ditch-and-gravity" era.
1724–26	*Conseil Supérieur* decrees landholders responsible for maintaining drainage ditches fronting properties; *inspecteur de police* is charged with enforcement.
1728	Gov. Étienne de Périer takes charge of drainage, as enslaved Africans are put to digging gutters, moats, and outfall canals.
1732	Engineer Ignace-François Broutin sketches sophisticated subsurface drainage system, but it is not installed.
1732	King Louis XV weighs in on how to drain New Orleans.
Mid-1700s	Gravitational systems drain New Orleans streets via gutters to moat and Lake Pontchartrain via Bayou St. John, while scores of French long-lot plantations are drained via network of ditches to backswamp.
1762–69	Dominion of Louisiana colony shifts from France to Spain.
1770s–1800	*Regidores* of Spanish Cabildo, fielding complaints from citizens, advise governor on drainage policy and taxation for services.
1776–94	Nine tropical storms and two major fires subject *Nueva Orleans* to semi-annual pace of disasters.

1788	In wake of 1788 fire, City Surveyor Carlos Laveau Trudeau lays out *Suburbio Santa Maria* in former Gravier Plantation, expanding ditch-and-gravity system into today's Central Business District.
1789	*Regidores* establish St. Louis No. 1 Cemetery with above-ground tombs, rather than subterranean burials used in old French graveyard, citing overcrowding, malodor, and public health as main motivations.
1794	Gov. Héctor Carondelet oversees excavation of new drainage outfall to Bayou St. John, later expanded for navigational use as Carondelet Canal (Old Basin Canal). Channel is filled in 1920s and is now Lafitte Greenway, still used for drainage and stormwater management.
1796	City's first yellow fever epidemic kills 600 out of 8,756 population; seven more plagues strike during 1799–1812. Miasmatic theory is leading explanation, indicting swamps as health problem.
1800–1803	Spain secretly retrocedes Louisiana to France; French administrators arrive just prior to learning that Napoleon has sold Louisiana to the Americans.
1803–20s	Shifts in river channel trigger formation of batture along St. Mary riverfront; adjacent owners extend levee to reclaim beach, triggering dispute with city over riparian landownership. Long legal case results in piecemeal reclamation of seven hundred acres, becoming today's Warehouse District and adjacent riverfront areas.
1806–10s	*Ingénieur géographe* Barthélemy Lafon becomes city's premier drainage expert, laying out Faubourg Marigny, today's Lower Garden District, and other subdivisions, each with well-designed gravitational ditches, water-retention systems, and Classical motifs.
1808	President Thomas Jefferson suggests methods to drain local parcel pertaining to Marquis de Lafayette; opines on St. Mary Batture dispute, arguing riparian lands ought to be public.
1810s	Saint-Domingue-born Jacques Tanesse, protégé of Lafon, lays out Faubourg Tremé as well as Canal Street, Rampart Street, and Esplanade Avenue corridors; becomes city's premier surveyor and drainage engineer.
1816	City ordinance makes it duty of homeowners to clean and maintain gutters along *banquettes* fronting properties.
1817, 1819	French-speaking physicians form *the Societé Médicale de la Nouvelle Orléans,* and English speakers form the Physico-Medical Society, each seeking to solve yellow fever.

1818–20s	Saint-Domingue-born Joseph Pilié, another Lafon protégé, follows Tanesse as city surveyor; oversees drainage of four urban polders extending from today's Lower Garden District to Bywater, each with ditch-and-gravity subsystems. Runoff is discharged into Bayou St. John or Bayou Bienvenue.
1828	William Gormley has canal excavated in what is now Central City; Gormley Canal is later used to drain parts of Uptown to Bayou St. John and Lake Pontchartrain.
1820s–30s	Syndics serve as drainage-system bosses, mostly by prodding residents to do work or billing them for it.
1820s–30s	Private waterworks draw water from river using steam pumps and deliver it gravitationally to subscribers for domestic use; surplus is used to flush gutters but also adds to water volume to be removed by drainage system.
1829	Francis Ogden's *City of New Orleans* map depicts dendritic network of now-gone tributaries flowing out Bayou St. John to Lake Pontchartrain, city's main outfall throughout 1700s and 1800s.
1830s	Most plantations in today's Uptown are subdivided by surveyors in collaboration with city engineers, who retrofit street grids and drainage gutters into preexisting French long lots. New subdivisions include Carrollton, Greenville, Hurstville, and the Faubourgs Bouligny, Plaisance, Delassize, Livaudais, Lafayette, and Nuns.
1830s	Louisiana Board of Public Works supports private infrastructure projects by granting legal powers and providing economic subsidies, including slave labor to dig navigation and drainage canals.
1832–38	New Basin Canal is excavated to today's West End, opening urban core to lake resources. Channel's guide levees split city's main watershed into two smaller polders, laying groundwork for mechanized drainage and "speculation mania" on new real estate.
1834	Charles Zimpel publishes *Topographic Map of New-Orleans and Its Vicinity,* showing neighborhood development and detailed geographical features.
1834	Seven local physicians launch Medical College of Louisiana to train doctors to study plagues associated with swamp miasmas and poor drainage; institution becomes University of Louisiana in 1847 and Tulane University of Louisiana in 1884.

1835   State charters New Orleans Drainage Company to drain swamps "on the same plan that is adopted by Holland, by hydraulic machines. The profits are derived from the increased value of the lands drained."[1] Effort marks end of purely gravitational drainage and opens steam-driven "polder-and-paddle" era.

1836   Journalist for *New Orleans Bee,* reporting on mechanized drainage and swampland speculation, predicts "proprietors of [said] lands will have to elevate the surface of their lots . . . if intended for building. The drainage cannot give substance to the spongy soil."[2] Comment foresees that drainage could lead to subsidence below sea level.

1836   State Engineer George T. Dunbar issues *Report on the Draining of the Back Lands Beyond Claiborne Street,* city's first comprehensive drainage plan.

1836   Ethnic infighting leads to division of New Orleans into three semiautonomous municipalities, further complicating drainage; inefficient "municipality system" lasts until 1852.

1840   Board of Health, predecessor of city's modern Health Department, is established.

1849   Sauvé's Crevasse triggers city's worst flood to date; leads to so-called "swamp buster acts," which eventually transfer sixty-four million acres of federal wetlands to states, to enable leveeing, reclamation, drainage, and economic development.

1853–55   City's worst-ever yellow fever epidemic in 1853 claims up to ten thousand lives; two more epidemics in 1854–55 kill another five thousand. Outbreaks inspire formation of sanitary commission, chaired by Dr. Edward Hall Barton, who spearheads groundbreaking medical research on yellow fever. Barton's report helps establish local ethos valorizing drainage and maligning water on landscape.

1855   As New Orleans Drainage Company's charter ends, all drainage apparatus—including steam engines, paddle wheels, and feeder ditches—transfers to city control. Twenty-year effort fails to reclaim Uptown backswamp, but nevertheless expels runoff faster than gravity, drying out polder's uppermost perimeter and allowing for some urban expansion.

1855   City passes ordinance to establish today's Health Department, funding it for staff and budget.

1857 City Surveyor Louis Pilié's *Report on Drainage* recommends building lakefront levee to polderize swamp to south, draining it through outfall canals rather than Bayou St. John, and using steam-operated lift pumps rather than push pumps.

1858 Pilié's report spurs state legislature to pass Act 165, creating three drainage districts, each staffed with knowledgeable professionals, and giving them taxation and other legal powers. But funding is limited, and staff service is uncompensated.

1858 Lewis DeRussy submits Raymond Thomassy's "colmates" plan to State Board of Commissioners. Sound in theory but impractical, colmates never get beyond envisioning stage.

1861–62 Secession of southern states and outbreak of Civil War relegate drainage and urban sanitation to low priorities under Confederate regime; systems clog and runoff filth accrues.

1862–65 Union takeover puts Maj. Gen. Benjamin Butler in charge of city; following President Lincoln's directive, Butler creates paid workforce to clean streets and restore drainage systems. Yellow fever deaths are negligible for remainder of war.

1864 City Surveyor George Willard Reed Bayley recommends turning Bayou St. John into drainage reservoir protected by artificial seawall; is first to envision future Lakefront Project, and to warn of drainage-caused subsidence.

1868 George Brott's New Orleans and Ship Island Canal Company proposes to bisect entire metropolis with navigational seaway dug to Mississippi Sound; outlandish plan eventually fails, but gains enough ground to reconfigure new drainage plans, with lasting consequences.

1868 City Surveyor Louis Surgi develops drainage plan to complement Brott's state-backed seaway scheme; envisions two super-polders with lakefront seawall, as well as first mention of future outfall canals in what are now Lakeview and Gentilly.

1869 *Map and Profile of the New Orleans and Ship Island Channel*, by A. F. Wrotnowski working for George Brott, is first to plot trajectories for the three outfall canals, the first of which (today's Seventeenth Street Canal) is labeled to imply navigational use. All three channels, dug in next few years, would bedevil drainage engineering for generations to come.

1869	In deference to Brott's scheme, state legislature repeals prior drainage acts and transfers drainage apparatus and funding mechanisms to aid Brott in construction of seaway.
1870–74	Amid controversy, City of New Orleans annexes Jefferson City, Algiers, and Carrollton, affecting political geography of drainage to this day.
1871	Crevasse at Bonnet Carré sends river water into Lake Pontchartrain and up New Basin Canal. Levee breach at Hagan Avenue leads to worst flooding since Sauvé's Crevasse; illustrates how artificial canals threaten population by bringing external water into city.
1872	City Surveyor William Bell, compelled to design drainage plan around Brott's seaway, builds upon ideas of Bayley and Surgi in sketching *Chart of Draining Sections of New Orleans, Showing Present Canals, With Protection Levees and Reservoir Canals,* envisioning outfall canals with pumps placed at lakefront perimeter.
1872	Brott's New Orleans and Ship Island Canal Company, nearing bankruptcy, transfers obligations to Warner Van Norden, who proceeds with drainage portion of project, using Bell's plans.
1873	Bell releases *Plan of Proposed Improvements for the Lake Shore Front,* reprising Bayley's 1864 lakeshore breakwater concept with reclaimed landmass protruding into Lake Pontchartrain, complete with landscaping, harbors, and drainage infrastructure. Idea comes to fruition fifty years later as Lakefront Project.
1872–75	Warner Van Norden makes substantial progress executing Bell's drainage plan, digging Seventeenth Street, Orleans Avenue, and London Avenue outfall canals, and establishing framework for future drainage system.
1876	Brott's caper ends as state legislature authorizes city to retake control of all drainage work from him and Van Norden. Brott's seaway, never built, nonetheless affects Bell's drainage design and Van Norden's construction, with permanent effects.
1876	Capt. James Eads builds jetties at South Pass to scour out sedimentation and reopen river to deep-draft shipping, sparing city of other risky seaway proposals.
1878	City's second-worse yellow fever epidemic claims 4,056 lives; prompts new wave of medical research, connecting doctors in New Orleans with colleagues in Cuba (1879) and eventually leading to breakthroughs in 1880s and 1890s.

1879   Edward Fontaine publishes "ekmuzesis" proposal, apex of quackery in drainage schemes.

1879   Establishment of Mississippi River Commission enables federal advising and financial support for building levees, a task traditionally left to local governments and private entities.

1879–80s   Citizens form New Orleans Auxiliary Sanitary Association, demonstrating what city ought to be doing to clean streets. Effort marks emergence of progressive push for civic improvements, namely drainage, and politicians soon heed demands.

1880s   City's drainage system, essentially Bell Plan from 1872, is capable of removing at most 0.125 inches of rainfall per hour, only 10 percent of what is needed; dewaters only a slender perimeter of backswamp.

1881   Following years of research, Cuban medical researcher Dr. Carlos Juan Finlay presents "The Mosquito Hypothetically Considered as the Agent of Transmission of Yellow Fever." Findings are ridiculed and ignored for nearly twenty years.

1888   State legislature creates Commission of Public Works for New Orleans; City Surveyor Maj. Benjamin Morgan Harrod pens *Report on Drainage, to the City Council of New Orleans,* assessing Bell Plan and outlining needed improvements. Public interest in drainage rises.

Late 1880s Drainage is talk of the town: city papers publish 154 drainage stories in 1886; 206 in 1887; 329 in 1888; and 326 in 1889. Around the time Harrod's report is released, drainage makes headline news 137 times. Public enthusiasm for drainage prompts various proposals proffered by civically engaged neighbors and armchair engineers.

1890   State legislature creates Orleans Parish Levee District and Board of Commissioners.

1892   John Fitzpatrick becomes mayor and makes drainage an executive priority.

1893   City council passes two ordinances to map city's topography and form Advisory Board on Drainage, led by City Engineer Linus W. Brown and tasked to prepare comprehensive new drainage plan.

1893   Advisory Board on Drainage carries out massive topographic survey and devises state-of-the-science drainage plan. Surveyors map evidence of first sections of city to drop slightly below sea level, but engineers express no concern that further drainage will exacerbate subsidence.

1895   Brown's advisory board releases influential *Report on the Drainage of the City of New Orleans;* decides to discharge runoff eastward into Bayou Bienvenue and Lake Borgne via pumps placed along main outfall canal on South Broad and Florida Avenue. Later changed to go northward to Lake Pontchartrain; initial decision to go eastward explains problematic location of pumps in interior of city, rather than along periphery.

1895   City council passes Ordinance No. 10991, making "the plan of drainage as submitted by the Advisory Board . . . approved and accepted and made the plan of drainage [for] New Orleans," watershed moment in city's history of dewatering.[3]

1895   Westinghouse adopts alternating current (AC) and 25 Hz frequency for electrical generating plant at Niagara Falls, helping set industry standard. Engineers in New Orleans later follow suit, designing generators, transformers, transmission wires, and pumps for 25 Hz. But situation later changes, leaving local pumping power on obsolete frequency.

1896   Drainage Commission is formed to execute Brown's advisory board plan, but is not yet reliably funded with tax revenue.

1896–99   Citizens' League forms to elect pro-drainage politicians, while other progressives create Sewerage, Water and Drainage Campaign Committee and Woman's Sewerage and Drainage League, led by suffragist Kate M. Gordon. Women's involvement focuses on drumming up vote for June 1899 public referendum to fund Drainage Commission with special tax millage on assessed real estate value.

1897   Contracts are let to start building new drainage system, even though long-term funding is still not yet secured.

1899   On June 6, voters, including property-owning women, approve millage in landslide referendum. On June 22, city council passes Ordinance 15,391, levying new tax and finally securing reliable funding stream for Drainage Commission.

1899   State legislature amends constitution "to establish therein public systems of sewerage and water,"[4] creating New Orleans Sewerage & Water Board (S&WB). Drainage is not part of original charge, being responsibility of Drainage Commission.

1899–1902   Over $3 million is spent building out 20 percent of new drainage system, as canals are dug, interconnected, and covered along principal arteries, and pumping stations are erected along main canal.

1899   Young prodigy Albert Baldwin Wood joins S&WB, where he makes engineering history repeatedly over next fifty-seven years.

1900   Maj. Walter Reed's US Army Fourth Yellow Fever Commission meets with Dr. Carlos Finlay in Havana; subsequent collaborations prove that *Aedes aegypti* mosquito is yellow fever vector. While poor urban water management produces most mosquito habitat, swamps are primarily impugned, further motivating drive for reclamation and drainage.

1902   State legislature merges Drainage Commission with S&WB.

1902   Board of Inquiry launched to investigate problems with drainage installation.

1904–9   Roughly half of new drainage system is completed; pace of water removal increases to five thousand cubic feet per second, and reclaimed expanse grows to twenty-two thousand acres. Over 20 miles of covered concrete-lined canals, 17 miles of open canals, and 3 miles of wood-lined canals crisscross cityscape; by 1909, another 106 miles of subsurface pipelines are installed to further lower soil water for urbanization.

1906   Baldwin Wood designs centrifugal pump with high-speed impeller, creating hydrodynamic energy to lift water from suction and expel it to external discharge basin with phenomenal speed.

1907–9   Workers dig Algiers Outfall Canal to connect to Bayou Barataria; sewerage and water lines are concurrently installed and activated in 1909.

1908–10   Frank Hayne forms New Orleans Lake Shore Land Company (1908) and acquires marshes in today's New Orleans East. Two years later, Warren Reed founds New Orleans Drainage Company and teams with Hayne to reclaim eastern marshes for leasing as truck farms.

1909–10s   Drainage progress triggers new real estate development in Gentilly and Lakeview.

Early 1900s   Electrical engineers nationwide shift from 25 Hz to 60 Hz frequency, but S&WB, having made massive investments at 25 Hz, misses chance to modernize, and soon becomes stuck with antiquated technology too costly to replace.

1910s–50s   Drainage makes New Orleans bigger, wealthier, drier, warmer, lower, riskier—and more racially segregated, as real estate industry instills discriminatory language in contracts for houses in new subdivisions built on reclaimed swamps.

1910	Investors led by L. B. Langworthy purchase fifty-nine hundred acres of Kenner backswamp with aims to reclaim it for leasing as truck farms.
1910s	Drainage success obliges additional capacity upgrades, as more impermeable surfaces increase runoff vis-à-vis diminishing natural storage areas. Cycle will repeat for remainder of century.
1912	State legislature enables police juries (parish governments) to form drainage districts, within which landowners may vote to tax themselves and issue bonds to install mechanized drainage.
1912–13	Bent on private drainage for real estate development, George Hero amasses thousands of acres of swampland on West Bank, and joins with other landholders in three parishes to form Orleans-Jefferson-Plaquemines Drainage District.
1912	Wood augments his 1906 centrifugal pump design to thirty-inch diameter; installed as constant-duty pump in 1912 and remains in service today.
1913	Wood ups his centrifugal pump design to twelve-foot diameter.
1913	Sub-Drainage District No. 1 of Fourth Jefferson Drainage District is created in Kenner backswamp, and Langworthy's team has it polderized with new levees along Lake Pontchartrain and LaBranche wetlands. Pumps are installed at lakefront for dewatering, and lattice of drainage canals is dug, forming framework for future Kenner urbanization.
1913	Cracks appear on wall of St. Louis Cathedral, drawing attention of architects and prompting public discussion of drainage-caused subsidence.
1914	Sewerage & Water Board has eleven of Wood's newly designed twelve-foot centrifugal screw pumps manufactured and installed citywide, increasing total system capacity by 250 percent, from 4,400 to 11,200 cubic feet per second.
1914	In apotheosis year of city's drainage success, Mayor Martin Behrman proudly addresses League of American Municipalities in Milwaukee with rousing speech titled "New Orleans—A History of Three Great Public Utilities, Sewerage, Water and Drainage, and Their Influences Upon the Health and Progress of a Big City." Behrman reports taxable property rose from $140 million in 1900 to $250 million by 1914, while typhoid cases had dropped by half, malaria by 95 percent, and yellow fever by 100 percent—all for a mere $12.5 million investment in drainage and $17.5 million for potable water and sanitation.

1914   Hero's engineers install drainage apparatus and Wood screw pumps in West Bank swamps, while also launching South New Orleans Realty Company and Development Company and proposing bridge to Gretna for East Bankers to buy soon-to-be-reclaimed swampland.

1915   Hero Day on Mardi Gras weekend: "Drainage King" George Hero is heralded in elaborate parade to Barataria swamp, where ceremony is staged involving President Wilson on wire to White House. Pumps are activated, and West Bank swamplands are dewatered in three hours.

1915   Jefferson Parish creates Sub-Drainage District No. 2 (Hoey's Basin) and No. 4 (greater Metairie), readying them for reclamation.

1915   Great Storm of 1915 causes extensive damage, though advocates later crow about city being "storm proof." Flooding of recently drained low-lands prompts calls for levee improvements and lakefront reclamation for surge protection.

1916   Wood patents Wood Trash Pump to remove debris from runoff be-fore it enters screw pumps; raking mechanisms and screens are subse-quently affixed to pumping stations, and remain in service today, elim-inating enormous cost of manual removal.

1916–47   Three-decade lull in tropical activity abets enthusiasm for real estate development in drained swamplands.

1916   Warren Reed's Drainage Company, working since 1910, dewaters east-ern marshes owned by Frank Hayne's Lake Shore Land Company. By 1912, over 600 acres are reclaimed in Little Woods; by 1914, over 5,400 acres are reclaimed and one-third are cultivated; by 1916, over 6,000 acres are reclaimed, most of them cultivated. Lattice of drainage canals becomes future New Orleans East street grid.

1916–18   US Weather Bureau forecaster Dr. Isaac Cline publishes evidence that drainage has made New Orleans lower in elevation and slightly hotter in microclimate.

1918–23   Excavation of Inner Harbor Navigation (Industrial) Canal radically re-works hydrology of downriver half of East Bank, complicating 1895 decision to discharge most city runoff eastward through Bayou Bien-venue to Lake Borgne.

1919–23   Frank Hayne's Lake Shore Land Company starts turning over its drain-age apparatus to S&WB, exhibiting process of private drainage exe-cuted to raise value of private land and then transferred to public domain. Colonel De Montluzin acquires thirty-five thousand acres in 1923, including Hayne's reclaimed lands, for Faubourg de Montluzin, said to be the world's largest metropolitan land tract under one ownership. Parts later become New Orleans East.

1919   *Times-Picayune* headlines read "New Orleans Is Sinking Slowly But Steadily[;] Subsidence is Caused by Drying Out of Earth by Drainage Canals," first prominent report about city sinking below sea level.

1920   Engineers install "underground river" siphon beneath Inner Harbor Navigation (Industrial) Canal to get drainage discharge from main canal on Florida Avenue to Bayou Bienvenue and Lake Borgne. But hydrological bottleneck does not bode well for 1895 decision to discharge eastward.

1920s   Baldwin Wood's screw pumps are installed between Amsterdam and Zwolle to drain Zuider Zee in the Netherlands.

1921   Ford Meter Box Company of Wabash, Indiana, works with S&WB to design special water meter to endure effects of soil subsidence, and puts on each cover its crescent-and-stars logo, probably influenced by Indiana state flag. Visible to every pedestrian, mystical design becomes iconic, and is later trademarked by S&WB, despite appearing in other cities.

1921   State constitution is revised to enable Orleans Parish Levee Board to acquire private lakefront properties and build landmass with dredged sediment to prevent flooding from Lake Pontchartrain.

1921–23 State legislature makes Metairie swamps into Sub-Drainage District No. 3, with power to levy taxes and install drainage system. Work begins in 1923 with construction of new lakeshore levee, excavation of gravity ditches, reservoir canal on what is now West Esplanade Avenue, and outfall canals to lake, where four pumping stations are built.

1925   Col. Marcel Garsaud unveils plans for Lakefront Improvement Project, calling for dredging of thirty-six million cubic feet of lake-bottom sediment to create two thousand acres of new land at five feet above lake level.

1926   Jefferson Parish's new East Bank system is in full operation to reclaim Metairie and Kenner backswamp, though subsurface drainage takes another thirty years.

1926–34 Lakefront Improvement Project is completed, radically altering city's geography while iterating its bowl-shaped topography with two thousand acres of new elevated land in what had previously been brackish water and saline marsh.

1927   Great Mississippi River Flood appears to threaten New Orleans proper, but local pluvial flash flooding unrelated to river threat inadvertently triggers rash decision to dynamite levees downriver from city, causing unnecessary damage and lingering resentment.

1927–36   Flash flood of 1927 leads to $9 million in drainage upgrades, mostly involving improvements to electrical power generation and transmission to pumps.

1928   Flood Control Act of 1928 replaces "levees-only" policy for Mississippi River with more robust strategy incorporating levees as well as spillways, floodways, and reservoirs. Law gives federal government responsibility for controlling river and tributaries, but immunizes itself from liability should systems fail.

1929   First of fourteen fourteen-foot-diameter Wood screw pumps is installed in Broadmoor, world's largest pump by a margin of 50 percent and largest casting ever made in local foundry.[5]

1930s   After $21.8 million in expenditures, New Orleans drainage system comprises ten pumping stations powered by 35,000-horsepower electrical generators capable of moving 27,000 cubic feet per second, through 73 miles of open and closed canals and 870 miles of pipes, and discharging to three outfalls. System drains 39,000 acres on East Bank and 11,000 acres on West Bank, making living space for nearly a half-million people. Systems in neighboring parishes with commensurate equipment serve another eighty thousand. Metro-area settlement now spans river to lake, with new development expanding westward, eastward, and southward across river.

1939   Evidencing relationship between drainage and real estate development with racist deeds, 100 percent of Lakeview's 206 populated blocks, all former swamps which by this time had subsided two to four feet below sea level, are 100 percent white. Demographics of new Gentilly and Old Metairie subdivisions are similar.

1939–45   Outbreak of World War II brings millions of people to or through New Orleans; S&WB's critical drainage, water, and sewerage systems enable nearly thirty thousand workers to manufacture over twenty thousand vessels at seven Higgins Industries plants, most of them built on or near former swamplands.

1941	Voters in New Orleans approve fifty-year renewal of 1899 millage, continuing critical revenue stream for drainage improvements and operations.
1943	Gulf Intracoastal Waterway is excavated in eastern New Orleans for domestic barge traffic to evade German submarines in Gulf. But shipping channel also creates new ingress for storm surge.
1946	Robert Moses's *Arterial Plan for New Orleans* calls for Tide Water Ship Canal and harbor in eastern New Orleans, latest in long line of speculative seaways—only this one will come to fruition in late 1960s.
1947	First major storm to strike city since 1915, September hurricane causes extensive flooding in East Jefferson Parish, Gentilly, and Ninth Ward. Ninety percent of Louisiana's $100 million of damage is flooding-related, nearly all of it on or near drained swamplands.
1947–69	Upswing in tropical activity illustrates risk of surge flooding to subdivisions built on drained, sinking swamplands; efforts ensue to shore up lakefront and lateral levees.
1947	Recent flooding motivates Jefferson Parish leaders to secure $5.1 million in Army Corps support to build barricading levee along lakeshore, marking new era of federal/local partnerships for surge protection.
1948	Funding is secured to dig alternative route for GIWW from Bayou Barataria to new lock at Cut Off in Lower Coast of Algiers.
1948–53	Lakefront levee is completed across East Jefferson Parish lakefront; now all that is needed for full-scale development of reclaimed swamps is installation of subsurface drainage, which comes in late 1950s.
1949	Jefferson Parish Sheriff Frank Clancy is one of few to understand and explain relationship among drainage, subsidence, development, and flood risk exposure to public.
1952	Army Corps adopts new internal policy to provide Standard Project Flood protection in riverine areas that are prone to hurricane surge.
1955	Congress authorizes Army Corps to study East and Gulf coasts toward preventing surge-related damage and loss of life, setting stage for federal commitment to coastal communities.
1956	President Eisenhower signs bill to fund excavation of seaway in eastern New Orleans, later named Mississippi River–Gulf Outlet (MR-GO) Canal.

1956   Seventy-seven-year-old Albert Baldwin Wood, "the man who made wa-
        ter run uphill," dies of heart attack aboard beloved *Nydia* sloop while
        sailing off Biloxi, his favorite weekend pastime for decades. Brilliant
        engineer, revered at S&WB, exemplifies civil servants' dedication to city
        services in drainage, water, and sewerage.[6]

1956   Hurricane Flossy causes some flooding, but not enough to slow devel-
        opment in drained lowlands.

1956   Work is completed on GIWW Alternative Route ("Algiers Canal"), giv-
        ing barge traffic additional access to the river, but radically reshuffling
        West Bank drainage strategy and incurring new storm-surge risk.

1958   Congress stipulates 30 percent of federal hurricane-protection costs
        must be paid by local beneficiaries, giving them fiscal burden but also
        leverage in design and decision-making.

1958–59 Work begins on MR-GO Canal, with excavation of twenty million cubic
        yards of marsh soil to enlarge GIWW, plus twenty-seven million cubic
        yards to clear channel to Breton Sound.

1959   Army Corps extends Project Flood and Standard Flood Protection con-
        cepts to coastal environments, furthering federal commitment to pro-
        tect from storm surge.

Late 1950s Regarding three drainage outfall canals, all open to Lake Pontchar-
        train, Army Corps seeks to reconcile its surge-protection priority with
        S&WB's drainage priority and parish levee boards' canal-maintenance
        duties. No one solution—installing gates, changing pumping capacity,
        widening channels—pleases all parties, much less the public.

1960–65 Barge-mounted dredges enlarge MR-GO Canal to width of five hun-
        dred feet and depth of thirty-six feet.

1960   Army Corps recommends installing gates with bypass pumps at mouths
        of drainage outfall canals. S&WB worries that proposed gates, being
        under authority of Orleans Parish Levee Board, might not function
        with proper regard to S&WB's drainage imperative.

1961   Meeting resistance to gating of drainage outfall canals, Army Corps in-
        stead proposes Barrier Plan, in which surge would be kept out of Lake
        Pontchartrain courtesy of two massive barricades across Rigolets Pass
        and Chef Menteur Pass. Barrier Plan, which many assume will eventu-
        ally come to fruition, silences further calls to gate outfall canals.

1965   Congress approves Barrier Plan.

1965 Hurricane Betsy causes extensive wind and water damage; surge intrudes via funnel formed by GIWW and MR-GO (still under construction), rupturing levees along Industrial Canal and flooding parts of Ninth Ward, Gentilly, and eastern New Orleans. Damages are much worse than those caused by 1947 hurricane, because of newly dug surge ingress and because of extensive residential development in low-lying areas over subsequent eighteen years.

1965 Betsy disaster prompts Congress to authorize Flood Control Act of 1965, including $56 million Lake Pontchartrain and Vicinity Hurricane Protection Project, next major step in federalization of storm-surge protection. But roster of projects is only partially built, increasing city's exposure to hazard in decades ahead.

1968 MR-GO Canal is completed and opened to ocean-going shipping, as Port of New Orleans prepares to shift facilities eastward to envisioned "Centroport, USA."

1968 Congress creates National Flood Insurance Program, offering government insurance as private firms pull out of market. NFIP spreads out financial risk of living in coastal, riverine, and deltaic zones, but also lures development into repeat-flood areas which marketplace would otherwise deem too risky—thus creating lobby for additional federal protection from storm surges.

1960s–70s Opponents of Barrier Plan propose alternative High-Level Plan, which aims to raise lakefront levees to prevent lake surge from entering city. Similarly, opponents of gating outfall canals propose raising levees and floodwalls along these drainage channels.

1969 Passage of National Environmental Protection Act puts new expectations on drainage and flood-protection authorities, as growing environmental movement brings new values and demands to water-management debates.

1972–77 Spate of house explosions brings dangers of subsidence to public attention, results in new building codes to mitigate effects of sinking soils on construction.

1975 Save Our Wetlands, Inc., files lawsuit against Army Corps's Barrier Plan on grounds that it had not properly considered environmental impacts on lake ecosystems.

1977 US Circuit Court of Appeals agrees with Save Our Wetlands, Inc., finding Army Corps failed to consider alternatives to Barrier Plan. Ruling nullifies key component of LP&VHPP; corps responds by reevaluating entire project, taking additional eight years.

1978   Intense rainfall on May 3 overwhelms drainage system, causing flash flooding.

1980   Intense rainfall on April 13 floods subdivision on West Bank.

1981   Intense rainfall on June 10 causes flash flooding.

1982   Intense rainfall on April 23 causes flash flooding.

1983   Intense rainfalls on April 7 and December 28 cause extensive flash flooding.

1983–91   Worldwide oil bust stifles Gulf Coast economy. Metro-area investments fold; government coffers empty; and civic malaise prevails amid rising crime and decaying infrastructure, including drainage. S&WB spends a quarter of a billion dollars on drainage upgrades during 1980s, but more are needed as funds run dry.

1985   Army Corps officially abandons Barrier Plan in favor of High-Level Plan, opting to raise levees around Lake Pontchartrain and floodwalls along outfall canals, rather than blocking surge where it enters.

1985   Hurricane Juan, only significant hurricane to strike metro area during long tropical lull, causes pluvial flooding in West Bank subdivisions.

1986   Twenty thousand acres of marshes slated for subsurface drainage and residential development by New Orleans East, Inc., are instead transferred to federal government to become Bayou Sauvage National Wildlife Refuge. For once, dewatering and peopling wetlands are deemed bad policy. Entire refuge is flooded by Hurricane Katrina in 2005.

1989   Intensive rainfall on November 7 causes flash flooding in St. Bernard Parish.

1991   Drainage millage, passed triumphantly in 1899 and now up for second fifty-year renewal, is rejected by voters; loss of funds forces S&WB to transfer authority of twelve hundred miles of subterranean pipes and seventy-two thousand catch basins to Department of Public Works.

1994   Intensive rainfall on May 9 dumps seven inches, five times more than pumps can handle.

1995–96   Torrential downpours on May 8, 1995, send ten to eighteen inches of rain upon metro area, worst of recent "date storms," causing extensive flash flooding. "May 8 Flood" precipitates inclusion of Southeastern Louisiana Urban Flood Control Project (SELA) in Energy and Water

Appropriations Act and Water Resources Development Act of 1996; $455 million budget represents largest-ever federal investment in local drainage. Effort aims to augment drainage capacity to protect against ten-year rainstorm, calculated at nine inches over twenty-four hours. SELA construction continues through late 2010s.

1995   Upswing in tropical activity in August ends lull ongoing since Hurricane Camille in 1969; persists into 2020s, amid growing evidence of changing global climate.

1996–2005 $823 million in federal dollars fund water-management projects in Louisiana, of which SELA's drainage efforts account for fully 53 percent, while LP&VHPP's surge protection projects are only 17 percent.

2005   Four decades after Flood Control Act of 1965, LP&VHPP remains up to 40 percent incomplete, due to underfunding, confusion, resistance, and incompetence—while soils continue to subside, coast erodes, and sea levels rise.

2005   S&WB drains over ninety-five square miles of Orleans Parish with annual discharge capacity of nearly thirteen billion cubic feet.

2005   Hurricane Katrina strikes east of city on August 29; surge enters via unbarricaded Lake Pontchartrain, unblocked MR-GO/GIWW "funnel," and ungated drainage outfall canals, leading to multiple breaches in federal levees and floodwalls. Catastrophic flooding ensues, deeply inundating urban polders drained of swamplands a century ago. More than three hundred S&WB personnel stay on job round-the-clock, and many are on front lines of deadly disaster, heroically saving colleagues and others. Worldwide news coverage puts geography of New Orleans, including histories of drainage, subsidence, and coastal erosion, into international discourse.

2005   Floodwaters reach S&WB's power-generating station in Carrollton on August 31; personnel manage to keep deluge at bay, saving vital turbines.

2005   Three days after Katrina, engineers at Carrollton Plant pull off something never done before: cold-starting turbines and restarting pumps after total system stoppage. In days ahead, S&WB personnel commence epic task of dewatering sunken city.

2005   S&WB removes most floodwaters from city's main polders in just eleven days, far faster than predicted. Assisted by National Guard, FEMA, Army Corps, and other recovery workers, S&WB restores rudimentary water and sewerage services to unflooded neighborhoods by late September, and works outwardly thereafter.

2005–6　Ambitious recovery plans and debates over "shrinking the urban foot-print" are for naught, as *status quo ante* prevails in Katrina aftermath. No neighborhoods are closed, and homeowners are greenlighted to re-turn to heavily flooded, low-lying subdivisions courtesy of Road Home grants to rebuild, else sell out. Most opt to rebuild, but process takes years, and much of population never returns, principally working-class renters.

2006　Sen. Mary Landrieu and diplomats in the Netherlands organize Louisi-ana delegation to tour Dutch flood-protection systems; one participant, architect David Waggonner, comes away impressed by stormwater-management systems, inspiring launch of Dutch Dialogues.

2006–10s　Army Corps fast-tracks design and construction of Hurricane and Storm Damage Risk Reduction System (HSDRRS), priced at $14.6 bil-lion and engineered against "1-percent storm." Envisioned as one inte-gral system held to single standard, HSDRRS puts most resources into gating or barricading drainage and navigation canals built from 1870s through 1960s.

2006–10s　S&WB sees major changes to governance structure, as its mission broadens from operations to emergency planning and "future proof-ing." Hundreds of millions of dollars are spent on long-needed up-grades, backup systems, and new command center at Carrollton Plant.

2008　Two Dutch Dialogues sessions initiate conversations between Nether-lands water experts and local counterparts; focus is on improving water management by storing runoff on landscape through green and blue rather than gray infrastructure.

2008　Hurricane Gustav strikes southeastern Louisiana, triggering full-scale regional evacuation.

2009　Army Corps begins work on East Bank linchpin of HSDRRS, Inner Har-bor Navigation Canal–Lake Borgne Surge Barrier, to barricade funnel formed by GIWW and MR-GO.

2010　Dutch Dialogues holds third session at national conference of Ameri-can Planning Association, whose theme is delta urbanism.

2011　Army Corps completes work on West Bank linchpin of HSDRRS, West Closure Complex, where North America's largest sector gates protect against Barataria Basin surge, while world's largest pumping station is capable of ejecting nineteen thousand cubic feet of runoff per second.

2012   Category 1 Hurricane Isaac strikes region, lingering to cause extensive wind, rain, and surge damage, along with days-long blackouts.

2013   Army Corps replaces post-Katrina temporary gates and bypass pumps with Permanent Canal Closures & Pumps at mouths of three drainage outfall canals in Lakeview and Gentilly.

2013   Waggonner & Ball, founder of Dutch Dialogues, releases Urban Water Management Plan for East Bank of metropolis. Basic principles of storing runoff on landscape and "living with water" resonate with city officials and nonprofits, which launch scores of micro-initiatives and pilot projects.

2014   The Water Institute, another outcome of Netherlands partnership traceable to 2006 junket, becomes Center of Excellence and builds riverside campus in Baton Rouge.

2015   City releases Resilient New Orleans Plan, incorporating recommendations of 2013 Urban Water Management Plan.

2016   US Department of Housing and Urban Development awards $141 million for Gentilly Resilience District, focused on implementing Mirabeau Water Garden as recreational park storing and circulating up to ten million gallons of runoff.

2017   Intense cloudburst on August 5 causes flash flooding in Mid-City, Seventh Ward, and Gentilly; S&WB initially reports all pumping stations operating at full capacity. Subsequent hearings find this and other statements to be faulty, leading to major shake-up in drainage leadership. Worst flash flood since 1995, "August 5 Flood" becomes rallying cry for sustainable urban water management.

2018   Violent rainstorm on May 18 overwhelms pump capacity and causes flash flooding. In subsequent press conference, Mayor LaToya Cantrell declares "We are a city that floods. *We are a city that floods.*"

2018   Engineer Ghassan Korban becomes executive director of S&WB, hired for his engineering acumen as well as support of green and blue approaches to urban water management.

2019   Internal S&WB investigation reports drainage to be "chronically underfunded" compared to fee-based water and sewerage service. "It has not gained a new revenue source since 1982, and in 1991 it lost a dedicated millage."[7]

2019   City council mandates businesses must use pervious pavements for new projects, such that runoff may filtrate into soils, rather than rushing into drainage system and further burdening pumps.

2019   After five years of construction, Army Corps completes $150 million SELA project known as Pump to the River, in which heavy runoff in Jefferson Parish's Hoey's Basin is pumped into nearby Mississippi River, rather than being last in line for discharge through Orleans Parish's Seventeenth Street Outfall Canal to distant Lake Pontchartrain.

2021   City announces S&WB and Entergy will split $74 million expenditure to construct new power station at Carrollton Plant to tie pumps to external municipal power, part of ongoing effort to wean S&WB off dependency on outdated in-house generation capacity.[8]

2021   Exactly sixteen years after Hurricane Katrina, Category 4 Hurricane Ida strikes southeastern Louisiana; storm's late intensification and eastward veer cripple Entergy's electrical transmission lines, leaving entire metropolis dark for many days. But feared stormwater flooding is mostly avoided, as S&WB manages to run drainage system exclusively on in-house generators, including recently repaired Turbines 4 and 5 at Carrollton Plant and backup generators. For days after landfall, "the New Orleans Sewerage & Water Board was producing more electricity in the city than the electric company." Episode reframes debate over municipal versus in-house power generation, even with obsolete 25 Hz frequency. "As Katrina Was to Levees," reads *Times-Picayune* headline, "Ida Is to Electricity."[9] Jefferson Parish fares worse, as loss of power hobbles water and sewerage services. "The crash of those basic utilities, especially in East Jefferson, amounted to an uneasy role reversal for a parish that prides itself on having its act together more than its notoriously dysfunctional neighbor, New Orleans."[10]

[1] Bowen, *American Almanac and Repository of Useful Knowledge for the Year 1837*, 324.

[2] "New Orleans, June 9," *New Orleans Bee*, rpt. in *Evening Post* (New York), June 22, 1836, 2.

[3] Article 583 of Ordinance No. 10991 of the City of New Orleans, July 10, 1895, *Flynn's Digest of the City Ordinances*, 268.

[4] Act No. 4, S.B. No. 3, "Joint Resolution Proposing an Amendment to the Constitution [for] Ratifying and Carrying Into Effect a Special Tax Levied in the City of New Orleans," 6–7.

[5] Ray M. Thompson, "Albert Baldwin Wood: The Man Who Made Water Run Uphill," article for *New Orleans Magazine* republished as booklet by Sewerage and Water Board of New Orleans (1999), 20.

[6] Thompson, "Albert Baldwin Wood," 28–29.

[7] Green et al., *Task Force on New Orleans Sewerage, Water, and Drainage Utilities: Report of Findings and Recommendations*, 2.

[8] Jeff Adelson, "S&WB to Switch to Entergy Power," *Times-Picayune/New Orleans Advocate*, June 17, 2021, 1.

[9] David Hammer, "S&WB Draws Praise in Storm Operations," *Times-Picayune/New Orleans Advocate*, September 14, 2021, 1B; Mark Ballard, "As Katrina Was to Levees, Ida Is to Electricity," *Times-Picayune/New Orleans Advocate*, September 6, 2021, 1.

[10] Faimon Roberts and Chad Calder, "Jefferson Systems Slammed by Storm, *Times-Picayune/New Orleans Advocate*, September 26, 2021, 7A.

Notes

PROLOGUE

1. "Largest Refinery in Louisiana," *Sugar: An English-Spanish Technical Journal Devoted to Sugar Production* 22 no. 4 (April 1920): 195.

2. "Largest Refinery in Louisiana," 195.

3. "'I'll Drain a Million Acres More!' George A. Hero to Celebration Dinner," *New Orleans Item,* February 14, 1915, 1.

4. "'I'll Drain a Million Acres More!'" 1.

INTRODUCTION: THESE OOZY AND MUDDY LANDS

1. Christina Rae Butler, *Lowcountry at High Tide: A History of Flooding, Drainage, and Reclamation in Charleston, South Carolina* (Columbia: University of South Carolina Press, 2020), 5.

2. "Tackling the World's Toughest Drainage Problem . . . With Success!" Community and Intergovernmental Relations, Sewerage & Water Board of New Orleans, undated agency flier. Engineers at the S&WB would probably agree their city is the toughest *urban* drainage problem.

3. "New Orleans [was once] entirely surrounded by water and with its soil saturated. . . . In late years, however, levees have prevented overflows, reclamation projects have [drained] swamps and, finally, sub-surface drainage . . . has eliminated surface water [and] ground moisture [by] eight or ten feet." Isaac Cline, qtd. in "New Orleans Is Getting Hotter—Increases in Temperature in Summer Attributed to the Drainage System," *Columbus [GA] Ledger-Enquirer,* June 13, 1918, p. 1, col. 3.

4. Dorota Z. Haman, "Pumps, Displacement," *Encyclopedia of Water Science,* ed. B.A. Stewart and Terry A. Howell (New York: Marcel Dekker, Inc. 2003), 759.

5. Mohammad Valipour et al., "The Evolution of Agricultural Drainage from the Earliest Times to the Present," *Sustainability* 12, no. 416 (2020): 22–23.

6. Shannon Lee Dawdy, *Building the Devil's Empire: French Colonial New Orleans* (Chicago: University of Chicago Press, 2008), 66. See also Karl August Wittfogel, *Oriental Despotism: A Comparative Study of Total Power* (New Haven, CT: Yale University Press, 1957), which explores how "hydraulic despotism" creates powerful centralized bureaucracies by exerting control over water, via irrigation, drainage, flood control, and potable water systems.

7. Steven J. Burian and Findlay G. Edwards, "Historical Perspectives of Urban Drainage," *Ninth International Conference on Urban Drainage,* April 2012, 5–6.

8. US Army Corps of Engineers, New Orleans District, *National Register Evaluation of New Orleans Drainage System, Orleans Parish, Louisiana,* prepared by Earth Search, Inc., December 1999, 3.

9. Fransje Hooimeijer, Han Meyer, and Arjan Nienhuis, *Atlas of Dutch Water Cities* (Netherlands: SUN Publishers, 2005), 155.

10. Hooimeijer, Meyer, and Nienhuis, *Atlas of Dutch Water Cities,* 21–41.

11. Paraphrased from *Land Subsidence in the United States,* ed. Devin Galloway, David R. Jones, and S. E. Ingebritsen (US Geological Survey, Circular 1182, 1999), 2.

12. Longshore currents, weak as they are, did sweep some of the river's main deltaic deposits westward, forming a derivative delta known as the Chenier Plain, named for its brushstroke-like east-to-west ridges upon which oak trees (*chêne*) grew. This is today's south-central and southwestern Louisiana.

13. Captain Philip Pittman, *The Present State of the European Settlements on the Mississippi, with a Geographical Description of That River* (London: J. Nourse, 1770), 5.

14. Elisée Reclus, "Fragment d'un voyage à la Nouvelle Orléans" (1855), trans. Camille Martin and John Clark, in *Mesechabe: The Journal of Surre(gion)alism,* New Orleans, vol. 12 (Winter 1993–94).

15. Letter from Father Vivier to the Society of Jesus, to a Father of the same Society (emphasis added), in *The Jesuit Relations and Allied Documents: Travels and Explorations of the Jesuit Missionaries in New France 1610–1791, Volume LXIX—All Missions, 1710–1756,* ed. Reuben Gold Thwaites (New York, 1959), 210–13.

16. "Those marshes which have not acquired a sufficient consistency to produce trees, and shake . . . when trodden on, are in Louisiana called *prairies tremlantes*" (trembling prairies). Edward Livingston, *An Answer to Mr. Jefferson's Justification of His Conduct in the Case of the New Orleans Batture* (Philadelphia, 1813), 6n.

17. Le Page du Pratz, *The History of Louisiana,* ed. Joseph G. Tregle Jr. (1758; rpt. Baton Rouge, 1976), 128.

18. *Benjamin Henry Boneval Latrobe, Impressions Respecting New Orleans: Dairy & Sketches 1818–1820,* ed. Samuel Wilson Jr. (New York: Columbia Press University, 1951), 68.

19. Adam Mandelman, *The Place with No Edge: An Intimate History of People, Technology, and the Mississippi River Delta* (Baton Rouge: Louisiana State University Press, 2020), 100.

I. ADAPT OR RETREAT

1. Daniele Coxe, *A Description of the English Province of Carolina, by the Spaniards Call'd Florida, and by the French La Louisiane* (1727; rpt. San Francisco: California State Library, 1940), 21; Marc-Antoine Caillot, *A Company Man: The Remarkable French-Atlantic Voyage of a Clerk for the Company of the Indies—A Memoir by Marc-Antoine Caillot,* ed. Erin M. Greenwald, trans. Teri F. Chalmers (New Orleans: Historic New Orleans Collection, 2013), 78; Pierre Clément de Laussat, *Memoirs of My Life* (1831; trans., Baton Rouge: Louisiana State University Press, 1978), 40.

2. Fred B. Kniffen, Hiram F. Gregory, and George A. Stokes, *The Historic Indian Tribes of Louisiana, from 1542 to the Present* (Baton Rouge: Louisiana State University Press, 1987), 20. See also "Indian Economies" in Fred B. Kniffen and Sam Bowers Hilliard, *Louisiana: Its Land and People* (Baton Rouge: Louisiana State University Press, 1988), 107.

3. Tristam Kidder, "Making the City Inevitable: Native Americans and the Geography of New

Orleans," in *Transforming New Orleans and Its Environs: Centuries of Change,* ed. Craig Colten (Pittsburgh: University of Pittsburgh Press, 2000), 13.

4. Spanish surveyor Carlos Trudeau, 1803, qtd. in Betsy Swanson, *Terre Haute de Barataria: An Historic Upland on an Old River Distributary Overtaken by Forest in the Barataria Unit of the Jean Lafitte National Historic Park and Preserve* (Harahan, LA: Jefferson Parish Historical Commission Monograph XI, 1991), 17.

5. Kidder, "Making the City Inevitable," 13–16.

6. Daniel H. Usner, "American Indians in New Orleans: Native Communities Were Integral to the City's Foundation," in *New Orleans and the World: The Tricentennial Anthology,* ed. Nancy Dixon (New Orleans: Louisiana Endowment for the Humanities, 2017), 13–16; Kniffen, Gregory, and Stokes, *The Historic Indian Tribes of Louisiana,* 52–57 and 123; presentations and personal communications at "Indigenous Spaces, French Expectations: Exploring Exchanges Between Native and Non-Native Peoples in Louisiana," New Orleans Center for the Gulf South Symposium, March 14, 2018.

7. Dawdy, *Building the Devil's Empire,* 65.

8. To get an idea of their relative worth to the company, the 1721–22 salaries of Bienville, La Tour, and Pauger, were 12,000 *livres,* 8,000 *livres,* and 5,000 *livres,* respectively, at a time when an enslaved person cost 660 *livres.* Jean-Baptiste Bénard de La Harpe, *The Historical Journal of the Establishment of the French in Louisiana,* trans. Joan Cain and Virginia Koenig, ed. Glenn R. Conrad (Lafayette: Center for Louisiana Studies, University of Southwestern Louisiana, 1971), 137–41.

9. Minutes of the Council of Commerce of Louisiana (Comparison between the old and the new Biloxi), November 25, 1720, *Mississippi Provincial Archives 1704–1743: French Dominion,* ed. Dunbar Rowland and Albert Godfrey Sanders (Jackson, MS, 1932), vol. 3: 298–99.

10. Le Blond de La Tour, *Plan des ouvrages projettés pour le nouveau establissement du Nouveau Biloxy,* January 8, 1721, French National Archives; Sébastien Le Prestre, Seigneur de Vauban, *The New Method of Fortification, As Practiced by Monsieur de Vauban, Engineer General of France, with an Explication of All Terms Appertaining to that Art* (London: Abell Swall, 1693), topics culled from table of contents of second edition.

11. Pierre Le Moyne, Sieur d'Iberville, *Iberville's Gulf Journals,* ed. Richebourg Gaillard McWilliams (University, AL, 1991 translation of 1700 journal), 53; Anonymous, "Historical Journal: or, Narrative of the Expedition Made by Order of Louis XIV, King of France, under Command of M. D'Iberville to Explore the Colbert (Mississippi) River and Establish a Colony in Louisiana," in *Historical Collections of Louisiana and Florida,* ed. B. F. French (New York, 1875; translation of 1699 journal), vol. 7: 57–58.

12. Antoine Crozat, qtd. by Charles L. Dufour, *Ten Flags in the Wind: The Story of Louisiana* (New York: Harper & Row, 1967), 75.

13. Qtd. by Baron Marc de Villiers du Terrage, "A History of the Foundation of New Orleans (1717–1722)," *Louisiana Historical Quarterly* 3, no. 2 (April 1920): 174 (emphasis in original). The date for September of 1717 is obscured in the register, but ancillary information indicates it was September 9.

14. Jonathan Darby, "New Orleans, The Capital of the Colony and the Seat of Government and the Courts of Justice," trans. Rev. Conrad M. Widman, SJ, and published in "Some Southern Cities (in the U.S.) about 1750," Records of the American Catholic Historical Society of Philadelphia, vol. 10 (1899): 202; Jonathan Darby, qtd. by Shannon Lee Dawdy, *Madame John's Legacy (160R51) Revisited: A Closer Look at the Archeology of Colonial New Orleans* (New Orleans, 1998), 26–29.

15. This information comes from the *Journal Historique Concernant l'Establissement de la Louisiane, tiré des Mémoires Originaux par le Chevalier de Beaurain, géographe ordinaire du Roy,* qtd. in the *Relation de Pénicaut, in* Pierre Margry, *Découvertes Et Établissements Des Français Dans L'Ouest Et Dans Le Sud L'Amérique Septentrionale, 1614-1754* (Paris: D. Jouaust, 1875), vol. 5: 549n.

16. Gilles-Antoine Langlois, "French Architect-Engineers of New Orleans, 1718-1730," in *New Orleans, the Founding Era / La Nouvelle-Orléans, les Années Fondatrices,* ed. Erin M. Greenwald, trans. Henry Colomer (New Orleans: Historic New Orleans Collection, 2018), 61.

17. Thomas Jefferys, *The Natural and Civil History of the French Dominions in North and South America* (London, 1760), 148-49.

18. Charles R. Maduell Jr., *The Census Tables for the French Colony of Louisiana from 1699 to 1732* (Baltimore, 1972), 16-22 and 81.

19. This line appears on Le Blond de La Tour's January 12, 1723, map, *Partie du Plan de la Nouvelle Orleans* (French National Archives), drawn in red and labeled "Alignement Suiuant le projet de Mr. de Bienville des premieres maisons." See also Samuel Wilson Jr., *The Vieux Carre, New Orleans: Its Plan, Its Growth, Its Architecture* (New Orleans, 1968), 4.

20. Marion Stange, "Governing the Swamp: Health and the Environment in Eighteenth-Century Nouvelle Orléans," *French Colonial History* 11 (2010): 7; Marcel Giraud, *A History of French Louisiana, Volume Five: The Company of the Indies, 1723-1731* (Baton Rouge: 1987), 498-99; Le Blond de La Tour, *Partie du Plan de la Nouvelle Orleans,* January 12, 1723, French National Archives; Pauger quote from Wilson, *Vieux Carre, New Orleans,* 12.

21. As quoted in Wilson, *Vieux Carre, New Orleans,* 11.

22. Villiers, "History of the Foundation of New Orleans," 223.

23. Villiers, "History of the Foundation of New Orleans," 222-23.

24. Villiers, "History of the Foundation of New Orleans," 229.

25. Qtd. by Villiers, "History of the Foundation of New Orleans," 226.

26. Pierre François Xavier de Charlevoix, *Journal of a Voyage to North-America Undertaken by Order of the French King* (London, 1761), vol. 2: 276.

27. Charlevoix, *Journal of a Voyage to North-America* 2: 271-73.

28. Letter, Bienville to the Council, February 1, 1723, *Mississippi Provincial Archives 1704-1743: French Dominion,* ed. Rowland and Sanders (Jackson, MS, 1932), vol. 3: 343-44.

29. A footnote in Dumont's journal, as well as a number of tertiary sources, date this hurricane to September 11, 1721, but 1722 is the more likely date.

30. Adrien de Pauger, qtd. by Wilson, *Vieux Carre, New Orleans,* 13.

31. M. Dumont, "History of Louisiana, Translated from the Historical Memoirs of M. Dumont," in *Historical Memoirs of Louisiana, From the First Settlement of the Colony to the Departure of Governor O'Reilly in 1770,* ed. B. F. French (New York, 1853), 24.

32. Adrien de Pauger, qtd. by Wilson, *Vieux Carre, New Orleans,* 13.

33. La Harpe, *Historical Journal of the Establishment of the French in Louisiana,* 156.

34. Adrien de Pauger, qtd. by Wilson, *Vieux Carre, New Orleans,* 13.

35. Adrien Pauger, in a letter sent to Paris and copied to Le Blond de La Tour in New Biloxi, April 14, 1721, qtd. in Wilson, *Vieux Carre, New Orleans,* 11.

36. "Pauger's Savvy Move: How an Assistant Engineer Repositioned New Orleans on Higher Ground," Richard Campanella, *Cityscapes of New Orleans* (Baton Rouge: Louisiana State University Press, 2017).

2. THE DITCH-AND-GRAVITY ERA, 1720S–1830S

1. Dumont, "History of Louisiana," 23–24.

2. M. Dumont, *History of Louisiana, Translated from the Historical Memoirs of M. Dumont* (chap. 18: *Arrival of the Royal Commissaries at New-Orleans—Establishment of a Council in That Capital*), trans. and rpt. in *Historical Memoirs of Louisiana,* ed. French, 41; Henry P. Dart, editorial introduction to "Allotment of Building Sites in New Orleans (1722)," *Louisiana Historical Quarterly* 7, no. 4, (October 1924): 564–65.

3. du Pratz, *The History of Louisiana,* 54.

4. Letter, Périer and De La Chaise to the Directors of the Company of the Indies, written April 22, 1727, received October 27, 1727, *Mississippi Provincial Archives 1701–1729: French Dominion,* ed. Rowland and Sanders (Jackson: Press of the Mississippi Department of Archives and History, 1929), vol. 3: 537 and 537n. See also Lawrence N. Powell, *The Accidental City: Improvising New Orleans* (Cambridge, MA: Harvard University Press, 2011), 67.

5. Stange, "Governing the Swamp," 2–3.

6. Anonymous, *New Orleans As It Is: Its Manners and Customs* ("By a Resident, Printed for the Publisher," 1850), 20.

7. Craig Colten, *An Unnatural Metropolis: Wresting New Orleans from Nature* (Baton Rouge: Louisiana State University Press, 2006), 33; Burian and Edwards, "Historical Perspectives of Urban Drainage," 3.

8. Dumont, "History of Louisiana," 23–24.

9. Letter, King to Bienville and Salmon, February 2, 1732, *Mississippi Provincial Archives 1704–1743: French Dominion,* ed. Rowland and Sanders, vol. 3: 563.

10. Letter, Périer and De La Chaise to the Directors of the Company of the Indies, written November 3, 1728, received June 22, 1729, *Mississippi Provincial Archives 1701–1729: French Dominion,* ed. Rowland and Sanders, vol. 2: 591–92.

11. Samuel Wilson Jr., "Bienville's New Orleans: A French Colonial Capital, 1718–1768, and La Nouvelle Orléans: Le Vieux Carré," *The Collected Essays of Samuel Wilson, Jr., F.A.I.A.,* ed. Jean M. Farnsworth and Ann M. Masson (Lafayette, LA, 1987), reproduced in *The Louisiana Purchase Bicentennial Series in Louisiana History, vol. 14: New Orleans and Urban Louisiana—Part A, Settlement to 1860,* ed. Samuel C. Shepherd Jr. (Lafayette: Center for Louisiana Studies, University of Louisiana, 2005), 42–43 and 47.

12. Letter, Bienville and Salmon to Maurepas, Reply to the King's Memoir, May 12, 1733, *Mississippi Provincial Archives 1704–1743: French Dominion,* ed. Rowland and Sanders, vol. 3: 594.

13. Langlois, "French Architect-Engineers of New Orleans, 1718–1730," 66.

14. Powell, *Accidental City,* 67; Stange, "Governing the Swamp," 8–9.

15. This according to Jacques Tanesse, as cited by François-Xavier Martin in *Orleans Term Reports, or Cases Argued and Determined in the Superior Court of the Territory of Orleans* (New Orleans: Roche, Brothers, 1813), vol. 2: 15.

16. Ignace-François Broutin, *Plans et Profils d'un Pont de Brique pour l'Ecoulement des eause de cette Ville a la Nouvelle Orleans,* April 3, 1732, and *Plan et Profil d'un Pont de Briques pour estre Construir Sur les Fossez de cette Ville a la Nouvelle Orleans,* April 2, 1732, Collection Moreau de Saint Méry, Archives Nationales de France.

17. *Partie du Plan de la Nouvelle Orleans Pour Faire,* 1734, Center for Louisiana Studies, University of Louisiana at Lafayette.

18. Brison, *Projet de fortification de la Nouvelle Orléans,* April 10, 1730, Dépôt des Fortifications des Colonies, Archives Nationales de France.

19. Burian and Edwards, "Historical Perspectives of Urban Drainage," 11–12.

20. A. Oakey Hall, *The Manhattaner in New Orleans; or Phases of "Crescent City" Life* (New York: J. S. Redfield, Clinton Hall, 1851), 51.

21. Letter, Bienville and Salmon to Maurepas, March 20, 1734, *Mississippi Provincial Archives 1704–1743: French Dominion,* ed. Rowland and Sanders, vol. 3: 637-38.

22. Stange, "Governing the Swamp," 9.

23. Burian and Edwards, "Historical Perspectives of Urban Drainage," 7

24. "Survey of the Plantation of the Company of the Indies, Made by the Sieur Lassus, February 8, 1728," and a subsequent 1731 inventory, reproduced in Samuel Wilson Jr.'s "The Plantation of the Company of the Indies," *Louisiana History* 31, no. 2 (Spring 1990): 170–74 and 179.

25. Wilson, "Plantation of the Company of the Indies," 179.

26. John H. B. Latrobe, *Southern Travels: Journal of John H. B. Latrobe, 1834,* ed. Samuel Wilson Jr. (New Orleans: Historic New Orleans Collection, 1986), 38.

27. Letter, King to Bienville and Salmon, February 2, 1732, *Mississippi Provincial Archives 1704–1743: French Dominion,* ed. Rowland and Sanders, vol. 3: 563–64.

28. As depicted in the *Plan de la Nouvelle Orleans* by Gonichon, 1731; see also Wilson, "Bienville's New Orleans," 42–43.

29. Letter, Périer and De La Chaise to the Directors of the Company of the Indies, written November 3, 1728, received June 22, 1729, *Mississippi Provincial Archives 1701–1729: French Dominion,* ed. Rowland and Sanders (Jackson: Press of the Mississippi Department of Archives and History, 1929), vol. 2: 591–92.

30. Letter, De La Chaise and the Four Councillors of Louisiana to the Council of the Company of the Indies, April 26, 1725, *Mississippi Provincial Archives 1704–1743: French Dominion,* ed. Rowland and Sanders, vol. 2: 464.

31. Letter, Périer and De La Chaise to the Directors of the Company of the Indies, written January 30, 1729, received June 22, 1729, *Mississippi Provincial Archives 1701–1729: French Dominion,* ed. Rowland and Sanders, vol. 2: 616–17.

32. James Pitot, *Observations on the Colony of Louisiana from 1796 to 1802* (New Orleans: The Historic New Orleans Collection, 1981), 101-2.

33. Letter, Périer to Maurepas, December 10, 1731, *Mississippi Provincial Archives: French Dominion, 1729–1748,* ed. Rowland and Sanders, rev. and ed. Patricia Kay Galloway (Baton Rouge: Louisiana State University Press, 1984), vol. 4: 107.

34. Letter, Périer to Maurepas, December 10, 1731, 107.

35. Letter, Michel to Rouillé, September 23, 1752, *Mississippi Provincial Archives: French Dominion, 1749–1763,* ed. Rowland and Sanders, rev. and ed. Galloway (Baton Rouge: Louisiana State University Press, 1984), vol. 5: 116.

36. Letter, Périer to Ory, December 18, 1730, *Mississippi Provincial Archives: French Dominion, 1729–1748,* ed. Rowland and Sanders, vol. 4: 44.

37. Louis Bouchereau, qtd. by Mandelman, *Place with No Edge,* 26.

38. de Laussat, *Memoirs of My Life,* 40; Letter, Périer and De La Chaise to the Directors of the Company of the Indies, written January 30, 1729, received June 22, 1729, *Mississippi Provincial Archives 1701–1729: French Dominion,* ed. Rowland and Sanders, vol. 2: 616–17.

39. Richard Campanella, *Time and Place in New Orleans: Past Geographies in the Present Day* (New Orleans: Pelican Publishing Co., 2002).

40. Memoir on Louisiana [by Bienville], 1726, *Mississippi Provincial Archives 1704–1743: French Dominion,* ed. Rowland and Sanders, vol. 3: 516.

41. Letter, Bienville and Salmon to Maurepas, Reply to the King's Memoir, May 12, 1733, *Mississippi Provincial Archives 1704–1743: French Dominion,* ed. Rowland and Sanders, vol. 3: 598.

42. The lakes of the Washas are today's Lake Cataouatche and Lake Salvador on the West Bank. Letter, Périer and De La Chaise to the Directors of the Company of the Indies, written January 30, 1729, received June 22, 1729, *Mississippi Provincial Archives 1701–1729: French Dominion,* ed. Rowland and Sanders, vol. 2: 617.

43. The Chickasaw War, a forerunner to the French and Indian War, was a series of battles fought by French colonists allied with Choctaw and Illinois tribes, against British colonists and their Chickasaw allies. Samuel Wilson Jr., "Seven Oaks Plantation (Petit Desert), Westwego, Jefferson Parish, Louisiana" (Washington, DC: National Park Service, Historic American Building Survey, 1953), HABS No. LA-1158, 2–3; Map, *A Plan of the Coast of Part of West Florida & Louisiana: Including the River Yazous,* by George Gauld and Julius Erasmus (1778), Library of Congress.

44. Pitot, *Observations on the Colony of Louisiana,* 68–69.

45. Thomas A. Becnel, *The Barrow Family and the Barataria and Lafourche Canal: The Transportation Revolution in Louisiana, 1829–1925* (Baton Rouge: Louisiana State University Press, 1989), 28; Betsy Swanson, *Historic Jefferson Parish, From Shore to Shore* (1975; rpt. Gretna, LA: Pelican Publishing Co., 2004), 87–88.

46. Friends of the Cabildo, *New Orleans Architecture, vol. 4: The Creole Faubourgs* (Gretna, LA: Pelican Publishing Co., 1974), 4–5.

47. Pittman, *The Present State of the European Settlements on the Mississippi,* 24.

48. "Survey of the Plantation of the Company of the Indies, Made by the Sieur Lassus, February 8, 1728," and subsequent 1731 inventory, reproduced in Wilson, "Plantation of the Company of the Indies," 170–74 and 179.

49. Gwendolyn Midlo Hall, *Africans in Colonial Louisiana: The Development of Afro-Creole Culture in the Eighteenth Century* (Baton Rouge: Louisiana State University Press, 1992), 137.

50. Letter. Louis XV to Kerlèrec, January 1762, and Act of Acceptance of Louisiana by Charles III of Spain, November 13, 1762, *Mississippi Provincial Archives: French Dominion, 1749–1763,* ed. Rowland and Sanders, vol. 5: 274–83.

51. Mandelman, *Place with No Edge,* 24.

52. Gilbert C. Din and John E. Harkins, *The New Orleans Cabildo: Colonial Louisiana's First City Government, 1769–1803* (Baton Rouge: Louisiana University Press, 1996), 42–58; Christina Vella, *Intimate Enemies: The Two Worlds of the Baroness de Pontalba* (Baton Rouge: Louisiana University Press, 1997), 22–32.

53. Din and Harkins, *New Orleans Cabildo,* 245.

54. Din and Harkins, *New Orleans Cabildo,* 246.

55. *Digest of the Acts and Deliberations of the Cabildo,* March 26, 1779, book 1, p. 312, Streets: Bridges and Gutters, as posted by the New Orleans Public Library.

56. *Records . . . of the Cabildo,* book 3, no. 2 (January 1, 1788, to May 18, 1792), February 24, 1792, 174–174-A, p. 195 of second microfilm roll; and book 4, no. 3 (January 1, 1799, to December 12, 1800),

February 15, 1799, 132A–133, p. 13 of third microfilm roll of Cabildo records, Historic New Orleans Collection.

57. *Records . . . of the Cabildo,* book 3, no. 2 (January 1, 1788, to May 18, 1792), February 24, 1792, 174-174-A, p. 195 of second microfilm roll; and book 4, no. 3 (January 1, 1799, to December 12, 1800), February 15, 1799, 132A–133, p. 13 of third microfilm roll of Cabildo records.

58. Din and Harkins, *New Orleans Cabildo,* 246.

59. *Digest of the Acts and Deliberations of the Cabildo,* February 15, 1799, book 4, vol. 3: 13, Drainage section, as posted by the New Orleans Public Library.

60. Samuel Wilson Jr., "Early History of the Lower Garden District, in Friends of the Cabildo, *New Orleans Architecture, vol. 1: The American Sector* (Gretna, LA: Pelican Publishing Co., 1971), 6.

61. Samuel Wilson, Jr., "Early History of Faubourg St. Mary," in Friends of the Cabildo, *New Orleans Architecture, vol. 2: The American Sector* (Gretna, LA: Pelican Publishing Co., 1972), 3–8.

62. *Records and Deliberations of the Cabildo, 1769–1803,* trans. Works Progress Administration, October 17, 1788, 107/107-A, 39–40 (emphasis added), second microfilm roll of Cabildo records.

63. *Report of the Sanitary Commission of New Orleans on the Epidemic Yellow Fever of 1853* (New Orleans: City Council, 1854), viii. This leper colony had existed at least since the 1790s, when the Carondelet Canal was cut near "the high lands of the Lepers," meaning somewhere along the Metairie or Gentilly Ridge by Bayou St. John. Martin, *Orleans Term Reports, vol. 2: 12.*

64. J. E. Alexander, *Transatlantic Sketches, Comprising Visits to the Most Interesting Scenes in North and South America, and the West Indies* (London, 1833), vol. 2: 32–34.

65. *Plano del Sector del Reducto de Borgoña,* Nueva Orleans, May 22, 1794, Archivo General de Indias, Sevilla, España; Friends of the Cabildo: Roulhac Toledano and Mary Louise Christovich, *New Orleans Architecture, vol. 6: Faubourg Tremé and the Bayou Road* (New Orleans: Friends of the Cabildo / Pelican Publishing Co., 1980), 6; Richard Campanella, "'Fort-Prints': Relics of New Orleans's Fortified Past," *New Orleans Times-Picayune,* May 12, 2017.

66. François-Xavier Martin, *The History of Louisiana From The Earliest Period* (New Orleans: A.T. Penniman & Co., 1829), vol. 2: 111–13.

67. Din and Harkins, *New Orleans Cabildo,* 246–49.

68. This curious curving canal—likely based on a natural bayou, given its shape—is shown on an 1804 map drafted by Carlos Laveau Trudeau based on Spanish mapmaker Vicente Sebastián Pintado's field survey of 1795–96. It is labeled "Antiguo Canal por donde se desagruaba la Ciudad de Nueva Orleans y terrano suburbano en tiempo de lluvia" ("Old Canal by which the City of Orleans and suburban terrain drains in times of rain"). Superimposed on a modern map, this drainage channel would have curved through today's Fourth Ward, starting behind St. Louis No. 1 Cemetery and joining a tributary of Bayou St. John around the intersection of Canal Street with South Lopez Street. Vicente Sebastián Pintado and Carlos Laveau Trudeau, *Map of New Orleans and Vicinity,* 1873 copy of original 1804 map depicting information compiled by Pintado in 1795–1796 and drafted by Trudeau in 1804, Library of Congress.

69. Martin, *Orleans Term Reports,* vol. 2: 10.

70. Martin, *Orleans Term Reports,* vol. 2: 10.

71. Jack D. L. Holmes, *A Guide to Spanish Louisiana, 1762–1806* ([A. F. Laborde], 1970), 19–28.

72. Martin, *Orleans Term Reports,* vol. 2: 10.

73. This according to Jacques Tanesse, cited by Martin, *Orleans Term Reports,* vol. 2: 15.

74. Martin, *Orleans Term Reports,* vol. 2: 11.

75. Martin, *Orleans Term Reports,* vol. 2: 13.

76. These quotations are drawn from statements made in 1795 as well as paraphrases made in 1811. Martin, *Orleans Term Reports,* vol. 2: *11–13.*

77. Din and Harkins, *New Orleans Cabildo,* 255–56; Charles Gayarré, *History of Louisiana: The Spanish Domination* (1903; rpt. 1965), vol. 3: 352–53; Friends of the Cabildo: Toledano and Christovich, *New Orleans Architecture, vol. 6: Faubourg Tremé and the Bayou Road,* 60; and James S. Janssen, "The Carondelet (or Old Basin) Canal," *Building New Orleans: The Engineer's Role* (New Orleans: Waldemar S. Nelson & Co, 1987), 63–65.

78. Qtd. in John G. Clark, *New Orleans, 1718–1812: An Economic History* (Baton Rouge: Louisiana State University Press, 1970), 295; de Laussat, *Memoirs of My Life,* 26.

79. de Laussat, *Memoirs of My Life,* 89.

80. Ina Fandrich with Jay Edwards, "Barthelemy Lafon (1769–1820): A Brief Biographical Overview," in *Barthélemy Lafon in New Orleans 1792–1820,* report by Jay D. Edwards, Ina Fandrich, and Gabriele Richardson (Baton Rouge: Louisiana Division of Historic Preservation, 2019), 50.

81. Gabriele Richardson with Jay Edwards, "Lafon as a Surveyor and City Planner," in *Barthélemy Lafon in New Orleans 1792–1820,* by Edwards, Fandrich, and Richardson, 108; as cited by Jay Edwards, "Louisiana and the Tempestuous Times of Barthelemy Lafon," in *Barthélemy Lafon in New Orleans,* by Edwards, Fandrich, and Richardson, 29–30.

82. As cited by Ina Fandrich with Jay Edwards, "Barthelemy Lafon (1769–1820): A Brief Biographical Overview," in *Barthélemy Lafon in New Orleans,* by Edwards, Fandrich, and Richardson, 52.

83. Fandrich with Edwards, "Barthelemy Lafon (1769–1820)," 45.

84. Fandrich with Edwards, "Barthelemy Lafon (1769–1820)," 45; Cameron B. Strang, *Frontiers of Science Imperialism and Natural Knowledge in the Gulf South Borderlands 1500–1850* (Chapel Hill: University of North Carolina Press, 2018), 193.

85. Fandrich with Edwards, "Barthelemy Lafon (1769–1820)," 58.

86. Strang, *Frontiers of Science Imperialism,* 197.

87. Benjamin Latrobe, qtd. by Strang, *Frontiers of Science Imperialism,* 202.

88. Letter, Thomas Jefferson to M. du Plantier, Monticello, September 4, 1808, Louisiana Research Collection, Tulane University Special Collections; Colten, *Unnatural Metropolis,* 39.

89. de Laussat, *Memoirs of My Life,* 17.

90. Carlos Trudeau, *Plano de la Ciudad de Nueva Orleans y de las habitaciones imediatas formado en virtud del decreto del Ill. Cabo,* December 24, 1798.

91. Friends of the Cabildo, *New Orleans Architecture, vol. 4: The Creole Faubourgs* (Gretna, LA: Pelican Publishing Co., 1974), 8.

92. Nicholás de Finiels, qtd. by Samuel Wilson Jr., "Early History," in Friends of the Cabildo, *New Orleans Architecture, vol. 4: The Creole Faubourgs,* 9; Richardson with Edwards, "Lafon as a Surveyor and City Planner," 151.

93. Qtd. by Richardson with Edwards, "Lafon as a Surveyor and City Planner," 121.

94. In time, that line would become Florida Avenue, and in 1895, engineers selected that trajectory as the main East Bank drainage discharge. That Lake Borgne decision would late prove ill-fated, when the discharge route was shifted to Lake Pontchartrain. But it made sense in 1895, and it originated in Lafon's 1806 work on the Faubourg Marigny, as well as the 1740s work of Dubriel and his slaves in digging the mill race on what is now Elysian Fields Avenue.

95. Wilson, "Early History of the Lower Garden District," 3–24.

96. These are the names and spellings that appear on the 1815 *Plan of the City and Suburbs of New Orleans from an Actual Survey,* by Jacques Tanesse, Historic New Orleans Collection. Many have since been changed.

97. Harnett T. Kane, *Place du Tivoli: A History of Lee Circle* (Boston: John Hancock Mutual Life Insurance Co., 1961), 5.

98. Wilson, "Early History of the Lower Garden District," 12–13.

99. Thomas K. Wharton, *Queen of the South—New Orleans, 1853–1862: The Journal of Thomas K. Wharton,* ed. Samuel Wilson Jr., Patricia Brady, and Lynn D. Adams (New Orleans: Historic New Orleans Collection, 1999), 84–86.

100. This section is drawn from Campanella, *Time and Place in New Orleans,* and Richard Campanella, "A River Ran Through It: How the Watery St. Mary Batture Evolved Into Convention Center Boulevard," *New Orleans Times-Picayune,* July 22, 2018.

101. Ari Kelman, "*A River and Its City: Critical Episodes in the Environmental History of New Orleans.*" PhD diss., Brown University, 1998, 19 and 25.

102. Campanella, *Time and Place in New Orleans.*

103. Campanella, *Time and Place in New Orleans.*

104. John Adems Paxton, *The New-Orleans Directory and Register* (New Orleans, 1822), 32–33.

105. Paxton, *New-Orleans Directory and Register,* 32–33.

106. Latrobe, *Southern Travels,* 52.

107. "New Marine Hospital Site," *Daily Picayune,* November 11, 1882, 2.

108. Richard Campanella, "Fragmented Front Street: Intermittent Relic of St. Mary Batture," *New Orleans Advocate,* June 9, 2019.

109. "Personal Visit to the Great Port of New Orleans; City's Gigantic Cotton Warehouse System," *Times-Picayune,* July 4, 1915.

110. Measurement by author of all land between North Peters Street / Tchoupitoulas Street and the river, from the middle of the French Quarter to the edge of Carrollton.

111. Powell, *Accidental City,* 101.

112. Robert C. Vogel, "The Patterson and Ross Raid on Barataria, September 1814," *Louisiana History* 33, no.2 (Spring 1992): 159.

113. Jay D. Edwards, "New Orleans Urban Landscapes and Its Architecture on the Eve of Americanization: Barthelemy Lafon's Role in Shaping their Character," in *Barthélemy Lafon in New Orleans 1792–1820,* by Edwards, Fandrich, and Richardson, 188.

114. Strang, *Frontiers of Science Imperialism,* 204–7.

115. As if to make his demise even more tragic, Lafon's elderly father, Pierre Lafon, ventured to New Orleans in 1822 to claim his son's estate (his grandchildren having been denied their inheritance on account of their mixed race), only to himself perish to yellow fever. Next came his brother and sister-in-law from France, on the same mission—and with the same fate, both ending up in St. Louis Cemetery. Their daughter, Lafon's niece, followed in their quest, and while she managed to evade the saffron scourge, her years-long inheritance battle ended with the revelation that her late uncle had, in fact, been "wholly insolvent and unable to pay the legacies and debts." Qtd. by Fandrich with Edwards, "Barthelemy Lafon (1769–1820)," 63–66.

116. Strang, *Frontiers of Science Imperialism,* 197.

117. Roulhac Toledano, *A Pattern Book of New Orleans Architecture* (Gretna, LA: Pelican Publishing Co., 2010), 170.

118. Map, *Plan of the City and Suburbs of New Orleans from an Actual Survey,* Tanesse. On this map, the canal on Orleans Avenue is labeled *"Canal Girod,"* and Bayou Road is *"Grand Chemin du Bayou St. Jean."*

119. Qtd. in Charles Gayarré, *History of Louisiana: The American Domination* (1866; rpt. 1965), vol. 4: 194.

120. "An Act Respecting Claims to Land in the Territories of Orleans and Louisiana" (March 3, 1807), as recorded in *The Debates and Proceedings in the Congress of the United States* (1852), 1283. The planned extension of the Carondelet Canal also accounts for the great width of Basin Street. Francis Burns, "The Spanish Land Laws of Louisiana," *Louisiana Historical Quarterly* 11, no. 4 (October 1928): 566.

121. Map, *Plan of the City and Suburbs of New Orleans from an Actual Survey,* Tanesse.

122. Latrobe, *Impressions Respecting New Orleans,* 67–68.

123. Toledano, *Pattern Book of New Orleans Architecture,* 23; S. Frederick Starr, *Une Belle Maison: The Lombard Plantation House in New Orleans's Bywater* (Jackson: University of Mississippi Press, 2013), 55–56.

124. Map, *City of New Orleans,* Francis P. Ogden, 1829, Historic New Orleans Collection.

125. Latrobe, *Southern Travels,* 47.

126. Map, Henry Möllhausen, *Plan of the City of New Orleans: A Guide for Citizens and Strangers and Shewing [sic] the Situation of the Fire-plugs and Pipes of the Commercial Waterworks* (1837), Historic New Orleans Collection, accession no. 2021.0116.

127. Map, Charles Zimpel, *Topographic Map of New-Orleans and Its Vicinity,* 1834.

128. Richard Campanella, "Little-Known Gormley Canal Once Brought Cypress Logs and Brick Clay into Today's Central City," *New Orleans Times-Picayune,* March 9, 2018.

129. Vella, *Intimate Enemies,* 6; "General Police—An Ordinance Containing Several Provisions of Police, for the City of New Orleans," *A General Digest of the Ordinances and Resolutions of the Corporation of New-Orleans* (New Orleans: Jerome Bayon, 1831), July 28, 1828, 303.

130. *Historical Sketch Book and Guide to New Orleans and Environs* (New York: Will H. Coleman, 1885), 277.

131. "Upper Liberties—An Ordinance Concerning the Police of the Liberties of New-Orleans, and the Appointment of Syndics for Said Liberties," *General Digest of the Ordinances and Resolutions,* July 28, 1815, 41.

132. "Public Health—An Ordinance to Prevent Nuisances, and to Provide for the Security of the Public Health of the City of New Orleans," *General Digest of the Ordinances and Resolutions,* March 19, 1816, 343–47.

133. "An Ordinance Concerning the Natural Drains in the City of New-Orleans," *General Digest of the Ordinances and Resolutions,* December 15, 1817, 297.

134. Kelly Birch, "Slavery and Origins of Louisiana's Prison Industry, 1803–1861," PhD diss., University of Adelaide, 2017, 144.

135. James Stuart, *Three Years in North America* (Edinburgh: Robert Cadell and Whittaker and Co., 1833), vol. 2: 235.

136. "An Ordinance Concerning the Police Jail for the Detention of Slaves," October 8, 1817, *General Digest of the Ordinances and Resolutions,* 127–29; "An Ordinance to Regulate the Service of Slaves Employed in the Works of the City," November 10, 1817, *General Digest of the Ordinances and Resolutions,* 141–45.

137. *Conseil de Ville,* Session of December 31, 1824, 160; Session of March 22, 1825, 211; Session of July 6, 1825, 317; Session of November 4, 1826, 279; Session of July 14, 1827, 27 of microfilm no. 90–223, AB301, NOPL-LC; Richard Wade, *Slavery in the Cities: The South 1820–1860* (London: Oxford University Press, 1964), 23, 37; Becnel, *Barrow Family and the Barataria and Lafourche Canal,* 1–3.

138. "Unincorporated Suburbs—An Ordinance Concerning the Unincorporated Boroughs and Suburbs within the City of New Orleans," *General Digest of the Ordinances and Resolutions,* December 15, 1817, 165.

139. This area pertained to Jefferson Parish at the time, but Jefferson's syndic system was nearly identical to that of New Orleans proper. Giquel, Syndic of the Second Ward, "Parish of Jefferson—Works To Be Made," *Louisiana Advertiser,* May 10, 1827, 3.

140. "To the Honorable Senate," *Louisiana Advertiser,* December 2, 1826.

141. Sally K. Evans Reeves and William Dale Reeves/Kitree Corp., *Management Summary Cultural Resources Survey—United States Public Health Service Hospital* (US Department of Health and Human Services, March 1981), 13.

142. Charles Gayarré, "A Louisiana Sugar Plantation of the Old Regime," *Harper's New Monthly Magazine* 74 (December 1886–May 1887): 618.

143. Gayarré, "Louisiana Sugar Plantation of the Old Regime," 618.

144. Richard Campanella, "A Lake in Broadmoor? Natural Paradise Graced Last Open Tract," *Times-Picayune–New Orleans Advocate,* November 1, 2020, 1.

145. Gayarré, "Louisiana Sugar Plantation of the Old Regime," 621.

146. *Report of the Sanitary Commission of New Orleans on the Epidemic Yellow Fever,* viii; City Planning and Zoning Commission, *Major Street Report* (New Orleans, 1927), 75.

147. M. Perrin Du Lac, *Travels Through the Two Louisianas . . . in 1801, 1802, & 1803* (London, 1807), 8.

148. Timothy Flint, *Recollections of the Last Ten Years* (Boston: Cummings, Hilliard, and Co., 1826), 305.

149. John Wilds, *Crisis, Clashes, and Cures: A Century of Medicine in New Orleans* (New Orleans Medical Society, 1978), 35–37; A. E. Fossier, "History of Medical Education in New Orleans," *Annals of Medical History* 6, no. 4, 1934 reprint of late-1860s article; Richard Campanella, *"Bio-Medical Innovation in New Orleans: A Timeline of Three Centuries of Care, Research, and Innovation,"* unpublished report for the Director of Business Development and Strategy for BioInnovation and Health Service Innovation, New Orleans Business Alliance, 2017; Joseph Jones, "Yellow Fever Epidemic of 1878 in New Orleans," *New Orleans Medical and Surgical Journal,* March 1879. The cited fumigation methods date mostly from the 1860s to 1890s.

150. "An Act To Incorporate the Faculty of the Medical College of Louisiana, and the Medical College of New Orleans," April 2, 1835, in *Acts Passed at the First Session of the Twelfth Legislature of the State of Louisiana* (New Orleans: Jerome Bayon, State Printer, 1835), 200–222; John P. Dyer, *Tulane: The Biography of a University, 1834–1965* (New York: Harper & Row, 1966), 18–19; *Daily Picayune,* August 19, 1840, 3.

151. Fossier, "History of Medical Education in New Orleans," 24; *Daily Picayune,* April 20, 1847

3. THE POLDER-AND-PADDLE ERA, 1830S–1850S

1. Becnel, *Barrow Family and the Barataria and Lafourche Canal,* 1–3.

2. T. P. Thompson, "Early Financing in New Orleans, 1831—Being the Story of the Canal

Bank—1915," *Publications of the Louisiana Historical Society,* vol. 7: *1913-1914* (1915): 24-25; Canal Bank and Trust Company, *Through Ninety-Five Years* (1926), 6-7; Richard Campanella, "185-Year-Old New Basin Canal Continues to Affect Thousands of New Orleanians Every Day," *New Orleans Times-Picayune,* December 8, 2017.

3. "New-Orleans, Sept. 6," *Newburyport Herald* (Newburyport, MA), September 24, 1833, 3.

4. Alexander, *Transatlantic Sketches* 2: 39-40.

5. Alexander, *Transatlantic Sketches* 2: 39-40.

6. Campanella, "185-Year-Old New Basin Canal Continues to Affect Thousands."

7. B. M. Harrod, City Surveyor, *Report on Drainage, to the City Council of New Orleans,* November 22, 1888, 3-4.

8. "Speculation—Swamp," *Daily Picayune,* June 25, 1839, 2.

9. Report of City Surveyor L. Surgi to the Common Council, 1868, as cited by Drainage Advisory Board members John Fitzpatrick et al., *Report on the Drainage of the City of New Orleans* (New Orleans: T. Fitzwilliam & Co., 1895), 48.

10. *New Orleans Canal and Navigation Company v. The City of New Orleans,* in A. N. Ogden, *Reports of Cases Argued and Determined in the Supreme Court of Louisiana, Volume XII. [sic] for the Year 1857* (New Orleans: Office of the State Courier, 1858), 364; Mary Lilla McLure, *Louisiana Leaders 1830-1860* (Shreveport, LA: Journal Printing Co., 1935), 68.

11. "New Orleans Draining Company," *Commercial Advertiser* (New York), June 3, 1835, 2; *Daily Picayune,* June 9, 1853, 2; *Annual Report of the State Engineer, to the Legislature of the State of Louisiana* (New Orleans: Commercial Bulletin Office, January 1843), 22.

12. Charles Bowen, *The American Almanac and Repository of Useful Knowledge for the Year 1837* (Boston: Charles Bowen, 1836), 324.

13. "Speculation—Swamp," *Daily Picayune,* June 25, 1839, 2.

14. Bowen, *American Almanac and Repository of Useful Knowledge for the Year 1837,* 324.

15. Erik F. Haites, James Mak, and Gary M. Walton, *Western River Transportation: The Era of Early Internal Development, 1810-1860* (Baltimore: Johns Hopkins University Press, 1975), 17-18 and 130-31; "Steam Saw Mill," advertisement in *Orleans Gazette And Commercial Advertiser,* October 29, 1807, 4; Robert H. Thurston, *A History of the Growth of the Steam-Engine* (New York: D. Appleton and Co., 1903), 284; Frank Haigh Dixon, "A Traffic History of the Mississippi River System," document 11, National Waterways Commission (Washington, DC: Government Printing Office, December 1909), 15.

16. "New Orleans and Richmond," *Richmond Enquirer,* December 22, 1836, 3.

17. "Draining Company," *New Orleans Bee,* December 6, 1836, as picked up by *Evening Post* (New York), December 24, 1836, 2; "New Orleans and Richmond," *Richmond Enquirer,* December 22, 1836, 3.

18. *New Orleans Canal and Navigation Company v. The City of New Orleans,* 364.

19. *New Orleans Bee,* qtd. in *Elyria Republican and Working Men's Advocate* (Elyria, OH), November 12, 1835, 1.

20. "New Orleans, June 9," *New Orleans Bee,* rpt. in *Evening Post* (New York), June 22, 1836, 2 (emphasis added).

21. US Army Corps of Engineers, *National Register Evaluation of New Orleans Drainage System,* 11; Colten, *Unnatural Metropolis,* 41.

22. Roger Baudier, "Sanitation in New Orleans: Further Drainage Efforts," *Southern Plumbing and Heating Register,* April 1955, 17-18.

23. "The Inundation. A Glimpse, from the Cupola of the Charles, *Daily Picayune,* June 3, 1849, 2.

24. "New Orleans Draining Company (communicated)," *Daily Picayune,* June 9, 1853, 2; "Sale Without Reserve of 31 Negroes," *Daily Picayune,* December 3, 1851, 3.

25. "New Orleans Draining Company," *Daily Picayune,* June 20, 1853, 2.

26. "Lake Poydras," *Daily Picayune,* November 22, 1844, 2.

27. Gilbert Fowler White, "Human Adjustment to Floods," PhD diss., University of Chicago, 1942, 2.

28. Paxton, *The New-Orleans Directory and Register* (New Orleans, 1823), 138.

29. Qtd. by Adam Hodgson, *Remarks During a Journey Through North America in the Years 1819, 1820, and 1821* (New York, 1823), 164.

30. "The Bell Crevasse," *Daily Picayune,* May 17, 1858, 5.

31. George E. Waring Jr., *Report on the Social Statistics of Cities, Part II: The Southern and the Western States* (Washington, DC, 1887), 261.

32. Wilton P. Ledet, "The History of the City of Carrollton," *Louisiana Historical Quarterly* 21, no. 1 (January 1938): 228.

33. Inserted table and graph by Edward H. Barton, "Report Upon the Sanitary Condition of New Orleans," in *Cause and Prevention of Yellow Fever in New Orleans* (1854), following p. 100; Stanford E. Chaillé, "Inundations of New Orleans and Their Influence on Its Health," *New Orleans Medical and Surgical Journal,* July 1882, excerpt in Tulane University Special Collections Vertical File, flooding folder, 5.

34. "The Inundation," *Daily Picayune,* June 4, 1849, Monday evening edition, 2.

35. "The Inundation," 2.

36. "An Act to Better Protect the Interest of the State in the Property Appurtenant to the Canal of the New Orleans Canal and Banking Company," March 10, 1858, *Acts Passed by the Fourth Legislature of the State of Louisiana* (Baton Rouge: J. M. Taylor, State Printer, 1858), 50.

37. Nelson M. Blake and Robert L. Izlar, "Flood Control and Drainage," in *The New Encyclopedia of Southern Culture, vol. 8: Environment,* ed. Martin Melosi and Wilson Charles Reagan (Chapel Hill: University of North Carolina Press, 2007), 70–73.

38. "An Act to Aid the State of Louisiana in Draining the Swamp Lands Therein," no. 166, March 2, 1849, in *Decisions of the Interior Department in Public Land Cases, and Land Laws Passed by the Congress of the United States,* ed. W. W. Lester (Philadelphia: H. P. & R. H. Small Publishers, 1860), 150.

39. Blake and Izlar, "Flood Control and Drainage," 70–73.

40. White, "Human Adjustment to Floods," 2.

41. "Division of the City of New-Orleans into Three Municipalities," March 8, 1836, *A New Digest of the Statute Laws of the State of Louisiana,* compiled by Henry A. Bullard and Thomas Curry (New Orleans: E. Johns & Co., 1842), vol. 1: 15.

42. "The undersigned have been duly appointed commissions," *Daily Picayune,* July 7, 1840, 3.

43. See, for example, "First Municipality Council," *Daily Picayune,* June 16, 1848, 2.

44. "The season has now arrived . . . ," *Picayune,* August 19, 1837, 2.

45. "There is a great disparity . . . ," *Picayune,* September 5, 1837, 2.

46. *Picayune,* October 21, 1837, 2.

47. Benjamin Moore Norman, *Norman's New Orleans and Environs* (New Orleans: B. M. Norman, 1845), 84–85.

48. Thomas Affleck, "Drainage of the City," *Daily Picayune,* April 26, 1850, 2.

4. PLAGUES AND PROGRESS, 1850S–1860S

1. Testimonies of Mr. Vanderlinden, Dr. M. M. Dowler, Mr. Clark, Mr. Pashley, and Mr. Ebbinger, 3–7 of section within *Report of the Sanitary Commission of New Orleans on the Epidemic Yellow Fever.*

2. Testimonies of Mr. Ebbinger and Dr. Lemonier, 4–9 of section within *Report of the Sanitary Commission of New Orleans on the Epidemic Yellow Fever.*

3. Henry Tudor, *Narrative of a Tour in North America* (London: James Duncan, 1834), vol. 2: 380; Hugh Murray, *Historical Account of Discoveries and Travels in North America* (London: Longman, Rees, Orme, Brown, & Green, 1829), 427.

4. Wrote Timothy Flint in 1826, "The inhabitants of New Orleans retire for the summer, either to the North or to the pine woods. The pine woods, in the ear of a Louisianian, is a synonyme [*sic*] with health." Flint, *Recollections of the Last Ten Years,* 329.

5. New Orleans had a population of 154,133 in 1853, not including outlying areas nor visitors and transients, who would have been interred locally if they died. Table R in Barton, "Report Upon the Sanitary Condition of New Orleans," 89.

6. Jones, "Yellow Fever Epidemic of 1878," 699.

7. Table G, "Monthly Returns from Each of the Cemeteries," *Report of the Sanitary Commission of New Orleans on the Epidemic Yellow Fever,* 246; Table R, Barton, "Report Upon the Sanitary Condition of New Orleans," 89; Colten, *Unnatural Metropolis,* 36.

8. "Sanitary Commission," *Daily Picayune,* September 30, 1853, 2; *Report of the Sanitary Commission of New Orleans on the Epidemic Yellow Fever,* iv.

9. William Dosité Postell, "Edward Hall Barton, Sanitarian," *Annals of Medical History,* 3rd ser., vol. 4 (1942): 370–75.

10. Postell, "Edward Hall Barton, Sanitarian," 375.

11. *Report of the Sanitary Commission of New Orleans on the Epidemic Yellow Fever,* 171, 223, and 233.

12. Barton, "Report Upon the Sanitary Condition of New Orleans," chart on 101. See also *Report of the Sanitary Commission of New Orleans on the Epidemic of Yellow Fever,* 314.

13. *Report of the Sanitary Commission of New Orleans on the Epidemic Yellow Fever,* ii–ix.

14. Barton, "Report Upon the Sanitary Condition of New Orleans," 207 and 216. Emphasis in original.

15. Barton, "Report Upon the Sanitary Condition of New Orleans," 207–8.

16. Barton, "Report Upon the Sanitary Condition of New Orleans," 209.

17. *Report of the Sanitary Commission of New Orleans on the Epidemic Yellow Fever,* 416.

18. Postell, "Edward Hall Barton, Sanitarian," 370–81.

19. *The Statues of the State of Louisiana, revised and prepared by U. B. Phillips* (New Orleans: Emile La Sere, State Printer, 1855), section 88, "Draining of Swamp Lands," 398–99.

20. "An Ordinance for the Establishment of a Health Department for the City of New Orleans," posted in *Daily Picayune,* January 12, 1855, 1; "The Council," *Daily Picayune,* February 14, 1855, 6.

21. Louis H. Pilié, City Surveyor, *Report on Drainage, Communicated to the Common Council October 5th, 1857* (New Orleans, 1857).

22. John Smith Kendall, *History of New Orleans* (Chicago: Lewis Publishing Co., 1922), vol. 2: 567.

23. Drainage Advisory Board, *Report on the Drainage of the City of New Orleans,* 47.

24. Kendall, *History of New Orleans* 2: 567; "An Act to Provide for Leveeing, Draining and Re-

claiming Swamp Lands in Certain Portions of Parishes of Orleans and Jefferson," no. 165, March 18, 1858, *Acts Passed by the Fourth Legislature of the State of Louisiana,* 114.

25. *Gardner's New Orleans Directory for 1861* (New Orleans: Charles Gardner, 1861), xiv, 32, 273.

26. "An Act to Provide for Leveeing, Draining and Reclaiming Swamp Lands in Certain Portions of Parishes of Orleans and Jefferson," no. 165, March 18, 1858, *Acts Passed by the Fourth Legislature of the State of Louisiana,* 115.

27. *Gardner's New Orleans Directory for 1861,* xiv, 32, 273; Kendall, *History of New Orleans 2:* 567; Jones, "Yellow Fever Epidemic of 1878," 699.

28. *Daily Picayune,* August 26, 1856, 2.

29. "An Act to Provide for Leveeing, Draining and Reclaiming Swamp Lands in Certain Portions of Parishes of Orleans and Jefferson," no. 165, March 18, 1858, *Acts Passed by the Fourth Legislature of the State of Louisiana,* 116.

30. US Army Corps of Engineers, *National Register Evaluation of New Orleans Drainage System,* 12.

31. Personal communications via email with New Orleans engineer Dennis G. Lambert, PE, New Orleans, July 21, 2016, and other emails during 2014–16. Lambert researched Raymond Thomassy's connection to Lambert's great-great-great-grandfather, Lucius Place, who served as president of the Second Draining District in 1858.

32. "Floral, Hydro and Terrene Dynamics," *New Orleans Medical and Surgical Journal,* ed. Bennet Dowler et al. (New Orleans: Bulletin Book and Job Office, 1859), vol. 16: 193.

33. Lewis G. DeRussy, *Special Report Relative to the Cost of Draining the Swamp Lands Bordering Lake Pontchartrain* (Baton Rouge: J. M. Taylor, 1859).

34. William W. Howe, "Municipal History of New Orleans," *Johns Hopkins University Studies in Historical and Political Science,* ed. Herbert B. Adams (Baltimore: N. Murray, Publication Agent), ser. 7, no. 4 (April 1889): 171–75 (emphases added).

35. William Howard Russell, *My Diary North and South* (Boston, 1863), 230–31.

36. Russell, *My Diary North and South,* 230–31.

37. Maj. Gen. Benjamin F. Butler, qtd. in James Parton, *General Butler in New Orleans: History of the Administration of the Department of the Gulf in the Year 1862* (New York: Mason Brothers, 1864), 400.

38. Letter, Benjamin F. Butler to Abraham Lincoln, May 8, 1862, Abraham Lincoln Papers, Library of Congress; *Historical Sketch Book and Guide to New Orleans,* 291.

39. Russell, *My Diary North and South,* 230–31.

40. Gerald M Capers Jr., "Confederates and Yankees in Occupied New Orleans," *Journal of Southern History* 30 (1964): 405–10; David Herbert Donald, *Lincoln* (New York: Simon & Schuster, 1995), 484–85.

41. Communiques from Military Commandant Geo. F. Shepley, Maj. Gen. Benj. F. Butler, and Mr. Leefe, published under "The City—City Council," *Daily Picayune,* June 6, 1862, 2.

42. Communiques, published under "The City—City Council," 2.

43. George Ingham, President of the Board of Commissioners, "Drainage Report," January 31, 1863, published in *Daily Delta,* February 7, 1863, 2.

44. Gayarré, "Louisiana Sugar Plantation of the Old Regime," 618. The four canals in place by 1866 were on Melpomene Street, Toledano Street, Peters (now Jefferson) Avenue, and Dublin Street, and they discharged into the New Basin Canal, via steam pump, at what is now the Carrollton Interchange at Interstate 10. This map, which appears on page 2 of the 1866 *Acts of the Legislature,* was

provided to the author by Dennis Lambert, whose great-great-great-grandfather was president of the Second Drainage District during the Civil War.

45. G. W. R. Bayley, "Drainage—Important Correspondence Between Mayor Hoyt and City Surveyor Bayley," *Daily True Delta,* October 4, 1864, 1.

46. Bayley, "Drainage—Important Correspondence," 1. For the 1836 *Bee* report, see "New Orleans, June 9," rpt. in *Evening Post* (New York), June 22, 1836, 2.

47. *Biographical and Historical Memoirs of Louisiana* (Chicago: Goodspeed Publishing Co., 1892), vol. 1: 121.

48. *Biographical and Historical Memoirs of Louisiana,* 121.

5. CAPERS AND CONSEQUENCES, 1860s–1870s

1. "A Ship Canal Through New Orleans," *Daily Picayune,* September 2, 1868, 1; A. F. Wrotnowski, *Map and Profile of the New Orleans and Ship Island Channel Compiled and Drawn from Actual Survey and Hydrographical Charts* (1869), in "Mapping Louisiana: Louisiana Maps from 1513–1900," Tulane University, Howard-Tilton Library.

2. "The Ship Island Canal Scheme," *Daily Picayune,* January 7, 1869, 4.

3. "A Ship Canal Through New Orleans," 1.

4. William B. Mitchell, *St. Cloud in the Territorial Period* (St. Paul: Minnesota Historical Society, 1908), 641.

5. "Ship Island Canal Scheme," 4. See also "Encore Ship Island," *Daily Picayune,* September 12, 1869, 8.

6. Louis Surgi, "Drainage Report" (1868), qtd. in "Protection Against Overflow," *Daily Picayune,* September 15, 1869, 1.

7. Kendall, *History of New Orleans 2.*

8. Thomas Sydenham Hardee, *Topographical and Drainage Map of New Orleans and Surroundings,* 1877, Historic New Orleans Collection.

9. Wrotnowski, *Map and Profile of the New Orleans and Ship Island Channel.*

10. Drainage Advisory Board, *Report on the Drainage of the City of New Orleans,* 48.

11. Act No. 51, "An Act to Repeal All Laws . . . Creating Drainage Districts in . . . the Parishes of Orleans and Jefferson," March 2, 1869, *Acts Passed by the General Assembly of the State of Louisiana, at the Second Session of the First Legislature* (New Orleans: A. B. Lee Printers, 1869), 49–53; *Official Journal of the Proceedings of the House of Representatives of the State of Louisiana, At the Session Begun and Held in New Orleans, January 2, 1871* (New Orleans: A. L. Lee-State Printers, 1871), 145.

12. Act No. 51, "An Act to Repeal All Laws . . . Creating Drainage Districts," 49–53. See also *Report of the Special House Committee To Whom He Was Referred, The Charges Preferred Against Governor Warmoth, by George Wickliffe* (New Orleans: A. L. Lee, State Printer, 1870), 13.

13. "The South—New Orleans Ship Canal," *Saint Paul Daily Press* (St. Paul, MN), October 5, 1869, 1.

14. *Report of the Special House Committee To Whom He Was Referred, The Charges Preferred Against Governor Warmoth,* 13.

15. Richard Campanella, "The Annexation of Carrollton," *Times-Picayune,* September 8, 2017, InsideOut section; "The Annexation of Algiers," *Times-Picayune,* August 11, 2017, InsideOut section; "How New Orleans Took Uptown from Jefferson Parish," *Times-Picayune,* July 14, 2017, InsideOut

section; and "When Lafayette City Became New Orleans," *Times-Picayune,* June 9, 2017, InsideOut section.

16. Howe, "Municipal History of New Orleans," 171–72.

17. Act No. 30, "An Act to Provide for the Drainage of New Orleans," *Acts Passed by the General Assembly of the State of Louisiana at the First Session of the Second Legislature* (New Orleans: Republican Office, 1871), 75–79.

18. *Peake v. New Orleans,* Appeal from the Circuit Court of the United States for the Eastern District of Louisiana, No. 852, March 9, 1891, *United States Reports, vol. 139: Cases Adjudged in the Supreme Court at October Term, 1890* (New York: Banks and Brothers, Law Publishers, 1891), 345.

19. "The Courts. The Ship Island Canal Controversy Mandamus upon Drainage Commissioners Made Peremptory," *Daily Picayune,* May 13, 1870, 2.

20. "Encore 'Ship Island,'" *Daily Picayune,* September 12, 1869, 8.

21. *Peake v. New Orleans,* 346.

22. Letter, W. H. Bell to Hon. John Cochran, Administrator of Improvements, June 17, 1871, rpt. in Drainage Advisory Board, *Report on the Drainage of the City of New Orleans,* 49–50.

23. Drainage Advisory Board, *Report on the Drainage of the City of New Orleans,* 49–50.

24. US Army Corps of Engineers, *National Register Evaluation of New Orleans Drainage System,* 14.

25. Frank L. Schneider, "Vision, Effort Effect Big Changes at Lakefront," *Times-Picayune,* January 24, 1965, section 2: 9.

26. Drainage Advisory Board, *Report on the Drainage of the City of New Orleans,* 51.

27. Computations by author based on data from *Historical Sketch Book and Guide to New Orleans,* 278, and Harrod, *Report on Drainage, to the City Council of New Orleans,* 10. Rainfall accumulation assumes six inches of rain over twenty-four hours falling on the improved and unimproved basins of the East Bank of Orleans Parish.

28. *Warner v. City of New Orleans,* 167 US 467, May 24, 1897, No. 282, *The Supreme Court Reporter, vol. 17: Cases Argued and Determined in the United States Supreme Court, 1896–1897* (St. Paul, MN: West Publishing Co., 1897), 893; *Peake v. New Orleans,* 346–48; Act No. 30, "An Act to Provide for the Drainage of New Orleans," 77.

29. *Peake v. New Orleans,* 348.

30. "A Railway Bicycle," *Philadelphia Inquirer,* June 7, 1896, 31.

31. Wrotnowski, *Map and Profile of the New Orleans and Ship Island Channel.*

32. "Alexander's Bill," *Daily Picayune,* March 2, 1870; "Amended Charter," *New Orleans Times,* March 15, 1870.

33. "Barber's Bill," *Daily Picayune,* February 28, 1874.

34. Meloncy C. Soniat, "The Faubourgs Forming the Upper Section of the City of New Orleans," *Louisiana Historical Quarterly* 20 (January–October 1937), 192–211.

35. "An *Act* to Authorize the Barataria Ship Canal Company," June 15, 1878, *Annual Report of the Chief of Engineers to the Secretary of War for the Year 1878, Part I* (Washington, DC: Government Printing Office, 1878), 168; "Harvey's Canal—It Is Purchased by the Barataria Ship Canal Company for $100,000," *Daily Picayune,* May 6, 1880, 2.

36. "Excursion To Grand Pass—The Barataria Ship Canal," *Daily Picayune,* November 11, 1877, 6.

37. Edward Fontaine, *Contributions to the Science of Engineering* (Washington, DC: Government Printing Office, 1879), 15; *American Architect and Building News* 6, no. 209 (December 27, 1879): 202–3.

6. THE PROGRESSIVE ERA, 1880s–1890s

1. Jones, "Yellow Fever Epidemic of 1878," 683, 698–99.

2. Jones, "Yellow Fever Epidemic of 1878," 698.

3. Rudolph Matas, qtd. by Juan A. Del Regato in "Carlos Juan Finlay (1833–1915)," *Journal of Public Health Policy* 22, no. 1 (2001): 99.

4. Del Regato, "Carlos Juan Finlay (1833–1915)," 98–104.

5. Carlos Juan Finlay, qtd. by Enrique Chaves-Carballo in "Carlos Finlay and Yellow Fever: Triumph over Adversity," *Military Medicine* 170, no. 10 (2005): 882.

6. Chaves-Carballo, "Carlos Finlay and Yellow Fever," 883.

7. Chaves-Carballo, "Carlos Finlay and Yellow Fever," 883.

8. Walter Reed et al., "The Etiology of Yellow Fever—A Preliminary Note," *Public Health Papers and Reports,* American Public Health Association, 1900 (Columbus, OH: Berlin Printing Co., 1902), vol. 26: 41–42.

9. Howe, "Municipal History of New Orleans," 185–86.

10. Howe, "Municipal History of New Orleans," 185–86.

11. "Report of the Flushing Committee, New Orleans Auxiliary Sanitation Association," December 17, 1881, Tulane University Special Collections, New Orleans Municipal Papers, Manuscript Collection 17, box 7 of 17, Public Works Folder.

12. Edward Fenner, qtd. in Dennis East II, "Health and Wealth: Goals of the New Orleans Public Health Movement, 1879–84," *Louisiana History* 9, no. 3 (Summer 1968): 250.

13. Jones, "Yellow Fever Epidemic of 1878," 693; Kendall, *History of New Orleans 2:* 763.

14. Ernst von Hesse-Wartegg, *Travels on the Lower Mississippi, 1879–1880: A Memoir by Ernst von Hesse-Wartegg,* ed. and trans. Frederic Trautmann (Columbia: University of Missouri Press, 1990), 161.

15. Howe, "Municipal History of New Orleans," 171–75.

16. "Two Worthies," *Harvard Graduates' Magazine,* 1913, 239; "Harrod," Willi H. Hager, *Hydraulicians in the USA 1800–2000: A Biographical Dictionary of Leaders in Hydraulic Engineering and Fluid Mechanics* (London: CRC Press, 2015), 2135.

17. Harrod, *Report on Drainage, to the City Council of New Orleans,* 10.

18. Harrod, *Report on Drainage, to the City Council of New Orleans,* 15–16.

19. Harrod, *Report on Drainage, to the City Council of New Orleans,* 3–4.

20. Harrod, *Report on Drainage, to the City Council of New Orleans,* 3–4.

21. Harrod, *Report on Drainage, to the City Council of New Orleans,* 4–5.

22. Harrod, *Report on Drainage, to the City Council of New Orleans,* 7.

23. Harrod, *Report on Drainage, to the City Council of New Orleans,* 15–16.

24. Harrod, *Report on Drainage, to the City Council of New Orleans,* 9.

25. Harrod, *Report on Drainage, to the City Council of New Orleans,* 13.

26. Based on keyword searches by author of the *New Orleans Item, Daily Picayune,* and *Weekly Picayune,* using America's Historic Newspapers database, accessed through Howard-Tilton Library at Tulane University.

27. Act 93, "An Act to Establish the Orleans Levee District," July 7, 1890, *Acts Passed by the General Assembly of the State of Louisiana* (New Orleans: Ernest Marchand Printers, 1890), 95–97.

28. "Beginnings of a Drainage System," *Daily Picayune,* January 4, 1891, 4.

29. Ed Eisenhauer, "Agitating for a Drainage System," *Daily Picayune,* January 4, 1891, 8.

30. Drainage Advisory Board, *Report on the Drainage of the City of New Orleans,* 50.

31. Richard Campanella, *Bienville's Dilemma: A Historical Geography of New Orleans* (Lafayette: University of Louisiana Press, 2008).

32. Drainage Advisory Board, *Report on the Drainage of the City of New Orleans,* 50.

33. Kendall, *History of New Orleans 2:* 506–7.

34. Drainage Advisory Board, *Report on the Drainage of the City of New Orleans,* 51.

35. Articles 574–76 of Ordinance No. 7170 of the City of New Orleans, January 31, 1893, *Flynn's Digest of the City Ordinances . . . of the City of New Orleans* (New Orleans: L. Graham & Son, Ltd., 1896), 266–67; Drainage Advisory Board, *Report on the Drainage of the City of New Orleans,* ix.

36. Tonja Koob, *Historic Civil Engineering Landmark Nomination Report* (American Society of Civil Engineering, April 2020, provided to author by Koob), 6–7.

37. Kendall, *History of New Orleans 2:* 507

38. "The Proposed Drainage System," *Daily Picayune,* February 2, 1893, 4.

39. "Proposed Drainage System," 4.

40. Drainage Advisory Board, *Report on the Drainage of the City of New Orleans,* 13–17.

41. Drainage Advisory Board, *Report on the Drainage of the City of New Orleans,* 53–60.

42. Drainage Advisory Board, *Report on the Drainage of the City of New Orleans,* 60.

43. W. C. Kirkland, *Topographical Map of New Orleans,* Historic New Orleans Collection, accession nos. 1987.116.1 through 1987.116.12.

44. M. A. Baccich, E. E. Lafaye, and R. E. E. DeMontluzin–Gentilly Terrace Company, "Gentilly Terrace: Here's Your Opportunity" (1909), pamphlet archived at Williams Research Center, The Historic New Orleans Collection.

45. Drainage Advisory Board, *Report on the Drainage of the City of New Orleans,* 23–24.

46. "The Drainage Plan Before the Council; The Advisory Board Making a Complete Official Report," *Daily Picayune,* May 1, 1895, 6.

47. Drainage Advisory Board, *Report on the Drainage of the City of New Orleans,* 27.

48. See table on p. 29, "Probable Run-Off in Cubic Feet Per Second," and dry-weather figures on p. 35, in Drainage Advisory Board, *Report on the Drainage of the City of New Orleans.*

49. Drainage Advisory Board, *Report on the Drainage of the City of New Orleans,* 26.

50. Drainage Advisory Board, *Report on the Drainage of the City of New Orleans,* 34.

51. Drainage Advisory Board, *Report on the Drainage of the City of New Orleans,* 38 (emphasis added).

52. Drainage Advisory Board, *Report on the Drainage of the City of New Orleans,* 22.

53. Drainage Advisory Board, *Report on the Drainage of the City of New Orleans,* 36–37; "Map of the City of New Orleans Showing System of Drainage," in *Fifteenth Semi-Annual Report of the Sewerage and Water Board of New Orleans, to the Honorable City Council, June 30, 1907* (New Orleans: Searcy & Pfaff, 1907), map insert.

54. Drainage Advisory Board, *Report on the Drainage of the City of New Orleans,* 34.

55. US Army Corps of Engineers, *National Register Evaluation of New Orleans Drainage System,* 20; Map, "Growth in Areas Served by Storm Water Drainage System, New Orleans, 1894–1926, Compiled from the Records of the Sewerage and Water Board," City Planning and Zoning Commission–Advisory Commission, *The Handbook to Comprehensive Zone Law for New Orleans, Louisiana* (New Orleans, 1929–33), Tulane University Special Collections, 976.31 (711) N469h 1933.

56. "Tackling the World's Toughest Drainage Problem . . . With Success!"

57. Drainage Advisory Board, *Report on the Drainage of the City of New Orleans,* 17.

58. "The Drainage Plan Before the Council; The Advisory Board Making a Complete Official Report," *Daily Picayune,* May 1, 1895, 6.

59. "Broad Street Active In Fighting the Removal of Tracks and the Drainage Canal Digging," *Daily Picayune,* May 21, 1895, 3.

60. "The Rain Makes A Plea For Drainage," *Daily Picayune,* May 25, 1895, 12; "The Drainage Report To Be Considered To-Day," *Daily Picayune,* May 25, 1895, 4; "Mr. Louque Makes His Final Fight Against the Drainage Plans of the Advisory Board . . . And the Scheme Passes by a Vote of 18 to 4," *Daily Picayune,* July 10, 1895, 3.

61. Article 583 of Ordinance No. 10991 of the City of New Orleans, July 10, 1895, *Flynn's Digest of the City Ordinances,* 268.

62. Orleans Levee Board, *Nature Changes from Moment to Moment* (New Orleans: Orleans Levee Board, September 8, 1972), 1; "The Drainage Commission," *Daily Picayune,* September 5, 1896, 4.

63. Kendall, *History of New Orleans 2:* 517–26.

64. Kendall, *History of New Orleans 2:* 525.

65. Abraham Brittin, qtd. in Kendall, *History of New Orleans 2:* 526; "Sewerage Closer To The Victory. More Than Enough Names on the Petitions. The Ordinance Drawn," *Daily Picayune,* April 18, 1899, 7.

66. "Sewerage Closer To The Victory," 7.

67. "Women's League To Aid Sewerage Meets With the Same Old Barrier of Indifference," *Daily Picayune,* March 5, 1899, 12.

68. Sebastian Junger, "The Pumps of New Orleans," *American Heritage's Invention and Technology Magazine* 8, no. 2 (Fall 1992), www.inventionandtech.com/content/pumps-new-orleans-1 (accessed May 31, 2020).

69. "All Special Tax Points Made Clear—How to Vote, Who Can Vote, and Other Questions," *Daily Picayune,* May 31, 1899, 3.

70. "Hats Off To Our Patriotic Women!" *Daily Picayune,* June 7, 1899, 1.

71. Act No. 4, S.B. No. 3, "Joint Resolution Proposing an Amendment to the Constitution [for] Ratifying and Carrying Into Effect a Special Tax Levied in the City of New Orleans," August 18, 1899, *Acts Passed by the General Assembly of the State of Louisiana, at the Extra Session, August 1899* (Baton Rouge: Advocate, 1899), 6–7.

72. Thomas Watkins Campbell, *Manual of the City of New Orleans* (New Orleans City Council, 1903), 263–64; "The Sewerage & Water Board of New Orleans: How It Began, The Problems It Faces, The Way It Works, the Job It Does," Sewerage & Water Board of New Orleans, undated agency pamphlet; "Tackling the World's Toughest Drainage Problem . . . With Success!"

73. Orleans Parish School Board, *The New Orleans Book* (New Orleans: Searcy & Pfaff, Ltd., 1919), 44.

7. DEWATERING NEW ORLEANS, 1900S–1910S

1. US Army Corps of Engineers, *National Register Evaluation of New Orleans Drainage System,* 81–87.

2. "Drainage Board To Give A Hearing To Property Owners Along the Toulouse Canal," *Daily Picayune,* February 16, 1900, 12.

3. Paper by L. W. Brown, rpt. in Rudolph Hering, Geo. H. Benzenberg, and Howard A. Carson, *Report on the Board of Inquiry of the Conduct and Character of the Drainage Works* (New Orleans: C. W. Corson, March 1902), 32; Hering, Benzenberg, and Carson, *Report on the Board of Inquiry of the Conduct and Character of the Drainage Works,* 6–7.

4. Drainage Advisory Board, *Report on the Drainage of the City of New Orleans,* 38, 28, 43. For the sake of clarity, irregular capitalization in the original document has been standardized in the quoted material.

5. Letter, L. W. Brown to Hon. R. M. Walmsley, August 2, 1901, rpt. in Hering, Benzenberg, and Carson, *Report on the Board of Inquiry of the Conduct and Character of the Drainage Works,* 22.

6. Hering, Benzenberg, and Carson, *Report on the Board of Inquiry of the Conduct and Character of the Drainage Works,* 9.

7. Hering, Benzenberg, and Carson, *Report on the Board of Inquiry of the Conduct and Character of the Drainage Works,* 12–13.

8. "Map of the City of New Orleans Showing System of Drainage," in *Fifteenth Semi-Annual Report of the Sewerage and Water Board of New Orleans, to the Honorable City Council, June 30, 1907,* map insert.

9. American Society of Mechanical Engineers and Sewerage and Water Board of New Orleans, *A National Engineering Landmark—The A. B. Wood Low Head High Volume Screw Pump, No. 1 Pumping Station, New Orleans, La.,* monograph, June 11, 1974, 9.

10. Letter, Charles Louque to Mayor Capdevielle, February 1899, qtd. in Hering, Benzenberg, and Carson, *Report on the Board of Inquiry of the Conduct and Character of the Drainage Works,* 29.

11. Hering, Benzenberg, and Carson, *Report on the Board of Inquiry of the Conduct and Character of the Drainage Works,* 9–10.

12. Hering, Benzenberg, and Carson, *Report on the Board of Inquiry of the Conduct and Character of the Drainage Works,* 63.

13. Hering, Benzenberg, and Carson, *Report on the Board of Inquiry of the Conduct and Character of the Drainage Works,* 91 (emphasis added).

14. Thomas Ewing Dabney, "New Orleans Builds Own Underground River," *New Orleans Item,* May 2, 1920, 1.

15. Why? Because the lifting of the discharge would be done at the very last moments of its transport across recently subsided neighborhoods, just prior to its ejection into the lake, rather than miles inland, upon which the raised discharge would have to remain raised above the ground level of sinking new subdivisions.

16. US Army Corps of Engineers, *National Register Evaluation of New Orleans Drainage System,* 21.

17. US Army Corps of Engineers, *National Register Evaluation of New Orleans Drainage System,* 24.

18. *Nineteenth Semi-Annual Report of The Sewerage and Water Board of New Orleans, to the Honorable City Council, June 30, 1909* (New Orleans: J'Alfrey-Rodd-Pursell, Co., 1909), 16; US Army Corps of Engineers, *National Register Evaluation of New Orleans Drainage System,* 22–23.

19. Isaac M. Cline, "Temperature Conditions at New Orleans, As Influenced by Subsurface Drainage," *Proceedings of the Second Pan American Scientific Congress:* Section II—Astronomy, Meteorology, and Seismology, December 1916-January 1917 (Washington, DC: Government Printing Office, 1917), 483.

20. "Algiers Canal Next Drainage Work, Sewerage Board Opening Bids for the Contract," *Daily Picayune,* March 28, 1907, 5; "Algiers Drainage Is Now Assured"; *Daily Picayune,* September 13, 1907, 4.

21. "Algiers System of Water and Sewerage To Begin Operation on the First of August," *Daily Picayune,* June 27, 1909, 32; "Over The River—Items Along the Line From Gretna to Algiers; Drainage and Other Improvements," *Daily Picayune,* December 11, 1910, 5; "Gretna Gossip," *Daily Picayune,* March 17, 1909, 5; "Drainage Scheme For Reclaiming 36,000 Acres Of Land In Three Parishes," *Daily Picayune,* October 11, 1911, 5.

22. US Army Corps of Engineers, *National Register Evaluation of New Orleans Drainage System,* 24.

23. Baccich, Lafaye, and DeMontluzin–Company, "Gentilly Terrace: Here's Your Opportunity." 20–21.

24. "Lakeview Has Arrived," *Daily Picayune,* June 13, 1909, 7.

25. R. Christopher Goodwin & Associates and US Army Corps of Engineers—New Orleans District, *National Register Assessment of the Broadmoor Neighborhood, New Orleans, Orleans Parish, Louisiana* (New Orleans, September 2003), 11.

26. "Greatest Building Activity Now Is In 'Little California,'" *Times-Picayune,* August 22, 1920, 33.

27. "From the Iron Age," *New Orleans Item,* August 15, 1903, 4.

28. George Washington Cable, "New Orleans Revisited," *Book News Monthly,* April 1909, 564 and 560, qtd. by Kelman, "*A River and Its City,*" 281.

29. Harrod, *Report on Drainage, to the City Council of New Orleans,* 4–5.

30. US Army Corps of Engineers, *National Register Evaluation of New Orleans Drainage System,* 24.

31. US Army Corps of Engineers, *National Register Evaluation of New Orleans Drainage System,* 69.

32. Most of what has been published about A. Baldwin Wood's early life comes from an undated manuscript by Ray M. Thompson archived at Tulane University's Louisiana Research Collection in Jones Hall.

33. "Honored City Servant—A. Baldwin Wood," *New Orleans States,* November 15, 1954, 14; American Society of Mechanical Engineers and Sewerage and Water Board of New Orleans, *National Engineering Landmark—The A.B. Wood Low Head High Volume Screw Pump, No. 1 Pumping Station,* 5; "Albert Baldwin Wood," www.findagrave.com/memorial/115233383/albert-baldwin-wood (accessed June 5, 2020).

34. American Society of Mechanical Engineers and Sewerage and Water Board of New Orleans, *National Engineering Landmark—The A.B. Wood Low Head High Volume Screw Pump, No. 1 Pumping Station,* 9.

35. Hering, Benzenberg, and Carson, *Report on the Board of Inquiry of the Conduct and Character of the Drainage Works,* 178.

36. *Twenty-Third Semi-Annual Report of The Sewerage and Water Board of New Orleans, to the Honorable City Council, June 30th, 1911* (New Orleans: America Printing Co., 1911), 13.

37. Sewerage and Water Board Report, 1915, qtd. in US Army Corps of Engineers, *National Register Evaluation of New Orleans Drainage System,* 24–27.

38. American Society of Mechanical Engineers and Sewerage and Water Board of New Orleans, *National Engineering Landmark—The A.B. Wood Low Head High Volume Screw Pump, No. 1 Pumping Station,* 9.

39. Based on descriptions by Ray M. Thompson and the 1915 Sewerage and Water Board Report, in US Army Corps of Engineers, *National Register Evaluation of New Orleans Drainage System,* 24–27.

40. Junger, "Pumps of New Orleans."

41. "Theard Stands by Board Engineers; Farrar Attacks Sewerage Pump Awards Before the Council, *Daily Picayune,* January 14, 1914, 4.

42. "Farrar, Angry, Retired from Pump Dispute," *New Orleans Item,* January 21, 1914, 1.

43. American Society of Mechanical Engineers and Sewerage and Water Board of New Orleans, *National Engineering Landmark—The A.B. Wood Low Head High Volume Screw Pump, No. 1 Pumping Station,* 10.

44. Based on descriptions by Mayor Martin Behrman, Ray M. Thompson, and the 1915 Sewerage and Water Board Report, in US Army Corps of Engineers, *National Register Evaluation of New Orleans Drainage System,* 27.

45. W. H. Creighton, qtd. in American Society of Mechanical Engineers and Sewerage and Water Board of New Orleans, *National Engineering Landmark—The A.B. Wood Low Head High Volume Screw Pump, No. 1 Pumping Station,* 10 (emphasis added).

46. US Army Corps of Engineers, *National Register Evaluation of New Orleans Drainage System,* 28.

47. US Army Corps of Engineers, *National Register Evaluation of New Orleans Drainage System,* 28–29; American Society of Mechanical Engineers and Sewerage and Water Board of New Orleans, *National Engineering Landmark—The A.B. Wood Low Head High Volume Screw Pump, No. 1 Pumping Station,* 8.

48. Hon. Martin Behrman, "New Orleans—A History of Three Great Public Utilities, Sewerage, Water and Drainage, and Their Influences Upon the Health and Progress of a Big City," paper presented at Convention of League of American Municipalities, September 29, 1914, Milwaukee.

49. Behrman, "New Orleans—A History of Three Great Public Utilities."

50. American Society of Mechanical Engineers and Sewerage and Water Board of New Orleans, *National Engineering Landmark—The A.B. Wood Low Head High Volume Screw Pump, No. 1 Pumping Station,* 6 and 11.

51. As the Heros were involved in the maritime industry, they may have split their time between the two port cities. "George Hero—Background," by George Hero III, Hero Family Papers, provided by family to author; Numa C. Hero Jr., "And I Saw It: A Prologue," Hero Family Papers.

52. Treasury Department, Bureau of Statistics, *Tables Showing Arrivals of Alien Passengers and Immigrants in the United States from 1820 to 1888* (Washington, DC: Government Printing Office, 1889), 108–9; Richard Campanella, *Geographies of New Orleans: Urban Fabrics Before the Storm* (Lafayette: University of Louisiana Press, 2006), 193.

53. "Andrew Hero Dead at Age of 74 Years; Was Distinguished Confederate Soldier and Officer and Lawyer and Notary," *Times-Picayune,* April 25, 1914.

54. George Hero III, "Why We Are Here—Background of Hero Lands: A Concise History," family pamphlet dated March 2012, provided by George Hero III to author, May 2019, 2–3.

55. Hero, "Why We Are Here—Background of Hero Lands," 2–3.

56. George Hero III, "Louisiana Historical Society Tour—South St. Bernard and Plaquemines," family pamphlet dated November 2010, provided by George Hero III to author, May 2019.

57. Hero, "Why We Are Here—Background of Hero Lands," 4.

58. Drawn from the author's prior publication, *The West Bank of Greater New Orleans—A Historical Geography* (Baton Rouge: Louisiana State University Press, 2020).

59. "Large Refinery for Louisiana," *Sugar: An English Spanish Technical Journal devoted to Sugar Production* 22, no. 1 (January 1920), 195.

60. "By George A. Hero, President, Jefferson-Plaquemines Drainage District," *New Orleans Item,* June 22, 1913, 26.

61. "Monumental Task—Mayor Says George A. Hero Carrying Heaviest Burden of All—Drainage in Jefferson-Plaquemines District Will Work Wonders for All," *Daily Picayune,* October 19, 1913 (emphasis added).

62. "George A. Hero," "*The South—The Nation's Greatest Asset,*" *Manufacturers Record* 63, no. 12 (March 27, 1913), pt. 2: 194; "By George A. Hero, President, Jefferson-Plaquemines Drainage District," *New Orleans Item,* June 22, 1913, 26.

63. Numa Hero, "Jefferson and Plaquemines Drainage District," *Jefferson Parish Yearbook* (Jefferson Parish, LA: Police Jury of Jefferson Parish, 1935), 121.

64. "Greatest Drainage Pumps in Operation in Louisiana," *Municipal Engineering* 48, no. 1 (January 1915): 216.

65. Historical Records Survey, *Transcription of Parish Records of Louisiana, no. 26, Jefferson Parish (Gretna): ser. 1: Police Jury Minutes, vol. 2, 1905–1912* (Division of Professional and Service Projects, Work Projects Administration, July 1940), 442, and *vol. 10, 1918–1924* (Division of Professional and Service Projects, Work Projects Administration, July 1940), 18 and 37.

66. So phenomenal was its capacity that some reporters got carried away and put the discharge at 45 million gallons per minute, even a billion gallons! "Large Refinery for Louisiana," 195; "New Orleans, La," *Western Contractor* 25, no. 666 (October 15, 1913): 30.

67. "The Panama Canal of New Orleans," *Daily Picayune,* March 8, 1914, real estate section: 62. This ad did not mention Hero's name, only that of the general agent of the South New Orleans Realty Company. But it had "Hero" written all over it, and the company's office, at 921 Gravier, adjoined Hero's office in the Cotton Exchange Building, on Gravier Street at Carondelet. "For Rent or Sale," *New Orleans Item,* June 1, 1913, 34.

68. "Greatest Drainage Pumps in Operation in Louisiana," 216.

69. "Great Pumping Station at Work Across the River," *Times-Picayune,* February 14, 1915, 1.

70. "Large Refinery for Louisiana," *Sugar: An English Spanish Technical Journal devoted to Sugar Production* 22, no. 1 (January 1920): 195.

71. "Great Pumping Station at Work Across the River," *Times-Picayune,* February 14, 1915, 1.

72. "'I'll Drain a Million Acres More!'" 1.

73. "Hero, George A." entry in *The National Cyclopædia of American Biography* (New York: James T. White & Co., 1916), vol. 15: 196; "Wilson Greets City as Giant Pumps Reclaim West Section," *New Orleans Item,* February 13, 1915 (afternoon edition), 1.

74. "Wilson Greets City as Giant Pumps Reclaim West Section," 1.

75. "Photo of Car Entered by South New Orleans Realty and Development Company in the Hero Day Celebration," *New Orleans Item,* February 14, 1915, real estate section: 62.

76. Modjeski and Masters, *Greater New Orleans Bridge Over the Mississippi River: Final Report to the Mississippi River Bridge Authority* (New Orleans: Modjeski and Masters–Mississippi River Bridge Authority, 1960), 7.

77. Map, "Lakes Pontchartrain and Maurepas" (Washington, DC: US Coast and Geodetic Survey, 1920).

78. "New Orleans Drainage Company," advertisement and article in *Commercial West* 17 no. 17 (April 23, 1910): 17. These passages are drawn from the author's earlier articles, "Lost Coastal Communities of Eastern New Orleans," *Times-Picayune/New Orleans Advocate,* January 5, 2020, and "Addressing New Orleans East's Core Problem," *Times-Picayune,* December 13, 2013.

79. Senate Bill No. 164, "An Act to Sell to the New Orleans Lake Shore Land Company," *Daily State* (Baton Rouge), June 18, 1908, 8; *Report of the Board of State Engineers of the State of Louisiana* (Baton Rouge: Daily State Press, 1908), 84.

80. "Frank B. Hayne, N.O. Leader, Is Dead; Outstanding Cotton Map and Citizen Succumbs," *New Orleans States,* August 3, 1935, 1–2.

81. "New Orleans Lake Shore Land Co.," in "The South—The Nation's Greatest Asset," 198.

82. "Three Hundred Ten Acre Farms," *New Orleans Item,* August 31, 1911, real estate section: 3.

83. Orleans Parish School Board, *New Orleans Book,* 93.

84. "Truck Farming a Great Industry," *Daily Picayune,* September 1, 1900, 4.

85. "Three Hundred Ten Acre Farms," *New Orleans Item,* August 31, 1911, real estate section: 3; "New Orleans Drainage Company," 16.

86. "Boothe's Death Removes Leader of the National Reclamation Cause; Warren B. Reed, of New Orleans, Becomes President; *Daily Picayune,* April 13, 1913, 13.

87. "Runoff from Drained Prairie—Little Woods Tract—Area, 6,943 Area," *Journal of Agricultural Research* 11, no. 6. (November 6, 1917): 257.

88. "New Orleans Drainage Company," 16.

89. "New Speedway to Cost $120,000," *New Orleans Item,* July 29, 1911, 5.

90. Names drawn from 1936 and 1938 Louisiana Spanish Fort and Chef Menteur quadrangle maps produced by the War Department–Corps of Engineers and distributed by the US Geological Survey.

91. Charles W. Okey, "New Orleans Lakeshore Land Co. Tract, New Orleans," in "The Wet Lands of Southern Louisiana and Their Drainage," *Bulletin 652,* US Dept. of Agriculture (Washington, DC: Government Printing Office, June 6, 1918), 28.

92. "Runoff from Drained Prairie—Little Woods Tract," 257–59.

93. Okey, "New Orleans Lakeshore Land Co. Tract, New Orleans," 30.

94. A. M. Shaw, "Largest Electric Drainage Pump," *Engineering News,* 74, no. 17, October 21, 1915, 805.

95. Okey, "New Orleans Lakeshore Land Co. Tract, New Orleans," 30.

96. See, for example, *Thirty-Fifth Semi-Annual Report of the Sewerage and Water Board of New Orleans, La.* (New Orleans: American Printing Co., 1917), 14.

97. *Thirty-Ninth Semi-Annual Report of the Sewerage and Water Board of New Orleans, La.* (New Orleans: American Printing Co., 1919), 77.

98. The purchase was in 1923. "Gentilly's Developer Honored at Banquet," *New Orleans States,* December 3, 1955, 25.

99. "To Reclaim Big Tract Near Kenner," *New Orleans Item,* July 10, 1910, 7.

100. These lagoons would have been near today's Esplanade Mall in Kenner. "Hunting Grounds and Fishing Camps," *Daily Picayune,* January 2, 1900, 3.

101. "Kenner Project—The Nation's Greatest Asset," *Manufacturer's Record,* March 27, 1913, 197.

102. "Ohio Bank Bought Kenner Bond Issue Bonds of Drainage and Development Project Popular With Northern Bank," *Daily Picayune,* February 19, 1914, 5.

103. Craig A. Bauer, "From Burnt Canes to Budding City: A History of the City of Kenner, Louisiana," *Louisiana History: The Journal of the Louisiana Historical Association* 23, no. 4 (Autumn 1982): 353–81.

104. Bauer, "From Burnt Canes to Budding City," 371.

105. "Kenner Project—The Nation's Greatest Asset," 197.

106. Based on 1932 and 1936 Bonnet Carré, Spanish Fort, Hahnville, and New Orleans quadrangle maps produced by the War Department–Corps of Engineers and distributed by the US Geological Survey.

107. Justin F. Bordenave, "Fourth Jefferson Drainage District, Sub-Drainage Districts 1-2-3-4," *1938 Jefferson Parish Yearly Review* (Jefferson Parish Police Jury, 1938), 187.

108. Bordenave, "Fourth Jefferson Drainage District," 187.

109. Bordenave, "Fourth Jefferson Drainage District," 187.

110. Map, M.D. McAlester, "Extract from Dept. of the Gulf Map No. 5" (1868), Historic New Orleans Collection, accession no. 1957.14.

111. Bordenave, "Fourth Jefferson Drainage District," 187–88.

112. Names drawn from 1936-52 Indian Beach and Spanish Fort quadrangle maps produced by the War Department–Corps of Engineers and distributed by the US Geological Survey.

113. Frank J. Clancy, "High Hopes for the Low Section," *Jefferson Parish Yearly Review 1949* (Kenner, LA: Jefferson Parish Police Jury, 1949), 85.

114. Bordenave, "Fourth Jefferson Drainage District," 188–89.

115. John J. Holtgreve, "Look at Us Now, Mr. Jefferson!" *Jefferson Parish Yearly Review 1953* (Kenner, LA: Jefferson Parish Police Jury, 1953), 17 (emphasis added).

116. Richard Campanella, "Bucktown and the Lost Bayous of East Jefferson," *Time-Picayune/New Orleans Advocate,* February 3, 2020, 1.

8. GEOGRAPHIES REARRANGED, 1910S–1920S

1. Cline, "Temperature Conditions at New Orleans," 483.

2. John Magill, "A Conspiracy of Complicity," *Louisiana Cultural Vistas* 17, no. 3 (Fall 2006): 43.

3. "Drainage Bond Sale at 95 Opens Way for Great Things," *New Orleans Item,* May 30, 1915, 34.

4. *Plan de la Ville de la Nouvelle Orleans projetteé en Mars 1721,* Historic New Orleans Collection; Letter, De La Chaise and the Four Councillors of Louisiana to the Council of the Company of the Indies, April 26, 1725, *Mississippi Provincial Archives 1704–1743: French Dominion,* ed. Rowland and Sanders, vol. 2: 464.

5. *Plan du Canal de jonction du Mississippi au Lac Pontchartrain,* Historic New Orleans Collection.

6. "An *Act* to Authorize the Barataria Ship Canal Company," 168; "Harvey's Canal—It Is Purchased by the Barataria Ship Canal Company," 2; "Excursion To Grand Pass—The Barataria Ship Canal," 6.

7. Thomas Ewing Dabney, *The Industrial Canal and Inner Harbor of New Orleans: History, Description and Economic Aspects of Giant Facility Created to Encourage Industrial Expansion and Develop Commerce* (1921), 10–12.

8. *Official Journal of the Proceedings of the Senate of the State of Louisiana* (Baton Rouge: Ramires-Jones Printing Co., 1916), 536; "A Good Foundation," *New Orleans Item,* December 31, 1914, 6.

9. "Industrial Canal Now Assured, Says Governor," *New Orleans Item,* January 16, 1916, second section, 5.

10. Campanella, *Time and Place in New Orleans.*

11. Dabney, "New Orleans Builds Own Underground River," 1.

12. Orleans Levee Board, *Nature Changes,* 4.

13. Dabney, "New Orleans Builds Own Underground River," 1.

14. Dabney, "New Orleans Builds Own Underground River," 1.

15. Joshua Lewis, "*Deltaic Dilemmas: Ecologies of Infrastructure in New Orleans*," PhD diss., Stockholm University, 2015, section 2, "The Disappearing River": 11.

16. Dabney, "New Orleans Builds Own Underground River," 1.

17. Dabney, *The Industrial Canal and Inner Harbor of New Orleans,* 17–30.

18. "Deepest Excavation in New Orleans," *New Orleans Item,* March 30, 1919, Magazine and New Orleans Life, 3.

19. Dabney, "New Orleans Builds Own Underground River," 1.

20. Qtd. in Gary A. Bolding, "The New Orleans Seaway Movement," *Louisiana History* 10 (1969): 53–54.

21. Lewis, "*Deltaic Dilemmas,*" section 2, "The Disappearing River": 13.

22. "Editorial Notes and Comment—The Industrial Canal," *Proceedings of the Louisiana Engineering Society* 7, no. 5 (October 1921): 185–87; Lewis, "*Deltaic Dilemmas,*" section 2, "The Disappearing River": 13–14.

23. Lewis, "*Deltaic Dilemmas,*" section 2, "The Disappearing River": 18–19.

24. Elisée Réclus, *A Voyage to New Orleans,* ed. John Clark and Camille Martin (1855; translation, Thetford, VT, 2004), 50.

25. B. M. Harrod, "The Topography of New Orleans, with Reference to a System of Drainage," *Papers Read Before the New Orleans Academy of Sciences,* 1887–188 (New Orleans: L. Graham & Son, 1888) , vol. 1, no. 1: 46 and 49; Harrod, *Report on Drainage, to the City Council of New Orleans,* 3.

26. Okey, "New Orleans Lakeshore Land Co. Tract, New Orleans," 30.

27. Captain Philip Pittman, *The Present State of the European Settlements on the Mississippi* (Gainesville, FL, 1973 reprint of 1770 original), 11.

28. Based on C. R. Kolb and Roger T. Saucier, qtd. by Independent Levee Investigation Team, *Investigation of the Performance of the New Orleans Flood Protection Systems in Hurricane Katrina on August 29, 2005, vol. 1: Main Text and Executive Summary* (Washington, DC: National Science Foundation, July 31, 2006), 3–19.

29. Roger T. Saucier, *Geomorphology and Quaternary Geologic History of the Lower Mississippi Valley* (Vicksburg, MS, 1994), 1–53.

30. *Land Subsidence in the United States,* ed. Galloway, Jones, and Ingebritsen, 1–10.

31. "New Orleans, June 9," *New Orleans Bee,* rpt. in *Evening Post* (New York), June 22, 1836, 2.

32. Bayley, "Drainage—Important Correspondence," 1.

33. Qtd. by Stanley C. Arthur, *A History of the U.S. Custom House at New Orleans* (Baton Rouge: Work Projects Administration of Louisiana, 1940), 45.

34. "Home and Foreign Gossip." *Harper's Weekly,* October 14, 1871, 963, col. 2.

35. Drainage Advisory Board, *Report on the Drainage of the City of New Orleans,* 17.

36. "Cathedral Causes Debate," *Daily Picayune,* February 9, 1913, 6.

37. "Climate at New Orleans," *Times-Picayune,* February 25, 1918, 4.

38. Cline, "Temperature Conditions at New Orleans," 494.

39. Cline, "Temperature Conditions at New Orleans," 485–88.

40. Halle Parker, "Feeling the Heat: City's 'Island Effect,' Worst in the Nation, Intensifying Temps," *Times-Picayune/New Orleans Advocate,* Monday July 26, 2021, Metro section: 1B.

41. W. S. Callender, "New Orleans Is Sinking Slowly But Steadily Down Toward China," *New Orleans Item,* August 3, 1919, internal section: 5.

42. Callender, "New Orleans Is Sinking Slowly But Steadily Down Toward China," 5.

43. "New Orleans All Primed and Fit for Convention," *Plumbers' Trade Journal, Steam and Hot Water Fitters' Review,* June 1, 1921, 829.

44. Sewerage & Water Board of New Orleans photograph of "Samples of soils taken at varying depths" (1927), shown by David Waggonner in "Living with Water" presentation, TEDxNOLA, November 22, 2010, www.youtube.com/watch?v=EyywQ04e7dc.

45. Richard Campanella, "Above-Sea-Level New Orleans: The Residential Capacity of Orleans Parish's Higher Ground," white paper, Center for Bioenvironmental Research at Tulane and Xavier Universities, April 2007.

46. Callender, "New Orleans Is Sinking Slowly But Steadily Down Toward China," 5.

47. Dabney, "New Orleans Builds Own Underground River," 1.

48. Isaac M. Cline, "The Tropical Hurricane of September 29, 1915," *Monthly Weather Review,* September 1915, 457.

49. George G. Earl, *The Hurricane of September 29, 1915 and Subsequent Heavy Rainfalls: Report of George G. Earl, Gen'l Supt. to Sewerage and Water Board of New Orleans* (New Orleans, October 14, 1915), 7.

50. Cline, "Tropical Hurricane of September 29, 1915," 463.

51. As recorded by Cline, "Tropical Hurricane of September 29, 1915," 462.

52. "City Cut Off from Rest of World," *New Orleans Item,* September 29, 1915, evening edition, 1.

53. "Storm Takes Toll in Property Loss and Human Life," *Times-Picayune,* September 30, 1915, 1.

54. Cline, "Tropical Hurricane of September 29, 1915," 456.

55. "'STORM PROOF!' The Record Shows New Orleans," *New Orleans Item,* September 30, 1915, 1 (banner headline) and 10.

56. Schneider, "Vision, Effort Effect Big Changes at Lakefront"; Independent Levee Investigation Team, *Investigation of the Performance of the New Orleans Flood Protection Systems in Hurricane Katrina on August 29, 2005, vol. 1: Main Text and Executive Summary,* 3–14, 3–20, and 4–20.

57. This feature is known as the Pine Island Trend, a sandy deposit of the Pearl River which remains buried beneath eastern New Orleans. Richard Campanella, "Beneath New Orleans, A Coastal Barrier Island," *Louisiana Cultural Vistas,* Winter 2017.

58. A hurricane did the same in 1860, sending surge up "the Old and New Basin drainage canals and . . . allowing the onrushing water to flood . . . the back side of New Orleans." Though the deluge was not as severe as in 1871, it showed the danger of ungated canals to the lake, be they for drainage or navigation. Independent Levee Investigation Team, *Investigation of the Performance of the New Orleans Flood Protection Systems in Hurricane Katrina on August 29, 2005, vol. 1: Main Text and Executive Summary,* 4–10.

59. "The Overflowed District and the Drainage Machines," *Daily Picayune,* June 15, 1871, 2; W. H. Bell qtd. by Independent Levee Investigation Team, *Investigation of the Performance of the New Orleans Flood Protection Systems in Hurricane Katrina on August 29, 2005, vol. 1: Main Text and Executive Summary,* 4–16.

60. Chaillé, "Inundations of New Orleans and Their Influence on Its Health," 13–16.

61. Orleans Levee Board, *Nature Changes,* 1; Richard Campanella, "West End: 'The Coney Island of New Orleans,'" *Times-Picayune/New Orleans Advocate,* November 3, 2019, 1.

62. Orleans Levee Board, *Nature Changes,* 4; Andrew W. Kahrl, *The Land Was Ours: How Black Beaches Became White Wealth in the Coastal South* (Chapel Hill: University of North Caroline Press, 2016), 119.

63. "Dock Board Engineering Consultant Dies," *Times-Picayune,* February 12, 1958, 1.

64. "Lakefront Program to Cost $27,000,000 Launched by Board," *Times-Picayune,* August 20, 1925, 1–3.

65. Qtd. from Kahrl, *The Land Was Ours,* 119. See also Judy A. Filipich and Lee Taylor, *Lakefront New Orleans: Planning and Development 1926–1971* (New Orleans: Louisiana State University in New Orleans, 1971), 7–13; Association of Levee Boards of Louisiana, *The System That Works to Serve Our State* (1990), 43.

66. "Lakefront Program to Cost $27,000,000 Launched by Board," *Times-Picayune,* August 20, 1925, 1–3; Schneider, "Vision, Effort Effect Big Changes at Lakefront," 9; Filipich and Taylor, *Lakefront New Orleans,* 7–13.

67. Schneider, "Vision, Effort Effect Big Changes at Lakefront," 9.

68. Peirce F. Lewis, *New Orleans: The Making of an Urban Landscape* (Cambridge, MA: Ballinger Publishing Co., 1976), 65.

69. Schneider, "Vision, Effort Effect Big Changes at Lakefront," 9.

70. Richard Campanella, "'From French Colonists to the Beginnings of Jazz,' Spanish Fort Traces Its History Across Three Centuries," *Times-Picayune/New Orleans Advocate,* September 9, 2019; Richard Campanella, "'Where I'll Do as I D—n Please: A Historical Geography of Milneburg," *Times-Picayune/New Orleans Advocate,* Oct 6, 2019; "Spanish Fort's Destiny," *Daily Picayune,* December 9, 1896, 12.

71. Lewis, *New Orleans,* 66.

72. John M. Barry, *Rising Tide: The Great Mississippi River Flood of 1927 and How It Changed America* (New York: Simon & Schuster, 1997), 228–55; "Documents Tell Steps to Save City," *Times-Picayune,* April 27, 1927, 1; "Dock Board Engineering Consultant Dies," *Times-Picayune,* February 12, 1958, 1; "Col. Garsaud Final Rites Tomorrow," *New Orleans States,* February 12, 1958, 6.

73. "Lakefront Program to Cost $27,000,000 Launched by Board," *Times-Picayune,* August 20, 1925, 3.

74. Douglas Woolley and Leonard Shabman, *Decision-Making Chronology for the Lake Pontchartrain & Vicinity Hurricane Protection Project—Final Report Submitted for the Headquarters* (US Army Corps of Engineers Institute for Water Resources, March 2008), 4–2.

75. H. W. Gilmore, *Some Basic Census Tract Maps of New Orleans* (New Orleans, 1937), map book, Tulane University Special Collections, C5-D10-F6.

76. Latrobe, *Southern Travels,* 38.

77. Richard Campanella, "'Two Centuries of Paradox': The Geography of New Orleans's African-American Population, from Antebellum to Postdiluvian Times," in *Hurricane Katrina in Transatlantic Perspective,* ed. Romain Huret and Randy J. Sparks (Baton Rouge: Louisiana State University Press, 2014).

78. *Daily Picayune,* "A Kaleidoscopic View of New Orleans," September 23, 1843, 2 (emphasis added).

79. Based on US Census data interpreted by Campanella in *Bienville's Dilemma,* graphical section.

80. Ten enumeration districts in this front-of-town section, including those of the Garden District, had black populations of 5 percent or less, and most of them were probably domestic employees.

81. Based on the author's GIS analysis of the 1910 Census, drawn from population schedules and mapped at the enumeration district level. Note that figures presented here as Black include those originally recorded as either "black" or "mulatto."

82. Quotes drawn from Historical Records Survey, "A Brief History of Jefferson Parish," *Jefferson Parish Yearly Review 1939* (Jefferson Parish Police Jury, 1939), 173, and Dr. Charles F. Gelbke, "City of Gretna," *Jefferson Parish Yearly Review 1944* (Jefferson Parish Police Jury, 1944), 173-74.

83. Martha B. Mallory, interviewed by Allyson Ward Neal in *Algiers—The Untold Story: The African American Experiences, 1929–1955* (New Orleans: Beautiful Zion Baptist Church, 2001), 35.

84. This was "Freetown," or the McDonogh section of Algiers. See Campanella, *West Bank of Greater New Orleans.*

85. Transcriptions of Historical Records Survey, *Parish Records of Louisiana,* no. 26, *Jefferson Parish (Gretna):* ser. 1: *Police Jury Minutes,* vol. 11, *1924–1929* (1940), entry in Police Jury notes of July 12, 1926, 149–50.

86. "Segregation Row to Get Hearing in High Court," *Times-Picayune,* March 1, 1927.

87. "Segregation by Co-operation of Civic Bodies," *Times-Picayune,* November 23, 1924, sec. 2, p. 1 (emphasis added); Campanella, "'Two Centuries of Paradox.'"

88. The *Mapping Prejudice* project of the Borchert Map Library at the University of Minnesota has documented over 21,000 racist deed covenants in the progressive northern city of Minneapolis. www.mappingprejudice.org/.

89. Baccich, Lafaye, and DeMontluzin–Gentilly Terrace Company, "Gentilly Terrace: Here's Your Opportunity," 20–21.

90. "Greatest Building Activity Now Is In 'Little California,'" *Times-Picayune,* August 22, 1920, 33.

91. Exhibit No. 19, Civil District Court for the Parish of Orleans, No. 239–741, Division A, Docket 5, *Mrs. Viola Livaudais, Wife of John Joseph Grosch, v. John Joseph Grosch,* in US Senate, *Hearings Before the Special Committee to Investigate Organized Crime in Interstate Commerce,* part 8: *Louisiana* (Washington, DC: US Printing Office, 1951), 443.

92. GIS analysis by author using block-level racial data from 1939, initially published by Sam R. Carter in *A Report on Survey of Metropolitan New Orleans Land Use, Real Property, and Low Income Housing Area* (New Orleans: Work Projects Administration, Louisiana State Department of Public Welfare, and Housing Authority of New Orleans, 1941), insert map on residency by race. See author's mapping analysis of these data in Campanella, *Geographies of New Orleans.*

93. Each space was reanalyzed using the 1939 data according to the same bounding streets used for the 1910 data analysis. GIS analysis by author using 1910 Census data at the enumeration-district level, and 1939 block-level data of percent Black residency.

94. Daphne Spain, "Race Relations and Residential Segregation in New Orleans: Two Centuries of Paradox," *Annals of The American Academy of Political and Social Science* 441 (January 1979): 89.

95. "How Much of City Is Below Sea Level?" *New Orleans Item,* April 13, 1948, 12. Although this article appeared in 1948, the data used in the map dates to 1935.

96. This spot was at Charity Hospital. Works Progress Administration, *Some Data in Regard to Foundations in New Orleans and Vicinity* (Works Progress Administration of Louisiana and Board of State Engineers of Louisiana, 1937), 1939 addendum, 10.

97. "Chapter 1: Historical and Geological," Works Progress Administration, *Some Data in Regard to Foundations in New Orleans and Vicinity,* 3.

98. This house was on Colbert Street. J. O. Snowden and James B. Rucker, "Subsidence in New Orleans: A Photographic Case History of the Lakeview Subdivision," *Gulf Coast Association of Geological Societies Transactions* 42 (1992), abstract on 857.

99. Thomas Ewing Dabney, "Pushing Back the Water Frontier," *Jefferson Parish Yearly Review 1948* (Kenner, LA: F. Bordenave, 1948), 81–83 (emphasis added).

100. Dabney, "Pushing Back the Water Frontier," 84–85.

9. DRAINAGE BECOMES A UTILITY, 1920S–1950S

1. These enumerations vary depending on what one considers to be a subsystem and a polder.

2. "Tackling the World's Toughest Drainage Problem . . . With Success!"

3. Works Progress Administration, "New Orleans Drainage System," report by Pugh (Louisiana Works Progress Administration, 1940), State Library of Louisiana.

4. Lewis, *New Orleans,* 61–62.

5. *Forty-Eighth Semi-Annual Report of the Sewerage and Water Board of New Orleans, La. 1923* (New Orleans: Crescent Printing Co., 1923), 15 and 100–101.

6. Thomas Ewing Dabney, "Sewerage and Water Board Has Made Enviable Record," *New Orleans States,* December 6, 1934, 22.

7. Works Progress Administration, "New Orleans Drainage System"; the $21 million figure is from Dabney, "Sewerage and Water Board Has Made Enviable Record," 22.

8. "Flooding of City Must Stop, Say Citizens," *New Orleans States,* April 17, 1927, 1–10.

9. "Flooding of City Must Stop, Says Citizens," 1–10.

10. Barry, *Rising Tide.*

11. Lynn Dinkins, qtd. in "Flooding of City Must Stop, Says Citizens," 1.

12. George Earl, qtd. in "Flooding of City Must Stop, Says Citizens," 10.

13. Fred Cumbus, "New Method Devised for Laying Underground Cable Supplying Pumping System, *Times-Picayune/New Orleans States,* March 8, 1936, section 2: 10–11

14. A. Baldin Wood, qtd. by Cumbus, "New Method Devised for Laying Underground Cable," 11

15. US Army Corps of Engineers, *National Register Evaluation of New Orleans Drainage System,* 30.

16. Clancy, "High Hopes for the Low Section," 85.

17. Robert D. Leighninger Jr., *Building Louisiana: The Legacy of the Public Works Administration* (Jackson: University of Mississippi Press, 2007), 174.

18. "545,041 New Population of New Orleans," *Old French Quarter News,* July 16, 1943, 1; "City Growing; Population [of Metro Area] is now 630,000," *Old French Quarter News,* November 30, 1945, 1; "Parade and Program to Dedicate New Canal Link," *New Orleans States,* July 27, 1943, 3.

19. Qtd. by Nicole Youngerman in "The Development of Manufactured Flood Risk: New Orleans's Mid-Century Growth Machine and the Hurricane of 1947," *Disasters* 39, no. 2 (September 2015), 166–87.

20. Jason Theriot, "Building America's First Offshore Oil Port: LOOP," *Journal of American History* 99, no. 1 (May 2012): 188–96; William Conner and John W. Day, *The Ecology of Barataria Basin, Louisiana: An Estuarine Profile* (US Dept. of the Interior—Fish and Wildlife Service, Biological Report 87[7.13], 1987), graphs on p. 135; B. R. Couvillion, Holly Beck, Donald Schoolmaster, and Michelle Fischer, *Land Area Change in Coastal Louisiana, 1932 to 2016: U.S. Geological Survey Scientific Investigations Map 3381* (2017), 4 and 11, *doi.org/10.3133/sim3381.*

21. Youngerman, "The Development of Manufactured Flood Risk," 166–87.

22. Quotations from Morrison and Wood cited by Youngerman in "The Development of Manufactured Flood Risk," 166–87.

23. Interview of Del Hall by Richard Campanella, July 2013, in Richard Campanella, *The Photojournalism of Del Hall: New Orleans and Beyond, 1950s–2000s* (Baton Rouge: Louisiana State University Press, 2015).

24. Dabney, "Pushing Back the Water Frontier," 84–85.

25. "Damage During 1947 Hurricane," Hearings Before the Subcommittee of the Committee on Appropriations, US Senate, 1st sess., HR 5376, *Civil Functions, Department of the Army Appropriations, 1954* (Washington, DC: US Senate Committee on Appropriations, 1954), 601.

26. "City Counts Damage; Hurricane Goes North," *New Orleans States,* September 19, 1947, 1.

27. "Flood Threatens Downtown Area; Rail Embankment Holding Bayou Overflows Breaks," *Times-Picayune,* September 21, 1947, 4; Youngerman, "The Development of Manufactured Flood Risk," 166–87.

28. H. C. Sumner, "North Atlantic Hurricanes and Tropical Disturbances of 1947, *Monthly Weather Review,* December 1947, 252–54; Richard Campanella, "Disaster and Response in an Experiment Called New Orleans, 1700s–2000s," *Natural Hazard Science: Oxford Research Encyclopedias,* March 2016.

29. Youngerman, "The Development of Manufactured Flood Risk," 166–87.

30. "City Survived Storm—News Ads Inform U.S.," *New Orleans States,* September 25, 1947, 10.

31. Youngerman, "The Development of Manufactured Flood Risk," 166–87.

32. Clancy, "High Hopes for the Low Section," 85.

33. "Damage During 1947 Hurricane," 601.

34. Clancy, "High Hopes for the Low Section," 85.

35. "Damage During 1947 Hurricane," 600–601; Woolley and Shabman, *Decision-Making Chronology for the Lake Pontchartrain & Vicinity Hurricane Protection Project,* 2–3.

36. Clancy, "High Hopes for the Low Section," 81–91.

37. Independent Levee Investigation Team, *Investigation of the Performance of the New Orleans Flood Protection Systems in Hurricane Katrina on August 29, 2005, vol. 1: Main Text and Executive Summary,* 4–22.

38. "Flood Refugees Return to Home," *New Orleans States,* September 25, 1956, 3; Youngerman, "The Development of Manufactured Flood Risk," 166–87.

10. THE FEDS WADE IN, 1960S–1990S

1. Albert E. Cowdrey, *Land's End: A History of the New Orleans District, and Its Lifelong Battle with the Lower Mississippi and Other Rivers Wending Their Way to the Sea* (New Orleans: US Army Corps of Engineers, 1977), 43–80.

2. US Army Corps of Engineers, Engineering Manual 1110-21411, March 26, 1952, qtd. by Woolley and Shabman, *Decision-Making Chronology for the Lake Pontchartrain & Vicinity Hurricane Protection Project,* 2–3.

3. Public Law 71, "An Act to Authorize an Examination and Survey of the Coastal and Tidal Areas of the Eastern and Southern United States," June 15, 1955, *United States Statutes at Large,* 1955, vol. 69, pt. 1 (Washington, DC: Government Printing Office, 1955), 132.

4. Public Law 85-500, July 3, 1958, and National Hurricane Research Project Report No. 33, November 1959, qtd. by Woolley and Shabman, *Decision-Making Chronology for the Lake Pontchartrain & Vicinity Hurricane Protection Project,* 2–3 and 2–19.

5. Woolley and Shabman, *Decision-Making Chronology for the Lake Pontchartrain & Vicinity Hurricane Protection Project,* 4–2.

6. Independent Levee Investigation Team, *Investigation of the Performance of the New Orleans Flood Protection Systems in Hurricane Katrina on August 29, 2005, vol. 1: Main Text and Executive Summary,* 4–22.

7. Independent Levee Investigation Team, *Investigation of the Performance of the New Orleans Flood Protection Systems in Hurricane Katrina on August 29, 2005, vol. 1: Main Text and Executive Summary,* 8–11–12.

8. Independent Levee Investigation Team, *Investigation of the Performance of the New Orleans Flood Protection Systems in Hurricane Katrina on August 29, 2005, vol. 1: Main Text and Executive Summary,* 4–22–23 and 8–12.

9. W. F. Minor, "Engineers See No Lake Change; Barriers Won't Affect Level, Salinity—View," *Times-Picayune,* July 21, 1961, 1–14.

10. "New Orleans Plays Waiting Game with Killer Hurricane," *Times-Picayune,* May 29, 1977, 1–14.

11. "Barrier Plan Receives Okay," *Times-Picayune,* September 2, 1965, 1.

12. "*Act* to Authorize the Barataria Ship Canal Company," 168; "Harvey's Canal—It Is Purchased by the Barataria Ship Canal Company," 2; "Excursion To Grand Pass—The Barataria Ship Canal," 6.

13. LeRoy L. Hall, "Final Report of the Jefferson Parish Police Jury," *Jefferson Parish Yearly Review* (Kenner, LA, 1954), 29; Thomas Ewing Dabney, "Go West-Side, New Orleans, to Your New Frontier," *Jefferson Parish Yearly Review* (Kenner, LA: Police Jury of Jefferson Parish, 1948), 38 and 41; "Parade and Program to Dedicate New Canal Link," 3; Lehde Is Elected Dock Board Head," *Times-Picayune,* June 24, 1943, 10; Arthur A. Grant, "The West . . . and Best Seaway to the Gulf," *Jefferson Parish Yearly Review* (Kenner, LA: Police Jury of Jefferson Parish, 1945), 27.

14. Robert Moses and Andrews & Clark, *Arterial Plan for New Orleans* (Baton Rouge: Department of Highways, 1946), map, 18–19.

15. Bolding, "New Orleans Seaway Movement," 49.

16. Orleans Levee Board, *Nature Changes,* 4.

17. Brent M. Johnson, "Development of the Mississippi River–Gulf Outlet," *Journal of the Waterways Division, Proceedings of the American Society of Civil Engineers,* 1969, 8–9.

18. US Army Corps of Engineers, New Orleans District, *Water Resources Development in Louisiana,* 1995, 89–90.

19. Campanella, *West Bank of Greater New Orleans.*

20. "$61,000,000 for Flood Control Work Assured," *New Orleans States,* June 11, 1948, 32; Daniel J. Hubbell, "*The Lower Mississippi River from Baton Rouge to Head of Passes: A Mariner's Handbook,*" unpublished manuscript provided to author by Hubbell, 13.

21. Hubbell, "*The Lower Mississippi River from Baton Rouge to Head of Passes,*" 13.

22. Campanella, *West Bank of Greater New Orleans.*

23. Data in this section are drawn from US Army Engineer District, New Orleans, *Hurricane Betsy, September 8–11, 1965, serial no. 1880* (New Orleans, November 1965), and US Army Engineer District, New Orleans, *Hurricane Betsy, September 8–11, 1965: After-Action Report* (New Orleans, July 1966).

24. US Army Engineer District, New Orleans, *Hurricane Betsy, September 8–11, 1965,* 13.

25. US Army Engineer District, New Orleans, *Hurricane Betsy, September 8–11, 1965,* 24.

26. Sewerage and Water Board of New Orleans, *Report on Hurricane "Betsy," September 9–10, 1965* (New Orleans, October 8, 1965), 32.

27. Don Lee Keith, "More Pumps to Speed Flood Water Removal," *Times-Picayune,* September 15, 1965, 1; "Dredges Pump Water from Flooded Area into Canal," *Times-Picayune,* September 16, 1965, 1–26; Statement of Col. Tom Bowen, Corps of Engineers, *Hearings Before the Special Subcommittee to Investigate Areas of Destruction of Hurricane Betsy,* Committee on Public Works, House of Representatives, Eighty-Ninth Congress, September 25, 1965 (Washington, DC: Government Printing Office, 1965), 44.

28. Andy Horowitz, *Katrina: A History, 1915–2015* (Cambridge, MA: Harvard University Press, 56–57.

29. Statement of Gov. John J. McKeithen of Louisiana, "Hurricane Betsy Disaster of September 1965," *Hearings Before the Special Subcommittee to Investigate Areas of Destruction of Hurricane Betsy,* 13.

30. US Army Corps of Engineers, New Orleans District, *Water Resources Development in Louisiana* (New Orleans, 1995), 90, and *Waterborne Commerce of the United States, part 2: Waterways and Harbors, Gulf Coast, Mississippi River System and Antilles* (Washington, DC, 1998), 165, 170, and 207.

31. Public Law 89–298, 89th Congress, S. 2300, Flood Control Act, General Projects, October 27, 1965.

32. Testimony of Maj. Gen. A. P. Rollins Jr. to Senator John C. Stennis, Department of the Army, Corps of Engineers, "Lake Pontchartrain Project," March 24, 1971, *Public Works Appropriations for Fiscal Year 1972, U.S. Senate Subcommittee of Committee on Appropriations* (Washington, DC: US Senate, 1972), 1432–35.

33. Independent Levee Investigation Team, *Investigation of the Performance of the New Orleans Flood Protection Systems in Hurricane Katrina on August 29, 2005, vol. 1: Main Text and Executive Summary,* 4–11 to 4–12.

34. Woolley and Shabman, *Decision-Making Chronology for the Lake Pontchartrain & Vicinity Hurricane Protection Project,* ES-8.

35. US Government Accountability Office, "Army Corps of Engineers Lake Pontchartrain and Vicinity Hurricane Protection Project," Statement of Anu Mittal, Director, Natural Resources and Environment, Testimony Before the Subcommittee on Energy and Water Development, Committee on Appropriations, House of Representatives, September 28, 2005, 3; "Barrier Plan Receives Okay," 1.

36. Testimony of Maj. Gen. Rollins, "Lake Pontchartrain Project," 1315 and 1437.

37. US Army Corps of Engineers–New Orleans District, *Water Resources Development in Louisiana 1998,* 106–7; Woolley and Shabman, *Decision-Making Chronology for the Lake Pontchartrain & Vicinity Hurricane Protection Project,* ES-6 to ES-7 and 2–4 to 2–9; "New Orleans Plays Waiting Game With Killer Hurricane," 1–14; Independent Levee Investigation Team, *Investigation of the Performance of the New Orleans Flood Protection Systems in Hurricane Katrina on August 29, 2005, vol. 1: Main Text and Executive Summary,* 4–23.

38. US Army Corps of Engineers–New Orleans District, *Water Resources Development in Louisiana 1998,* 106.

39. US Government Accountability Office, "Army Corps of Engineers Lake Pontchartrain and Vicinity Hurricane Protection Project," Statement of Anu Mittal, September 28, 2005, 5.

40. Testimony of Maj. Gen. Rollins, "Lake Pontchartrain Project," 1313.

41. Cowdrey, *Land's End,* 80–81.

42. US Army Corps of Engineers–New Orleans District, *Water Resources Development in Louisiana 1998*, 106–7.

43. US Army Corps of Engineers–New Orleans District, *Water Resources Development in Louisiana 1998*, 107.

44. US Government Accountability Office, "Army Corps of Engineers Lake Pontchartrain and Vicinity Hurricane Protection Project," Statement of Anu Mittal, September 28, 2005, 6; Independent Levee Investigation Team, *Investigation of the Performance of the New Orleans Flood Protection Systems in Hurricane Katrina on August 29, 2005, vol. 1: Main Text and Executive Summary*, 4–23.

45. US Government Accountability Office, "Army Corps of Engineers Lake Pontchartrain and Vicinity Hurricane Protection Project," Statement of Anu Mittal, September 28, 2005, 1.

46. "Eleven Persons Are Injured as Metairie House Blows Apart," *Times-Picayune*, September 2, 1975, 1; Bill Mongelluzzo, "Jeff Goal: Prevent More Explosions," *Times-Picayune*, September 2, 1975, 2; "Gas Line Link Discovered at Littles'," *Times-Picayune*, September 3, 1975, 1.

47. "Gas Line Link Discovered at Littles'," 1; Map, "Explosions, 1972–1976," author's collection.

48. Fred Barry, "State, Federal Help Sought to Avoid Jeff Explosion," *Times-Picayune*, January 21, 1977, 6.

49. The 1-to-3-foot figure comes from the map made of the incidents in 1977, titled "Explosions, 1972–1976," in the author's collection; the 8-to-9-foot figure comes from FEMA/State of Louisiana LIDAR elevation data captured in 1999–2000, which show the affected area to have been 7.5 to 8.5 feet below sea level at that time, probably lower today. Given that the swamp surface had originally been slightly above sea level, we may surmise this area had subsided by about 8 to 9 feet within a century.

50. "Gas Line Link Discovered at Littles'," continuation on p. 6.

51. List of headlines culled by Christine Moe, *Soil Subsidence in the New Orleans Area* (Monticello, IL: Vance Bibliographies, December 1979), and revised by the author.

52. Moe, *Soil Subsidence in the New Orleans Area*.

53. Colten, *Unnatural Metropolis*, 5.

54. Campanella, "'Two Centuries of Paradox.'"

55. Population figures computed by author using 1910 US Census Bureau data at the enumeration district level; 1960, 1970, 1980 Census data at the census tract level, and 2000 Census data at the block level. Campanella, "'Two Centuries of Paradox.'"

56. Okey, "New Orleans Lakeshore Land Co. Tract, New Orleans," 30.

57. Colten, *Unnatural Metropolis*, 180–81.

58. "Hats Off To Our Patriotic Women!" 1.

59. Sewerage & Water Board of New Orleans, "Flood Protection? It's Your Call March 23," *Times-Picayune*, March 14, 1991, A-29

60. Dawn Ruth and Rebecca Theim, "Barthelemy Plays It Close to Vest in Top-Cop Search," *Times-Picayune*, February 16, 1991, B-3.

61. Sewerage & Water Board of New Orleans, "Flood Protection? It's Your Call March 23," A-29.

62. Dawn Ruth, "Vote Strategy for Drainage Fails," *Times-Picayune*, March 26, 1991, B-4; Frank Donze, "Drainage Tax Dies in Sparse Voting," *Times-Picayune*, March 24, 1991, B-2.

63. "Hats Off To Our Patriotic Women!" 1; Donze, "Drainage Tax Dies in Sparse Voting," B-2.

64. Based on information provided by Janet Howard, formerly of the Bureau of Government

Research, to the authors of *Task Force on New Orleans Sewerage, Water, and Drainage Utilities: Report of Findings and Recommendations* (New Orleans Sewerage and Water Board, January 25, 2019), 2.

65. Ramsey Green et al., *Task Force on New Orleans Sewerage, Water, and Drainage Utilities: Report of Findings and Recommendations*, 2.

66. "A Rain of Biblical Proportions," *Times-Picayune*, May 10, 1995, 1; *Times-Picayune*, May 7, 1995, weather section: B-8; qtd. by Wayne Knabb, *Times-Picayune*, May 8, 1995, A-1 and A-12.

67. Qtd. by Wayne Knabb, *Times-Picayune*, May 8, 1995, A-1 and A-12.

68. "Floods Cripple Orleans Area," *Times-Picayune*, May 4, 1978, 1; Joe Darby, "May 3 Yardstick in Jeff Replaced by April 13, 1980," *Times-Picayune*, April 14, 1980, 1; "Spring Storm Floods, Isolated Metro Area," *Times-Picayune*, April 8, 1983, 1; "10-Inch Rain Covers St. Bernard," *Times-Picayune*, November 8, 1989, 1; "Swamped: Flooding Comes as No Surprise," *Times-Picayune*, December 4, 1990, 1 and 8; Mark Schleifstein, "Storm Dumps 7 Inches of Rain on City," *Times-Picayune*, May 10, 1994, 1; Sewerage & Water Board of New Orleans, "Flood Protection? It's Your Call March 23," A-29.

69. Bob Ross, "After the Flood," *Times-Picayune*, May 8, 2000, 1.

70. W. Scott Lincoln, *Rainfall Analysis for the August 5, 2017 New Orleans Flash Flood Event* (National Weather Service Lower Mississippi River Forecast Center, August 11, 2017, updated May 18, 2018), 9.

71. Qtd. by Ross, "After the Flood," 1.

72. Ross, "After the Flood," 1; Martha Carr, "Infrastructure Poses Demands on Parish," *Times-Picayune*, January 8, 2000, Kenner edition: A-1 and A-8; US Army Corps of Engineers–New Orleans District, *Water Resources Development in Louisiana 1998*, 111; Ross, "After the Flood," 1.

73. Ross, "After the Flood," 3; Drainage Advisory Board, *Report on the Drainage of the City of New Orleans*, 34–39.

74. Qtd. by Carr, "Infrastructure Poses Demands on Parish," A-1 and A-8.

75. Woolley and Shabman, *Decision-Making Chronology for the Lake Pontchartrain & Vicinity Hurricane Protection Project*, 5–8.

11. THE ONLY THING THAT CAN MESS US UP, EARLY 2000S

1. Junger, "Pumps of New Orleans."

2. Interview of Marcia St. Martin by Norman Robinson, 2015, New Orleans Sewerage & Water Board, "My Story" video, www2.swbno.org/form_video.asp?s=history&yt=WqESti-O-Qk.

3. St. Martin interviewed by Robinson.

4. Marcia St. Martin, "Sewerage & Water Board's Emergency Response to Katrina Shared with International Water Utility Professionals," in *Katrina 5 Years Later: A Brief Glimpse Back with Full Focus on the Future* (New Orleans: Sewerage & Water Board, August 29, 2010), 8; Independent Levee Investigation Team, *Investigation of the Performance of the New Orleans Flood Protection Systems in Hurricane Katrina on August 29, 2005, vol. 1: Main Text and Executive Summary*, 4–19; W. Bernard Carlson, "Pushing Alternating Current in America (1892–1893)," in *Tesla: Inventor of the Electrical Age* (Princeton, NJ: Princeton University Press, 2013), 158–75; Allison Lantero, "The War of the Currents: AC vs. DC Power," US Dept. of Energy, November 18, 2014, www.energy.gov/articles/war-currents-ac-vs -dc-power (accessed July 22, 2020); "History of Power Frequency," *Electrical Science*, electrical-science .blogspot.com/2009/12/history-of-power-frequency.html (accessed July 22, 2020).

5. Qtd. by Richard F. Snow, "Low and Dry," *American Heritage's Invention and Technology Magazine* 8, no. 2 (Fall 1992), www.inventionandtech.com/content/low-and-dry-1 (accessed July 22, 2020).

6. Sewerage & Water Board of New Orleans, "Power Generation and Drainage Systems—Frequently Asked Questions," August 14, 2017, www.google.com/url?sa=t&rct=j&q=&esrc=s&source=web&cd=&ved=2ahUKEwjuud762a_3AhU8l2oFHRLGDLQQFnoECAwQAQ&url=https%3A%2F%2Fwww.swbno.org%2Fdocuments%2Frainevent%2FDPS_PowerGeneration_DrainageSystem_FAQ.pdf&usg=AOvVaw37Z13VcHt-CcuG-Hi9zB6G.

7. Independent Levee Investigation Team, *Investigation of the Performance of the New Orleans Flood Protection Systems in Hurricane Katrina on August 29, 2005, vol. 1: Main Text and Executive Summary,* 4–19.

8. St. Martin, "Sewerage & Water Board's Emergency Response to Katrina," 8.

9. Joseph Sullivan, qtd. by Snow, "Low and Dry."

10. St. Martin by Robinson.

11. Interview of Gerald R. Elwood by Norman Robinson, 2015, New Orleans Sewerage & Water Board, "My Story" video, www2.swbno.org/form_video.asp?s=history&yt=YZVqoz92Ywk; Independent Levee Investigation Team, *Investigation of the Performance of the New Orleans Flood Protection Systems in Hurricane Katrina on August 29, 2005, vol. 1: Main Text and Executive Summary,* 8–5.

12. The exact chronology of the flooding varies according to several analyses. This chronology is the author's synthesis of published or presented studies done by Delft University of Technology in the Netherlands and the US Army Corps of Engineers, based largely on the findings of National Science Foundation–funded Independent Levee Investigative Team. They are supplemented by the author's own research and eyewitness experiences published in *Geographies of New Orleans* and *Bienville's Dilemma.* See also M. Kok et al., *Polder Flood Simulations for Greater New Orleans—Hurricane Katrina August 2005* (Netherlands: Delft University of Technology, 2007).

13. Interview of Richard Reese by Norman Robinson, 2015, New Orleans Sewerage & Water Board, "My Story" video, www2.swbno.org/form_video.asp?s=history&yt=sFuAexGLRxY.

14. Independent Levee Investigation Team, *Investigation of the Performance of the New Orleans Flood Protection Systems in Hurricane Katrina on August 29, 2005, vol. 1: Main Text and Executive Summary,* 3–14 to 3–17.

15. Independent Levee Investigation Team, *Investigation of the Performance of the New Orleans Flood Protection Systems in Hurricane Katrina on August 29, 2005, vol. 1: Main Text and Executive Summary,* 3–11 to 3–37.

16. Interview of Richard Alexander by Norman Robinson, 2015, New Orleans Sewerage & Water Board, "My Story" video, www2.swbno.org/form_video.asp?s=history&yt=jRm_oz-HtWg.

17. Interview of Reynaldo Robertson by Norman Robinson, 2015, New Orleans Sewerage & Water Board, "My Story" video, www2.swbno.org/form_video.asp?s=history&yt=AKrn37wwHnE.

18. Interview of Bob Moenian by Norman Robinson, 2015, New Orleans Sewerage & Water Board, "My Story" video, www2.swbno.org/form_video.asp?s=history&yt=90SCX6bcGQM.

19. Magill, "Conspiracy of Complicity," 43.

20. Author's eyewitness and time-stamped photographs, August 29–31, 2005.

21. Interview of Gabe Signorelli by Norman Robinson, 2015, New Orleans Sewerage & Water Board, "My Story" video, posted online https://www2.swbno.org/form_video.asp?s=history&yt=DgBJ815NruU.

22. Interview of Damon Adams by Norman Robinson, 2015, New Orleans Sewerage & Water

Board, "My Story" video, posted online https://www2.swbno.org/form_video.asp?s=history&yt=GNH 2ml9kgDI.

23. The author and his wife were among those people.

24. Interview of Jason Higginbotham by Norman Robinson, 2015, New Orleans Sewerage & Water Board, "My Story" video, www2.swbno.org/form_video.asp?s=history&yt=8Srt2gZkFb8.

25. Higginbotham interviewed by Robinson.

26. Interview of Rudy August by Norman Robinson, 2015, New Orleans Sewerage & Water Board, "My Story" video, www2.swbno.org/form_video.asp?s=history&yt=DltcYdXV6J8.

27. Higginbotham interviewed by Robinson.

28. Interview of Robert Jackson by Norman Robinson, 2015, New Orleans Sewerage & Water Board, "My Story" video, www2.swbno.org/form_video.asp?s=history&yt=_ibAIktiQGo.

29. Independent Levee Investigation Team, *Investigation of the Performance of the New Orleans Flood Protection Systems in Hurricane Katrina on August 29, 2005, vol. 1: Main Text and Executive Summary,* 8–11.

30. St. Martin interviewed by Robinson; Jackson interviewed by Robinson.

31. Adams interviewed by Robinson; interview of Vincent Fouchi by Norman Robinson, 2015, New Orleans Sewerage & Water Board, "My Story" video, www2.swbno.org/form_video.asp?s=history &yt=-VReXiCAT4s; Signorelli interviewed by Robinson.

32. Elwood interviewed by Robinson.

33. Adams interviewed by Robinson.

34. St. Martin, "Sewerage & Water Board's Emergency Response to Katrina," 7.

35. Jackson interviewed by Robinson.

36. Emphasis added. The death toll in Louisiana has since been revised to approximately 1,600, with another 220 in Mississippi. Independent Levee Investigation Team, *Investigation of the Performance of the New Orleans Flood Protection Systems in Hurricane Katrina on August 29, 2005, vol. 1: Main Text and Executive Summary,* 8–11.

37. Independent Levee Investigation Team, *Investigation of the Performance of the New Orleans Flood Protection Systems in Hurricane Katrina on August 29, 2005, vol. 1: Main Text and Executive Summary,* 4–47.

38. Adams interviewed by Robinson.

39. Nicolai Ouroussoff, "How the City Sank," *New York Times,* October 9, 2005.

40. Signorelli interviewed by Robinson.

41. Reese interviewed by Robinson.

42. St. Martin interviewed by Robinson.

43. St. Martin, "Sewerage & Water Board's Emergency Response to Katrina," 12.

44. Missy Wilkinson, "Winds of Change," *Times-Picayune/New Orleans Advocate,* August 29, 2020, Inside/Out section: 15–17.

45. US Army Corps of Engineers, *Comprehensive Environmental Document Phase 1—Greater New Orleans Hurricane and Storm Damage Risk Reduction System* (New Orleans, May 2013), vol. 1: 1.

46. US Government Accountability Office, *Army Corps of Engineers—Known Performance Issues with New Orleans Drainage Canal Pumps Have Been Addressed, But Guidance on Future Contracts Is Needed* (Washington, DC: Report to the Chairman, Ad Hoc Subcommittee on Disaster Recovery, Committee on Homeland Security and Governmental Affairs, December 2007), 9.

47. US Army Corps of Engineers, New Orleans District, *Elevations for Design of Hurricane Protec-*

tion Levees and Structures, Lake Pontchartrain and Vicinity, West Bank and Vicinity, and New Orleans to Venice, Louisiana Projects, Report Version 2.0, December 2014, 10; Southeast Louisiana Flood Protection Authority–West / West Jefferson Levee District / Algiers Levee District, "West Bank Hurricane Protection: Before and After," September 14, 2015, *slfpaw.org/west-bank-hurricane-protection-before-and-after/* (accessed November 16, 2018); "The West Closure Complex: How It Works," NOLA.com | *Times-Picayune,* October 19, 2015, *www.nola.com/environment/index.ssf/2015/10/the_west_closure_com plex_how_i.html* (accessed November 16, 2018).

48. Mark Schleifstein, "Stronger Levees a Key Part of Katrina's Legacy," *Times-Picayune/New Orleans Advocate,* August 24, 2020, 1–4.

49. US Army Corps of Engineers, *Comprehensive Environmental Document Phase 1—Greater New Orleans Hurricane and Storm Damage Risk Reduction System* 1: ES-9 to ES-11, and US Army Corps of Engineers, "IHNC-Lake Borgne Surge Barrier Fact Sheet" (May 2015).

50. US Army Corps of Engineers, "Outfall Canal Closure Structures Fact Sheet," May 2013.

51. Southeastern Louisiana Flood Protection Authority–East, "The Permanent Canal Closures and Pumps (PCCP)," 2018.

52. US Army Corps of Engineers, "Permanent Canal Closures & Pumps Fact Sheet," March 2017; Southeastern Louisiana Flood Protection Authority–East, "The Permanent Canal Closures and Pumps."

53. US Army Corps of Engineers, *Comprehensive Environmental Document Phase 1—Greater New Orleans Hurricane and Storm Damage Risk Reduction System* 1, and "IHNC-Lake Borgne Surge Barrier Fact Sheet."

54. US Army Corps of Engineers, New Orleans District, *Elevations for Design of Hurricane Protection Levees and Structures, Lake Pontchartrain and Vicinity, West Bank and Vicinity, and New Orleans to Venice, Louisiana Projects,* Report Version 2.0, 10; Southeast Louisiana Flood Protection Authority–West / West Jefferson Levee District / Algiers Levee District, "West Bank Hurricane Protection: Before and After"; "The West Closure Complex: How It Works."

55. Campanella, *West Bank of Greater New Orleans;* US Army Corps of Engineers, *Comprehensive Environmental Document Phase 1—Greater New Orleans Hurricane and Storm Damage Risk Reduction System* 1, and "West Closure Project Fact Sheet" (August 2015).

56. Green et al., *Task Force on New Orleans Sewerage, Water, and Drainage Utilities: Report of Findings and Recommendations,* 2.

57. Interview of Cedric Grant by Norman Robinson, 2015, New Orleans Sewerage & Water Board, "My Story" video, www2.swbno.org/form_video.asp?s=history&yt=atw27dv_rJw.

58. Green et al., *Task Force on New Orleans Sewerage, Water, and Drainage Utilities: Report of Findings and Recommendations,* 2.

59. Jeff Adelson, "S&WB General Superintendent Bob Turner, Who Joined Agency After 2017 Floods, To Retire," *Times-Picayune/New Orleans Advocate,* February 24, 2021, and "President Joe Biden to Tour S&WB's Carrollton Plant This Week," *Times-Picayune/New Orleans Advocate,* May 3, 2021.

60. Interviews of Cedric Grant, Jason Higginbotham, and Elwood by Norman Robinson, 2015, New Orleans Sewerage & Water Board, "My Story" video, www2.swbno.org/history_hurricaneka trina.asp.

61. Louisiana State Climatologist Barry Keim, qtd. by David J. Mitchell, "Researchers: La. Seeing More Bursts of Hard Rain," *Times-Picayune/New Orleans Advocate,* August 2, 2021, 1A–4A.

62. Katie Moore, "An Inch the First Hour? How Much Can the Pump System Handle?" WWL-TV

News, "Down the Drain" investigation, November 15, 2017, www.wwltv.com/article/news/local/down-the-drain/an-inch-the-first-hour-how-much-can-the-pump-system-handle/289–491650863.

63. New Orleans City Council, Special City Council Meeting on the August 5 Flood, held August 8, 2017 in the City Council Chambers, www.nola.com/news/article_3e0808a0-91b2-53e7-8f2a-91e36ddc8dce.html; Lincoln, *Rainfall Analysis for the August 5, 2017 New Orleans Flash Flood Event*, 11 and 16. The city official who referred to the storm as a "no-notice rain event" was Ryan Berni, spokesman for the Mayor's Office, August 5, 2017, www.youtube.com/watch?v=Orz4mZdz_b8.

64. Councilman Jason Williams, speaking at City Council Press Conference, New Orleans, August 6, 2017, www.youtube.com/watch?v=La9_HLIIbOY.

65. Cedric Grant, speaking at a press conference outside city hall, August 5, 2017, www.youtube.com/watch?v=Orz4mZdz_b8; New Orleans City Council, Special City Council Meeting on the August 5 Flood.

66. New Orleans City Council, Special City Council Meeting on the August 5 Flood.

67. David Hammer, "S&WB Pump Reverses to Flood Streets," *Times-Picayune/New Orleans Advocate,* March 27, 2021, Metro Section: 1; ABS Group, *City of New Orleans Stormwater Drainage System Root Cause Analysis Final Report* (City of New Orleans, October 2018), 3; *New Orleans Advocate* staff reports, "New Orleans Flood, Pumps Debacle: Recap of New Information Released Thursday," August 10, 2017, www.nola.com/news/article_c6517162-9df3-513f-9d60-1675d2fbf87f.html.

68. ABS Group, *City of New Orleans Stormwater Drainage System Root Cause Analysis Final Report,* 1–13 and 89.

69. Mayor LaToya Cantrell, speaking to WWL-TV anchor Thanh Truong, May 18, 2018, www.youtube.com/watch?v=9qO-_GkbZ8w (accessed September 16, 2020).

12. REWATERING NEW ORLEANS, 2010S–2020S

1. Barton, "Report Upon the Sanitary Condition of New Orleans," 207–16.

2. Interview of David Waggonner by Richard Campanella, September 10, 2020.

3. Waggonner & Ball Architects, *Greater New Orleans Urban Water Plan—Vision,* November 2013, 71.

4. "Report of the Flushing Committee, New Orleans Auxiliary Sanitation Association."

5. Lorena O'Neil, "Why Doesn't New Orleans Look More Like Amsterdam?" *Atlantic,* September 2, 2015, www.theatlantic.com/technology/archive/2015/09/why-doesnt-new-orleans-look-like-amsterdam/402322/ (accessed September 15, 2020).

6. Interview of Ramiro Diaz by WaterLoop, "Living with Water Approach to Architecture," June 3, 2020, www.youtube.com/watch?v=GjoTmXfYyiQ.

7. Waggonner interviewed by Campanella.

8. First quote from Waggonner interviewed by Campanella; second quote from interview of David Waggonner by O'Neil, "Why Doesn't New Orleans Look More Like Amsterdam?"

9. Richard Rainey, "N.O. Region Takes New Look at Drainage," *Times-Picayune,* March 24, 2011, B1–B2.

10. Officials with local governments, including the Sewerage & Water Board, did have seats on the project's advisory council. Waggonner & Ball, "Greater New Orleans Urban Water Plan: Principles: Adapting the Flow" and Project Team, livingwithwater.com/blog/urban_water_plan/solutions/ (accessed September 14, 2020).

11. Richard Rainey, "Water Plan Has Flood of Support; New Approach to Protection Touted," *Times-Picayune,* September 8, 2013, B1-B2.

12. Waggonner & Ball Architects, "Report Organization," *Greater New Orleans Urban Water Plan—Vision,* November 2013, 17.

13. Waggonner & Ball, *Greater New Orleans Urban Water Plan—Vision,* 111–19.

14. Waggonner & Ball, *Greater New Orleans Urban Water Plan—Vision,* 34–37 and 83–85.

15. Waggonner & Ball, *Greater New Orleans Urban Water Plan—Vision,* 34–37, 83–85, and 136–43.

16. Waggonner interviewed by Campanella.

17. O'Neil, "Why Doesn't New Orleans Look More Like Amsterdam?"; Waggonner interviewed by Campanella.

18. Qtd. by O'Neil, "Why Doesn't New Orleans Look More Like Amsterdam?

19. Mandelman, *Place with No Edge,* 8–10.

20. Waggonner interviewed by Campanella.

21. "New, Two-Story Mirabeau Structure of Sisters of St. Joseph," *Times-Picayune,* March 22, 1952, 11.

22. Qtd. by Matthew Teague, "Convent Inundated by Hurricane Katrina May Help Save New Orleans," *New York Times,* August 27, 2020, A22.

23. Orleans Parish Assessor's Office property records database; Teague, "Convent Inundated by Hurricane Katrina May Help Save New Orleans," A22.

24. City of New Orleans Resilience + Sustainability, "Gentilly Resilience District Overview and Background" and "Gentilly Resilience District Fact Sheet," August 2018, www.nola.gov/resilience-sustainability/areas-of-focus/green-infrastructure/national-disaster-resilience-competition/gentilly-resilience-district/ (accessed September 16, 2020).

25. Interview of David Waggonner by Martin C. Pedersen, "David Waggonner on New Orleans and the Way Forward After Ida," September 23, 2021, *Common Edge,* commonedge.org/ (accessed September 25, 2021).

26. O'Neil, "Why Doesn't New Orleans Look More Like Amsterdam?"

27. Waggonner interviewed by Campanella.

Index